BIRDS *of the*
PACIFIC NORTHWEST

John Shewey & Tim Blount   with Hendrik Herlyn, editor

TIMBER PRESS FIELD GUIDE

Frontispiece: Hooded Merganser

The Haseltine Building
133 S.W. Second Avenue, Suite 450
Portland, Oregon 97204-3527

timberpress.com
Printed in China
Book design by Adrianna Sutton
Cover design by Patrick Barber
Series design by Susan Applegate

Library of Congress Cataloging-in-Publication Data

Names: Shewey, John. | Blount, Tim.
Title: Birds of the Pacific Northwest / John Shewey and Tim Blount ;
    with Hendrik Herlyn, editor.
Description: Portland, Oregon: Timber Press, 2017. | Series: Timber Press field
    guide | Includes bibliographical references and index.
Identifiers: LCCN 2016027189 (print) | LCCN 2016041401 (ebook) | ISBN
    9781604696653 (flexibind: alk. paper) | ISBN 9781604697858 (e-book)
Subjects: LCSH: Birds—Northwest, Pacific—Identification.
Classification: LCC QL683.N75 S54 2017 (print) | LCC QL683.N75 (ebook) | DDC
    598.09795—dc23
LC record available at https://lccn.loc.gov/2016027189
ISBN 13: 978-1-60469-665-3

A catalog record for this book is also available from the British Library.

Great Egret

CONTENTS

Golden-crowned Sparrow

BIRDWATCHING IN THE PACIFIC NORTHWEST

The corner of the world we know as the Pacific Northwest is home to an amazing array of bird life. From the southern borders of Oregon and Idaho, north to southern British Columbia, to the Pacific Ocean in the west, and east to the Rocky Mountains, almost 400 bird species occur. This includes nearly 300 breeding species, thanks to the region's incredibly diverse habitat types: open ocean, beaches and bays, coniferous forest, deciduous woodlands, mixed forests, high mountains, deserts, shrub steppe, rivers and lakes big and small, just to name a few. Such a rich and varied assemblage of habitats assures an equally varied avian population, from bird species that occupy very small enclaves of the region, to widespread generalists that occur in many habitats. Birding strategies here are diverse, as well—from a casual interest to the near-addictive pursuit of certain species. Indeed, birdwatching enthusiasts from the far-flung corners of the United States and beyond routinely travel to the Pacific Northwest to pursue their passion—and for good reason: they can see birds found few other places and they can observe substantial numbers of species that are new to them.

We ourselves are Pacific Northwest lifers; we've lived in various parts of both Oregon and Idaho, and traveled substantially throughout the region. This is our home, and the amazing diversity of birds and habitats in the Northwest is a big reason why. This guide is the culmination of our passion for birds, the Pacific Northwest, and in particular, Pacific Northwest birds.

Our hope for *Birds of the Pacific Northwest* is to serve all levels of birdwatchers, whether you simply enjoy feeding songbirds in your backyard or are an ardent enthusiast who considers birding your primary hobby. Carry the book with you in the field, keep it handy on your bookshelf, study it and learn more about your favorite species and birds that are new to you. Above all, enjoy it and embrace the wonder of Northwest birds.

Basic equipment for birdwatching is simple: a pair of binoculars and this book.

TOOLS OF THE TRADE

Birdwatching is remarkably low tech compared to many other hobbies, encumbered by only 2 essential accoutrements: a pair of binoculars and a bird book. Binoculars come in myriad brands, models, and prices, and generally, you get what you pay for—quality, meaning both image quality and product durability, comes at higher prices. Importantly, binoculars vary in magnification power, field of view, and light gathering ability; binoculars or their packaging are marked with a code that provides those details, so binoculars marked "7×35" (popular with birders) offer magnification 7 times greater than what you see with the naked eye, and a front lens diameter of 35 millimeters. Also popular among birders, 8×42 binoculars have a magnification factor of 8, and a front lens diameter of 42 millimeters. The diameter of the front lens dictates the lens's ability to gather light; the more light gathered, the brighter and more detailed your view. However, the larger the lens diameter, the smaller the field of view.

The highest magnification and the largest front lens diameter possible might seem ideal, but binoculars more powerful than 10×50 enter the realm of diminishing marginal returns. For starters, smaller fields of view with larger-diameter lenses make finding the object you want to look at more difficult because the binoculars see a smaller area. A 50-millimeter lens views a smaller area than a 35-millimeter lens. A small field of view plus lots of magnification can make tracking a small bird at close range very difficult. Also, more powerful binoculars tend to be bigger and heavier, and they exaggerate the consequence of even slightly shaky hands into substantial image shakiness, although binoculars with image stabilization technology are also available. In essence, powerful 10×50 binoculars are best for distant birds that don't move quickly, and for stable hands (or simply rest your elbows on a solid surface when you use them). For magnification greater than a factor of 10, spotting scopes (tripod mounted or vehicle-window mounted) are a better option than binoculars and can be very useful in identifying birds at extreme distances.

You'll spend most of your birdwatching time outdoors, under a variety of conditions, so dress for the weather and the landscape. A sturdy, comfortable pair of hiking shoes or boots makes birding by foot much less fatiguing, and layered clothing for cool or cold weather is essential, including a rain parka. When you venture outdoors in the Northwest, remember that in many areas mosquitos abound in spring and summer, and ticks are often common. The sun can be intense, as can the cold of winter and early spring. Plan ahead and protect yourself accordingly.

Birdwatching by vehicle is also very popular—you can see many different species just by keeping your eyes open on a drive through the countryside or by parking at places that provide excellent viewing opportunities. Slowly driving a forest road deep in the mountains, windows down so you can hear the birds, is a decidedly pleasant way to spend an early morning, but even commuting along ever-busy Interstate 5 (or virtually any other freeway in the region) offers ample opportunity to see a wide variety of interesting birds. Safety first, however—the driver needs to do the driving, not the watching.

The Pacific Northwest offers adventurous birders a chance to visit remote regions, from the deserts of southeast Oregon to the mountains of central Idaho. Off-the-beaten-path birding mandates a few extra precautions. Backcountry roads in the Northwest can be rugged and sometimes downright treacherous. If you will be driving remote desert or mountain roads, consider carrying 2 spare tires, rather than one, and make sure you have all the tools and know-how you need to change a tire in the middle of nowhere. Be aware that

Spotting scopes mounted on tripods (or vehicle windows) are especially useful for viewing distant birds.

many backcountry roads, especially in the desert, can quickly turn to unpassable goo in the rain. Also consider carrying 5 to 10 gallons of extra gasoline (on the exterior of the vehicle), and give a spare vehicle key to a member of the party or stash a spare key on the vehicle where any member of the party can get to it if needed. Remember a couple of valuable adages we've learned to embrace: use four-wheel drive to get out of trouble, not to get into trouble, and if you can walk a road faster than you can drive it, you probably ought to walk it.

Carry basic emergency supplies to get you through a few extra days should you happen to be stranded by weather, road conditions, vehicle breakdown, or any other form of just plain bad luck. Essentials include plenty of extra drinking water, extra food, and a sleeping bag or two for emergency warmth.

In remote areas of the Northwest, don't count on cell phone service, and by all means don't navigate backroads by GPS. Get good old-fashioned paper maps, the most up-to-date available for the areas in which you will be traveling, and familiarize yourself with these maps. Good information and map sources include the Bureau of Land Management, U.S. Forest Service, and U.S. Geological Survey (as well as the U.S. Fish and Wildlife Service for federal refuges, and state fish and wildlife agencies for state wildlife areas). Plan your trip by checking road conditions and weather forecasts, and if you are counting on fueling up or getting supplies in tiny, remote towns, call ahead to check hours of operation and available services. Armed with good maps, proper equipment, and common sense, you can find amazing birds in remarkable untrodden places in the Pacific Northwest.

BIRD IDENTIFICATION

Bird identification begins with observation, and the more acute your observation skills, the easier it becomes to identify birds. Many birds are easily identified even by novice birders because they are either familiar and iconic—the American Robin and adult Bald Eagle, for example—or because they are easily separable from all other species and thus easy to find in a bird book—a Brown Pelican or a male Yellow-headed Blackbird, for instance.

Others require more scrutiny, and therein lies the challenge (and much of the fun) in birdwatching. Learning to properly identify birds requires familiarity with the characteristics and behaviors that aid in

While many birds are easy to identify, others, such as this juvenile Brewer's Sparrow, can be very challenging, requiring close scrutiny not only of appearance, but also consideration of the location, habitat, and time of year.

distinguishing one species from another, as well as a general understanding of the birds likely to occur in a specific place at a particular time. Physical characteristics include body shape and size, plumage color and pattern, physical attitude and posture; behaviors that aid in identification include flight pattern, feeding habits, and songs and calls.

Physical traits that help identify birds begin with general characteristics that provide clues to a bird's taxonomy and thereby help you to look for that species in the correct part of this book, which is organized in taxonomic order (except in a few cases where a slightly altered order aids in comparing similar species). Is the bird small or large? Slender or compact? Does a small songbird have a thin, sharp bill or a thick, conical bill? Does a bird of prey have a long, slender tail or a short, fanlike tail? Identifying birds requires that you learn what to look for—color and pattern isn't always enough. Practice looking carefully at general body shape and head shape, bill shape and size, tail length and shape. Also learn to carefully study plumage colors and patterns—sometimes the slightest variation separates similar species.

Beware, though, that for many species, plumage colors and patterns change with the seasons, the bird's sex, and with age (see the section on "Plumage Variation in Birds"). Body shape can change depending on the bird's posture and activity; size is relative and often difficult to judge. For example, a

Accurately judging the size of a bird can be difficult, but is easier when nearby objects or other species lend comparisons, such as with this Black-bellied Plover (right), which is clearly larger than the dowitcher (left) standing nearby.

small sandpiper might appear a bit larger than it really is until a blackbird lands nearby and you realize that the tiny sandpiper—perhaps a Least Sandpiper—is no bigger than a sparrow and certainly not as large as the blackbird.

At the same time, note the bird's habits and habitat: does that conical-billed songbird stick to the ground or does it remain up in the trees? Did you see it in arid southeast Oregon, temperate western Washington, or montane central Idaho? Does that thin-billed songbird slowly feed among the tree leaves or dash about frenetically catching insects? Did you find it high in the mountains or along a lowland stream? Does that hawk soar over open country or fly expertly through the forest? Does that sandpiper feed in flocks on the beach or by itself in a plowed field?

Vocalizations, too, are often helpful—and sometimes critical—in differentiating between similar species; learning different bird songs and calls can add immeasurably to the joy of birdwatching. After all, the remarkable variety of sounds made by birds is intriguing and frequently mesmerizing. Who doesn't enjoy the cheery musical songs of skilled feathered vocalists, the haunting calls of an owl or a loon, or the bizarre sounds produced by a Virginia Rail or American Bittern?

Finally, consider the time of year and specific location. Many birds are migratory and occur in the Northwest only in specific seasons, while others are year-round residents. If you find a hummingbird in the Northwest during January, for example, it's a very safe bet you are seeing an Anna's Hummingbird; the other species that live here

Learning to identify birds by their songs and calls is fun and rewarding; here, an aptly named Warbling Vireo delivers its distinctive and musical song.

migrate to warmer climes for the winter. Ranges—the areas within the Northwest where each species is normally found—likewise provide a significant clue to identification. Even relative abundance can be an important clue to identifying many birds simply because all things being otherwise equal, rare birds are less likely to be seen than common birds.

Learn to not only look carefully at birds, but also to consider the totality of evidence in identifying challenging species: habits, habitat, range, time of year, vocalizations, and relative abundance. That totality of evidence is what really helps discern the differences between similar-looking species.

Plumage Variation in Birds

Birds can exhibit a wide range of plumage variations within a species. In some cases,

males look different than females; this is called sexual dimorphism, and it can vary from extreme (such as in most ducks) to minor. The male Downy Woodpecker has a bright red patch on the back of his crown, whereas the female does not—the only difference between the two. In a few cases—notably the phalaropes—the female wears a brighter breeding-season plumage than the male (reverse sexual dimorphism).

Also, birds swap out feathers seasonally through a process called molting, shedding old, worn feathers for new plumage. The extent to which the molt changes a bird's appearance varies with the species. Molt generally occurs in late summer and fall, when some birds transition to a nonbreeding (winter) plumage, and others (ducks, for example) begin regrowing their breeding-season plumage in preparation for

Bird Topography   These terms are used frequently in the species descriptions in this book.

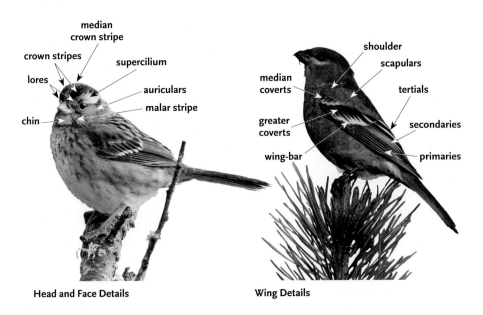

Head and Face Details

Wing Details

Surf Scoters are sexually dimorphic, meaning the male (top) and female (bottom) differ from one another in color. Sexual dimorphism can range from fairly extreme, such as in ducks like these, to minor variations in plumage between the male and female of a species.

the following year. Birds that transition to a distinct nonbreeding plumage molt again in late winter or spring, transitioning back to breeding plumage. The difference between breeding and nonbreeding plumages can be dramatic.

The timing of the molt varies both from species to species, and within species. A flock of sandpipers migrating through the Northwest in midspring, for example, often has individuals ranging from winter (nonbreeding) plumage to full breeding plumage; members of a flock of Golden-crowned Sparrows wintering in the Northwest acquire their beautiful black-and-yellow crowns at different rates, so that some individuals in the flock attain breeding plumage in March and others a month or more later. Moreover, many species retain their breeding plumage only for a scant 3 months or so. Because of the comparatively short duration of full breeding plumage in many birds,

ornithologists use the term "basic plumage" to describe the nonbreeding plumage, and "alternate plumage" in reference to the breeding-season plumage. In this book, however, we have opted for the more comfortable and familiar labels of "breeding" and "nonbreeding" plumage.

Also, while many birds, especially songbirds, acquire their adult plumage before one year of age, others require more than a full year, and in a few species—the Bald Eagle, for example—as many as 5 years are required to reach maturity and full adult plumage.

Because feathers can fade and become damaged with age, many birds can appear substantially different with worn plumage than with fresh plumage, creating further identification challenges. Typically such worn plumage occurs after the breeding season, so birders need to be especially aware of such possibilities in late summer and fall.

Many birds change with the seasons, transitioning between breeding and nonbreeding plumages, and the timing of such changes, called molts or molting, varies with the species and within the species. These Dunlins are in the process of molting from nonbreeding plumage to breeding plumage.

Bear in mind also that light plays a significant role in how colors appear, or fail to appear. This is especially true in birds with iridescent feathers. Both the male Brewer's Blackbird and the Tree Swallow, for example, are extensively iridescent, but when seen in poor light—on a cloudy day or near sunset—the colorful, shiny feathers may simply appear black. Hummingbirds can be especially tricky; the iridescent gorgets—the layered platelike feathers on the throat of male hummingbirds—can appear dark and colorless or can flash like a strobe light in brilliant colors, and the color can vary based on light conditions. For instance, the male Anna's Hummingbird's gorget, when perfectly illuminated, appears brilliant fuchsia, but depending on the angle of the light, can vary from deep blood red to coppery.

Identification Challenges: Lookalikes

With *Birds of the Pacific Northwest* at your side, you'll find most of the birds in the Northwest are recognizable and identifiable through observation and scrutiny of key field marks. But in some cases, similar or even virtually identical-looking species are very difficult to distinguish, even for expert birders. Among the most problematic are gulls, *Empidonax* flycatchers, *Calidris* sandpipers, and a handful of others.

Gulls are especially challenging because many species require several years to reach maturity and thus adult plumage, and their plumage is different each successive year leading to adulthood. Moreover, most adult gulls look somewhat different in winter than during the breeding season of spring and summer. Lastly, hybridization between

This male Anna's Hummingbird exhibits just a hint of bright color on his throat and crown when viewed from the side (left), but when he turns his head at just the right angle (right), his gorget—the shiny bib—and his crown flash like neon lightbulbs.

some species of gulls is common, leading to individuals with intermediate physical characteristics that are not easily reconciled with the features that identify either of the parent species. For these reasons, identifying gulls in their various age-determined plumage stages frequently requires attention to relatively minute details, which are described in this book for each species.

Empidonax flycatchers—a genus that includes 7 very similar-looking species in the Northwest—require acute attention not only to plumage details, but also to habitat, range, and voice. In fact, successfully identifying one of these lookalike flycatchers—separating a Dusky Flycatcher from a Hammond's Flycatcher, for example—usually mandates that observers consider the totality of evidence, especially if the bird is not singing or calling. Photos can help immeasurably, allowing for scrutiny of field marks that are very difficult to study in the field, such as relative wingtip length and bill length. But in the absence

Empidonax flycatchers, such as this Dusky Flycatcher, are notoriously difficult to identify, and in many cases, totality of evidence—plumage, song, habitat, location—is key to determining the species.

of photographic support, identification typically hinges on considering a collage of details in addition to appearance. A few other species are equally inseparable where their ranges overlap: the Juniper Titmouse and Oak Titmouse, for example; they co-occur in Oregon's Klamath Basin and even their various vocalizations overlap enough to render calls and songs unreliable in separating the two. Similarly, the female Rufous and Allen's Hummingbirds, which co-occur along the southwest Oregon coast, are effectively indistinguishable in the field.

The smaller *Calidris* sandpipers, colloquially and collectively referred to as "peeps," are likewise challenging to identify in the field and require close study. Some prior knowledge of what to look for helps a great deal when you see peeps in the field—a concept that holds true for all birds, but especially lookalikes. Peeps wear a breeding plumage in spring and part of summer, but then molt to a drabber non-breeding plumage. In most cases—again, as with many birds—they are easier to identify in breeding plumage, but that only accounts for a few months during the year. Juvenile peeps account for a third plumage variation during late summer and autumn, in which they are often easier to identify. In addition to basic colors and patterns, key field marks to differentiate peeps include leg color and length; bill color, length, and shape; head and body structure and posture; and in-flight color and pattern, especially of the dorsal surface.

Colloquially known as peeps, tiny sandpipers of the genus *Calidris* are represented by several very similar species in the Northwest, and identifying them often requires considering all the evidence: plumage (which varies with the age of the bird and with the season), as well as location, habitat, behavior, posture, shape, leg color, and relative abundance.

The Lesser Scaup (left) and Greater Scaup (right) resemble each other so closely that careful study is required to tell them apart.

Lesser and Greater Scaups, female Common Goldeneyes and Barrow's Goldeneyes, female hummingbirds in general, various sparrows in nonbreeding and/or juvenile plumages—these are just a few examples of the many birds that are so similar, identifying them requires careful study. In almost all cases, however, proper identification is not only possible, but fun and challenging, adding an enjoyable dynamic to birdwatching.

All told, correctly identifying lookalike species begins before you ever see them. First, familiarize yourself with the species most likely to be seen in specific places at specific times of year. Next, study and try to remember at least a few key species-specific field marks, including plumage, voice, and behaviors. Finally, realize that sometimes precise identification

Cooper's Hawks (above right) and Sharp-shinned Hawks (right) are very similar and closely related; adults can be identified by considering a variety of characteristics, including the head pattern—a dark hood on the Sharp-shinned Hawk and a dark cap strongly contrasting with the light gray nape on the Cooper's Hawk.

of a certain bird may not be possible for a variety of reasons, including the conditions under which you see the bird. After all, a stationary bird in good light at close range is far easier to study than a frenetic speedster or a shy skulking bird or an individual seen in poor light or at long range. And don't worry—veteran birders make identification mistakes, too, often with very challenging birds such as the *Empidonax* flycatchers, but also for other reasons. Once, while driving along Brownlee Reservoir on the Snake River just past dusk on an early summer evening, we stopped suddenly at the sight of a small owl perched atop a wooden fencepost, silhouetted against the hillside above the road. The first stars had appeared in the east, and darkness was upon us, but we could plainly see the owl in profile, and viewing it through binoculars, we were perplexed: a chubby little short-tailed owl with no ear tufts. We ruled out Northern

Pygmy-Owl and Burrowing Owl; Western Screech-Owl was the only logical choice, but despite swiveling its head around frequently and altering its perch slightly, this owl showed no indication whatsoever of having ear tufts.

We considered: a Northern Saw-whet Owl in a treeless desert environment? A Flammulated Owl far from its normal pine tree habitat and steadfastly refusing to raise its ear tufts? We groped through the cargo box for a powerful flashlight, but none was to be found; we thumbed frantically through a bird book. And then an adult Western Screech-Owl appeared like a ghost, and fluttered down to feed its fledgling, which we had not recognized as such. Of course. We had simply not considered all the possibilities, temporarily forgetting that this was the time of year that fledgling owls are out of the nest. We laughed about it and thoroughly appreciated the rare treat of watching an adult owl feed its offspring.

Female Common Mergansers (front) and female Red-breasted Mergansers (behind) are quite similar; the differences between such similar birds are clearly explained in the bird species accounts.

Alternate plumage Breeding plumage (see also "Basic plumage")

Auricular Area around the ear

Basic plumage Nonbreeding plumage (see also "Alternate plumage")

Cere A waxlike casing on the upper mandible, including the nostrils

Chin The small area immediately below the bill

Crest Tuft of feathers protruding from a bird's crown; most species can raise and lower the crest

Crown Top of the head (sometimes called "cap")

Culmen Center ridgeline of the upper mandible (jaw) of a bird's bill

Decurved bill Downward-curved bill

Eclipse plumage A plumage molt stage, best known in ducks, during which the bird acquires a drab post-breeding plumage (late summer) before beginning to regrow breeding plumage for fall and winter courtship. Eclipse phase is a complete molt, including flight feathers, so ducks become temporarily flightless at this time.

Eyebrow stripe or supercilium A stripe of contrasting shade running laterally above the eye

Eye ring A feathered ring of contrasting shade encircling or partially encircling the eye

Eye stripe or eye line A stripe of contrasting shade running laterally through the eyes on the sides of the face

Flanks The area along the bird's sides beneath the wings and above the base of the legs

Forehead The front of the crown, directly above the bill

Form A regional variation of a species with slightly different color or markings, but not one that has taxonomic status as a subspecies

Gorget A patch of iridescent feathers on the throat of male hummingbirds (some females and immature males have partial gorgets)

Lores The space between the eye and the base of the bill

GLOSSARY

Malar The cheek area

Malar stripe Cheek stripe

Mandible, lower Lower bill

Mandible, upper Upper bill (also correcty termed "maxilla")

Mantle Upper back; often used to describe a bird's back, tertials, and scapulars taken together

Moustache stripe or moustachial stripe A stripe of a contrasting shade resembling a moustache on the sides of a bird's throat

Nape Back of the neck

Necklace A band of contrasting shade running across the upper breast, as in the Varied Thrush

Orbital ring A ring of unfeathered skin surrounding the eye

Primaries Outer flight feathers, projecting from the outer half of the wing

Primary coverts The row of feathers that veils the basal portion of the primary flight feathers on the dorsal surface of the wing

Scapulars Feathers at the base of the dorsal wing surface along each side of the bird's back

Secondaries Inner flight feathers, projecting from the inner half of the wing

Secondary coverts Feathers covering the dorsal surface of the wing, exclusive of the secondary flight feathers; these comprise 3 rows: lesser secondary coverts, median secondary coverts, and greater secondary coverts, respectively, starting at the leading edge of the wing

Semipalmated Partially webbed toes

Sides The area on the sides of a bird forward of the shoulders

Spectacles White- or light-colored lore and eye ring combination

Speculum The row of colorful, often iridescent, feathers on the dorsal wing surface of many ducks

Subterminal band A band of contrasting shade across the tail, just above the tip

Supercilium See "Eyebrow stripe"

Supraloral The area on the face above the lores

Tail coverts and upper tail coverts Small dorsal feathers covering the base of the tail

Tertials Inner flight feathers at the base of the wings

Throat The area below the bill, between the chin and the upper breast

Transient A species found in an area between its breeding and nonbreeding ranges

Undertail coverts Ventral feathers covering the base of the tail feathers

Vagrant A bird that has strayed from its normal breeding range, wintering range, or migration route

Wingbars Stripes on the wing surface usually produced by differently colored wing covert tips

Wingtips The outer ends of the primary flight feathers

FINDING BIRDS

Trite as it may sound, birds truly are where you find them, and such a broad statement hints at an important strategy in bird-watching: if you want to see a certain kind of bird, go to where that bird is most likely to be found, at the time it is most likely to be there. For finding unusual and/or less-than-common species, it's best not to leave your fate to chance. Use this guide to study species you may want to see, then go to the right places at the correct times, armed with some knowledge about that bird's habits and habitat.

Conversely, many people simply enjoy visiting interesting places to see a variety of species, and perhaps even keep a trip record of all species found on each outing. Such field trips often lead to extraordinary places in the Pacific Northwest—a bay or estuary on the Oregon coast, a national wildlife refuge in the arid interior, a verdant woodland in Washington, a ponderosa pine forest in British Columbia, or a mountain summit in Idaho. Still others prefer a more passive strategy of adding birdwatching and bird identification as another layer to other outdoor activities. Whether you enjoy hiking, biking, boating, fishing, country driving, or other activities, identifying species seen along the way adds immeasurably to the intrigue.

When you visit new places in search of birds, familiarize yourself ahead of time with some of the key species that might occur there. This book and other resources can provide helpful information such as species lists generated by previous visitors and posted online, as well as checklists for some locations, especially wildlife refuges.

Throughout the Northwest, the mix of species changes with the time of year. Many migratory birds depart the region for the winter or return in spring, others arrive here to spend the winter, and yet others pass through along their migration routes between breeding ranges and wintering ranges outside the Northwest. Generally speaking, the diversity of species is greatest during spring and fall migration, when birds that breed in the Northwest arrive or depart at the same time that transient species are passing through.

The peak of spring and fall migration varies annually to some extent, and also varies with species and location. Some species that arrive in southwest Oregon in early March may not appear in northeast Washington until late May. As a generalization, however, spring migration—especially important to birdwatchers because many species arrive in prime breeding plumage—lasts from late February until early June, with much depending on species and location. Fall migration, when birds head south (or to the coast) for winter, begins as early as late July for some species and lasts into early winter for others.

In most of the Northwest, species diversity is at its lowest during winter, but winter also brings species that are uncommon or absent at other times of year. On coastal waters and shores, however, the diversity of species is often higher in winter than in summer because many birds winter on Northwest bays, estuaries, beaches, and nearshore seas.

Birds tend to be most active at specific times of day. For songbirds, morning is the time of frenetic activity, so birders should be in the field early to maximize opportunity. In warm spring and summer weather, the difference between early morning songbird activity and afternoon activity tends to be dramatic—the same location that was buzzing with activity at 8 a.m. may seem devoid of bird life at 1 p.m. Many other species are likewise most active in the morning; in fact entire wildlife communities, in spring and summer, tend to be liveliest in the morning.

Other birds might be more active at midday, such as raptors (birds of prey) that hunt by soaring and rely on thermals in the warming air to make for easy flight. Most owls and Common Poorwills are nocturnal, while birds such as nighthawks and swifts (which capture insects in flight) tend to be most active when flying insects are most abundant—often around dusk. In winter, when birds conserve body heat and energy, they may be most active at the warmest time of day.

Frequently, finding birds means entering their habitat as unobtrusively as you can. Some species are very tolerant of humans, even seemingly curious at times and thus easy to approach. But many birds are secretive and retiring by nature and may go undetected by birders who are not stealthy. So walk and talk quietly, stopping often to look and listen—birds often reveal their presence by sounds even if they remain well hidden. Consider other clues as well: bird tracks in the dust, droppings, feathers on the ground.

From well-known hotspots to places rarely visited by humans (let alone birders), the Northwest offers countless places to see a wide array of bird species.

Best of the Northwest

Following are sites throughout our region that offer particularly rich opportunities for birding. We make no promises, but if you travel to these areas and practice sound birdwatching behavior, the chances are high that your efforts will be rewarded.

Idaho

Camas National Wildlife Refuge and Market Lake Wildlife Area Diverse, expansive wetlands on the upper Snake River Plain provide nesting and migration habitat for a surprisingly robust list of birds. What birders don't find on the federal refuge, they'll likely find a short distance away at the state wildlife area.

Swans stage at Camas National Wildlife Refuge in eastern Idaho.

Coeur d'Alene Chain Lakes and Lane Marsh Extensive wetlands bordering forested uplands create a tremendous variety of species for birders by foot, car, or canoe.

Deer Flat National Wildlife Refuge The premier birding site in western Idaho, with expansive wetlands and waterways, riparian woodlands, and associated prairie and sagebrush uplands, Deer Flat serves as a critical migration stopover for myriad species, and provides nesting habitat for a variety of intriguing birds.

Grays Lake National Wildlife Refuge A rare and bird-rich example of a high-elevation wetlands and meadow ecosystem, Grays Lake is the world's largest hardstem bulrush marsh and supports the world's largest concentration of nesting Sandhill Cranes.

Deer Flat National Wildlife Refuge in southwest Idaho hosts a substantial population of nesting Clark's Grebes, as well as many other species.

Sandhill Cranes abound at eastern Idaho's Grays Lake National Wildlife Refuge.

Morley Nelson Snake River Birds of Prey National Conservation Area The massive Snake River canyon provides prime nesting habitat for a number of raptor species, while the surrounding riparian margins and sagebrush uplands are home to an impressive array of other birds.

Oregon

Klamath Basin Here you'll find extensive wetlands, including units of the Klamath Basin National Wildlife Refuge Complex as well as the serenely bucolic Wood River Valley, plus diverse forestlands—plan on 2 or 3 days to explore this extensive region.

Malheur National Wildlife Refuge One of the nation's premier birding sites, and an important staging area and breeding habitat for myriad waterbirds, Malheur is also a Northwest hotspot for rare vagrant species.

Metolius River Basin A ponderosa pine and mixed-species conifer forest, most of it within national forest lands, provides habitat for a wide variety of montane species, including 11 of the Northwest's 12 species of woodpeckers.

Upper Rogue River Valley From Ashland to Shady Cove to Grants Pass, diverse geology and habitat types support bird species absent from or difficult to locate elsewhere in the Northwest. Ample enticing public access sites make the birds easy to find.

Willamette Valley National Wildlife Refuge Complex Three separate refuges—Finley, Ankeny, and Baskett Slough—preserve a variety of wetland, prairie, and woodland habitats in the central Willamette Valley, creating superb habitat for an extensive mix of birds throughout the year.

The overlook at Vista Station provides a panoramic view of one small corner of southeast Oregon's Malheur National Wildlife Refuge.

Central Oregon's bird-rich Metolius Basin is home to 11 species of woodpecker, including the White-headed Woodpecker.

The Willamette Valley National Wildlife Refuge comprises 3 refuges that protect habitat for geese, waterfowl, and a diverse array of other species.

Yaquina Bay and Tillamook Bay Two bays, an hour apart on the Oregon coast, host an impressive quantity and diversity of migrating and wintering species, along with excellent access to prime habitat, including mudflats, sand flats, jetties, beaches, and more.

Washington

Blaine area and Drayton Harbor Incredible concentrations of migrating and wintering waterbirds and shorebirds use this expansive saltwater wetlands just south of the Canadian border.

Desert Unit, Columbia Basin Wildlife Area One of the premier sites east of the Washington Cascades, with wetlands amid desert-scrub and sagebrush steppe habitats, the colloquially named "Desert Wildlife

An intriguing land where water meets sagebrush steppe, east-central Washington's Desert Unit (Desert Wildlife Area) provides habitat for species such as the Yellow-headed Blackbird.

Oregon's bays are great places to find many different waterbirds; among the best bay locales are Tillamook Bay and Yaquina Bay (pictured).

The Grays Harbor area is renowned for prodigious numbers of migrating shorebirds.

Area" is home to an intriguing mix of resident and migrating birds.

Grays Harbor This massive bay complex attracts tremendous numbers and varieties of waterbirds and shorebirds (a term used for sandpipers, plovers, and related species), with excellent access for birders. Shorebirds are such a major attraction that the area is home to an annual shorebird festival.

Hurricane Ridge, Olympic National Park Drive from low-elevation, old-growth conifer forests all the way up to subalpine environs, and experience the changing mix of woodland species along the way.

Little Pend Oreille National Wildlife Refuge This little-known national refuge ranges from lowland lakes, ponds, and marshes up into forested mountains, meaning a

tremendous variety of birds, including species seldom seen in other regions of the Northwest.

Ridgefield National Wildlife Refuge Vast managed wetlands on the Columbia River plain host staggering numbers of waterbirds, making it a favorite among regional birders, especially from autumn through late spring.

British Columbia

Boundary Bay One of the great birding sites of the entire West, this sprawling estuary provides critical stopover and wintering grounds for countless birds, from tiny sandpipers to Snowy Owls.

Cathedral Provincial Park These intoxicating highlands rank among the best and most scenic locations in southern British

Hurricane Ridge provides both excellent birding and stunning views.

A male Hooded Merganser stretches its wings on a pond at Little Pend Oreille National Wildlife Refuge in northeast Washington.

A Great Blue Heron huddles in its roost at southwest Washington's Ridgefield National Wildlife Refuge.

Boundary Bay near Vancouver is one of the Northwest's premier birdwatching locations.

Scenic Cathedral Provincial Park offers birders a chance to find high-elevation species.

Columbia for finding high-elevation specialties, such as Spruce Grouse, White-tailed Ptarmigan, White-winged Crossbill, Boreal Owl, and Boreal Chickadee—among a host of other mountain species.

Creston Valley Wildlife Management Area More than 250 species have been tallied at this expansive site in southeastern British Columbia, and birders can also visit other prime sites nearby: Moberly Marsh, the Columbia River wetlands, and the Wasa Sloughs.

Okanagan Valley A verdant swath of wetlands, pine forests, quiet towns, and agrarian lands squeezed between rugged, forested uplands creates so much habitat diversity that birders need several days to sample this unique area.

Southern British Columbia's Okanagan Valley is home to many uncommon species, such as Pine Grosbeaks.

Fertile and rich in water, the Creston Valley Wildlife Management Area is one of British Columbia's top birding sites.

Victoria area waterfront and Saanich Peninsula Especially during migration and winter, incredible numbers and a variety of waterbirds and shorebirds abound in the rich habitat where southwest Vancouver Island meets the Strait of Georgia.

Ethics of Birdwatching

Enjoying birds in their natural habitats carries with it the essential responsibility to do no harm. Many birds are, to varying degrees, tolerant of human intrusion on their habitat and their daily activities, but birdwatching ethics mandate that we always do our best to limit our intrusion so as not to disturb birds unnecessarily. This is a fluid concept that takes on different meanings and parameters for different situations: a hummingbird feeding amid the flowers and foliage of a blooming tree is not easily put off and close approach won't likely disturb the bird. If the little hummer decides you are a bit too close, it can easily fly a few feet away to other flowers. But a raptor that has made a kill and is sitting on the ground or on a perch with its hard-won meal may fly off and leave its prey behind if you approach too closely, necessitating the hunt to begin anew, perhaps while hungry chicks wait back at the nest.

Seeing birds, however, often does require that we intrude upon their habitat, so it falls to common sense to manage our intrusion so as not to unduly disturb their activities. Just because an owl on a day roost, or a grouse feeding along the edge of a forest road, or a flock of terns on the beach might allow close approach does not mean birders should necessarily keep trying to inch closer and closer. Whenever possible, enjoy birds from a respectful distance. Luckily, many prime birding sites,

In some places, special restrictions limit access so that birds are not unduly disturbed. No matter where birders pursue their hobby, they must place the well-being of birds ahead of all else.

especially wildlife refuges, include observation stations and even photography blinds from which you can scan important locations without being intrusive.

Throughout North America (and the world), many bird species are critically imperiled, and others are in serious population decline. Birdwatchers can be important allies to scientists by providing empirical data. Events such as Christmas bird counts, breeding bird surveys, migration watches, and other opportunities to systematically quantify particular species help ornithologists build databases that show population trends. Additionally, birders can enjoy the opportunity to work directly with scientists on projects such as bird banding and site monitoring, and also to work with conservation organizations on projects ranging from observation and recording to habitat restoration and nesting box installation.

Various federal and state agencies are deeply involved with bird population monitoring, species protection, and habitat preservation and enhancement. As well, many private national and regional conservation organizations play important roles in habitat preservation and improvement, ornithology, and even species-specific population monitoring and conservation. See the Recommended Resources section at the end of this book for names and websites of government agencies and private organizations germane to birders.

Birds in the Backyard

Bird-friendly yards are great places to watch and study a variety of species, and attracting birds to your yard can be as simple as hanging a feeder, or as all-inclusive as installing a variety of bird feeders, watering stations, birdbaths, and bird houses. You can even landscape with specific plants that attract birds.

Many birds have specific food preferences, but high-quality seed mixes that include a high percentage of black-oil sunflower (in the shell, or shelled for a cleaner yard) attract a variety of songbirds. Certain species are especially fond of thistle seed, which can be contained in special thistle feeders; peanuts and other nuts can also be popular. Suet cakes and suet cylinders attract many species that don't normally frequent seed stations, including woodpeckers, and fruit stands can attract striking Bullock's Orioles.

The Northwest as a whole is breeding range for at least 5 species of hummingbirds, and all are easily attracted to special hummingbird feeders filled with a simple mix of sugar diluted in water (one cup of

American Goldfinches (left) and a Pine Siskin—two of the many species of birds that frequent bird feeders in the Northwest.

sugar per every 3 to 4 cups of water). Many hummingbird feeders are built so that bees and wasps cannot reach the nectar inside, and these are advantageous in areas with lots of such insects. Change the water often, especially in hot weather; it will last longer in shade than in direct sun. And keep those hummingbird feeders up and maintained during winter if you live in areas home to Anna's Hummingbirds, the only species that routinely stays year-round in the Northwest. Also, most hummingbirds, especially males, are territorial and will aggressively defend a feeder they have chosen as their own. So if you have more than one feeder, space them apart, and on different sides of the house.

Water is critical for birds throughout the year, but especially at times when little is available—during summer, and also during subfreezing conditions in winter, when a heated water source can provide

drinking water. A birdbath kept full of clean water is a great way to attract birds in summer and provides often-humorous entertainment, as various kinds of birds show up to splash about and take baths.

Be forewarned that backyard bird-feeding stations can attract more than just birds. Entire product lines are dedicated to strategies for keeping squirrels out of bird feeders; chipmunks, mice, rats, woodrats, and other rodents can be equally persistent; even deer frequent birdfeeding stations in many places. Rodents aren't there to eat birds, but rather to enjoy the bounty you provide. But bird feeders can also attract neighborhood housecats, and to counter the feline menace, you'll need to remain vigilant. If you own an outdoor cat, consider placing a little bell on its collar to

Birdhouses can provide nesting homes for a wide variety of species, and in some cases, boxes provide valuable habitat where natural cavities have been lost to land-use changes and invasive species.

Throughout the Northwest, hummingbirds are easily attracted to sugar-water feeders, and are always a joy to watch.

help warn birds away—many cats quickly learn to operate effectively even with the bell, so watchfulness on your part is the best protection for birds.

Throughout the Pacific Northwest, whether you live in the city or country, certain bird-eating raptors, especially Sharp-shinned Hawks and Cooper's Hawks, may include your backyard in their hunting grounds, and while a hawk materializing out of nowhere and snatching a songbird from your feeders in a puff of feathers comes as a sudden and disconcerting surprise, it also comes with the territory. In backyard bird feeding, sometimes you must simply take what nature gives you.

Other birds can likewise become problematic for other reasons. Flocks of starlings, blackbirds, and various other species can mob bird-feeding stations, rapidly eating everything in sight and aggressively driving off other species. If these species become too bothersome, you may have to stop feeding the birds for a week or two, although even then the pest birds tend to return quickly once you resume feeding. Also, woodpeckers—especially Northern Flickers—can damage buildings. Often they and other woodpeckers drum on buildings, including metal drainpipes and gutters, as a territorial announcement, but they may also drum into wood structures in search of food or nesting sites. Swallows, American Robins, Starlings, House Sparrows, and other species often build nests in crevices and platforms on buildings, and these nest sites can quickly become messy.

Insects can become a problem as well, ranging from ants to bees and even

Bird-feeding stations often attract Sharp-shinned Hawks such as this beautiful adult, as well as Cooper's Hawks, both of which feed on smaller birds.

notoriously aggressive yellow jackets. If oriole feeders or suet feeders begin attracting insects, you may simply need to remove them, at least for a time. In some areas, you can turn the tables on insects by erecting houses for insectivorous swallows and Purple Martins, and even bats for that matter.

All told, though, the good usually outweighs the bad in creating a bird-friendly yard, and your property can become an attractive refuge for migrating, wintering, roosting, and nesting birds if you provide for some or all of their basic needs: food, water, and cover.

USING THIS BOOK TO IDENTIFY BIRDS

Birds of the Pacific Northwest provides the essential information needed to help you identify each of the bird species found within this region, including the vagrant species (birds occurring outside of their normal ranges) found with the most regularity. The photos, range maps, and information provided in the species accounts all work in unison to help you identify the birds you see.

Like all living things, birds are organized scientifically through the system called taxonomy, which seeks to categorize organisms based on their natural relationships. Taxonomy classifies organisms on an ever-narrowing scale, beginning with their broadest relationships (for example, animal or plant?) to their closest relationships (species and then subspecies, for instance). The scale begins with kingdom (animal kingdom and plant kingdom are examples). Next, respectively, come phylum, class, order, family, genus, species, and subspecies. This book adheres to taxonomic classification in the grouping of species included herein, and generally this book is also organized to follow the order in which groups and species of birds are presented by the American Ornithologists' Union (AOU) *Checklist of North and Middle American Birds*. We have deviated from that order in a few cases. For example, while the AOU lists falcons well removed from the hawks and other raptors, we list them in the same section of this book for the simple reason that it seems logical for a field guide to group all the birds of prey together, so readers can more easily compare species that might have similar appearances in the field.

In all cases, *Birds of the Pacific Northwest* adheres to the accepted North American common names for all species, but birders should be aware that many species are known by other names, some of which are simply nicknames, such as "Hungarian partridge" or "Hun" for the Gray Partridge. Others are cases of pure colloquialism based not so much on misidentification as on uninformed but harmless local or regional tradition, such as both the California Scrub-Jay and Steller's Jay often being labeled simply "blue jays" by nonbirders, even though the Blue Jay is a separate species altogether. The surest way to avoid confusion over common names is to also include Latin scientific names in a field guide such as this—so while scientific names may not be particularly useful to most birdwatchers, they can be critical to scientists who might otherwise need to laboriously grapple with multiple monikers for a single species. Scientific names appear in parentheses after the common name for each species in the Bird Species Accounts chapter. To separate bird groups for easier referencing, we have color coded page edges. However, colors are not unique to individual bird groups and are repeated.

Photos Throughout this book, we have chosen photos that best depict the key features needed to identify each species. Beyond normal photo editing for light and color detail, we have opted not to alter the images in any way that misrepresents the

The distinctive pattern of this sandpiper's back, tail, and wings readily identifies it as a Surfbird.

photographer's work and vision. In a very few cases, we cloned out objects (such as limbs) that might obscure key parts of the bird in a photo; otherwise we have left natural elements unaltered. Many birds change little throughout the year, but others molt into different seasonal plumages; so too, sexual dimorphism (males and females differ) is common among birds, and juvenile plumages can differ markedly from adult plumages. Also, some birds have distinct color variations within the species, independent of age or sex. In all cases, we have included photography to represent the plumage stages and variations likely to be seen. Within each species account, the photos and text are intended to be used together to allow you to confidently identify the birds you see.

Range maps The geographical area in which a particular species occurs is known as its range. Range provides valuable infor-

mation for identifying birds, and in a few cases, range is essentially the only way to distinguish between lookalike species. The Juniper Titmouse and Oak Titmouse, for example, are identical, and probably only co-occur in Oregon's Klamath Basin. So if you see a Titmouse in southern Idaho or southeast Oregon, you can safely identify it as a Juniper Titmouse, and if you see one in Oregon's Rogue River Valley, you can rest assured that it is an Oak Titmouse.

The area in the Northwest occupied by each species can be quite static over time, changing little with the seasons or the decades, or be substantially dynamic. Some birds—Crossbills, for example—are notoriously nomadic; others reliably do not stray from their ranges in the region. Some species are expanding their ranges rapidly in the Northwest, such as the Eurasian Collared-Dove and Barred Owl; others are dwindling in population, and consequently, range. The maps in this book, color coded for seasonality (see Range Maps Key) are as accurate and up to date as possible, though birders should expect to occasionally find various birds outside their normal ranges—indeed, such sightings frequently stir excitement in the birdwatching community and are often rapidly shared via websites and internet forums. They are part of what makes birdwatching fun.

Birds that occur well outside their normal ranges are termed vagrants, and in this book, we have chosen not to include maps for the vagrant species described and shown, simply because in most cases they can and do show up nearly anywhere in the region, but only very rarely. Of course some vagrants are more likely to occur in broad general areas. For example, pelagic (open ocean) vagrant species are generally only seen out at sea, and certain vagrant songbirds are perhaps more likely to be found in specific general areas, such as the northernmost coverage area of this book, or west of the Cascades, or well inland. These tendencies are based on decades of sightings, but it pays to remember that sightings of rare birds occur most often in places visited most frequently by birders—in other words, having lots of eyes out there increases the chance that someone will find an oddity. But vagrants can and do show up in locations visited by few people, so of course by extension, they must also show up in many places where no one sees them at all.

Length, Wingspan, Sexes similar, Adult, Male, Female, Juvenile These headings and other variations describe the size, color, pattern, and shape of the bird. Measurements are given in length, and where useful, the wingspan measurement is provided (wingtip to wingtip). "Sexes similar" indicates that both sexes are largely identical, varying only slightly in brightness of plumage or size, but otherwise sharing the key physical attributes described. When male, female, and juvenile physical characteristics differ, they are called out as such. Plumage phases are also noted where relevant, such as breeding (spring and summer) and nonbreeding (fall and winter) plumage, juvenile plumage, and plumage variations. These descriptions focus on key field marks for identifying each species (and supspecies where appropriate) and call attention to details that help separate similar-looking species.

Voice Describing bird songs phonetically is both subjective and challenging. To arrive at our descriptions, we listened to as many recorded samples as possible, as well as live examples in the field. Certainly other listeners might translate songs and calls into different verbiage than we have done, but our hope is to provide concise enough descriptions of vocalizations (and some nonvocal sounds) to allow you to compare what you hear in the field with our descriptions in an effort to aid in identification. Punctuation in our bird song and call descriptions is related to cadence. For example, *chip chip chip* conveys a song with a distinct pause between notes, *chip-chip-chip* describes a song with indistinct pauses between notes, and *chip-chipchip* describes a rapidly delivered song with no discernible pause between notes. Many birds vary their cadence so that a song might be described as *chip-chip chip-chipchip*. Additionally, we describe emphasis in some songs by using capital letters, such as the song of the House Sparrow, which we describe as "...slightly wavering *cheeee-up cheeeeup* often punctuated by a screechy *terweeeeEEE* twitter call."

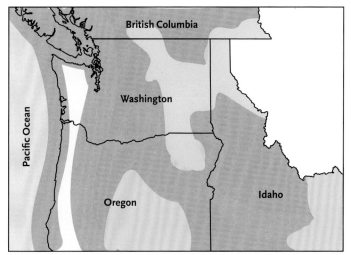

Summer Range

Winter Range

Year-round Range

Migration Range

The area covered by this book—Pacific Ocean nearshore, southern British Columbia, Oregon, Washington, and Idaho—with sample seasonal and migration ranges. (Montana and southern Alberta are not included.)

Summer Range Summer range generally indicates the area where birds breed and rear young, so in this sense, summer means the timeframe spanning from the arrival of the species in question in spring to its departure in fall. This timeframe often ranges from mid-spring to late summer, but varies by species and location.

Winter Range Many birds spend only the winter in the Northwest, typically dispersing southward from breeding ranges farther north. Some winter visitors consistently and predictably occupy the same areas at the same time each year, while others are highly variable and unpredictable.

Year-round Range A variety of bird species live in the Northwest year-round. In some cases this represents stationary populations, and in other cases it fully or partially represents replacement populations, wherein breeding birds of the species in question depart the area but are replaced by individuals of the same species migrating or dispersing in from other places.

Migration Range Migration patterns vary among species, but in most cases they are at least fairly predictable and consistent. These consistent migratory routes are mapped for species that occur as migrants in the Northwest. However, birds that occur only as summer residents or winter residents are also, by definition, migrants, as they migrate in and out of the Northwest seasonally. So, for example, if a bird winters along the Columbia River but breeds in the Far North, it must spend time in transit between those places, so appropriate habitat could attract it to locations not shown on the maps herein so long as the locations sit within the assumed migration route.

Behaviors Some bird behaviors are diagnostic, allowing certain or near-certain identification even if you don't benefit from a close look at the bird's color and pattern. Examples include the way a Wilson's Snipe flushes, the way a Belted Kingfisher hovers and then plunges into the water, and the way a Vaux's Swift flies on stiff, rapidly beating wings unlike the flap-and-glide style of a swallow. In the bird accounts herein, we strive to offer behavioral traits that can aid in identification, especially flight characteristics and feeding habits.

Habitat While some birds are habitat generalists, able to thrive in a variety of places, others are quite specific in their habitat requirements, offering not only a clue to identification based on where you see them, but also allowing you to know where to look if you want to see a certain species. The Northwest features numerous specific habitat categories; in fact, the Northwest Habitat Institute recognizes 32 different "wildlife-habitat types," with just a few examples including "montane mixed conifer forest," "ponderosa pine forest and woodlands," "shrub steppe," "herbaceous wetlands," "coastal dunes and beaches," and "marine nearshore." This book uses some of those same categories in describing habitat that supports certain birds, but also relies on more general as well as more specific habitat descriptions, depending on the species.

Status This heading describes the relative abundance of a bird species, ranging from abundant to very rare. Additional descriptors include the words "local" or "locally," which, when combined with a term of relative abundance, indicates that a species is rarely found outside a specific area. A good example in the Northwest is the Scott's Oriole, which is a rare summer resident that appears only in a single small, isolated mountain range in southern Idaho; hence, the bird is described herein as "very rare and local." As previously noted, some species are described as "vagrant," which means they are outside their normal ranges when seen in the Pacific Northwest, including outside of their normal migration paths. By definition, vagrant species are rare, and in choosing which vagrant species to include in this field guide, we opted to leave out species that have only been recorded a handful of times in the region. Status also describes the general time of year you can expect to find each species in the Northwest. Some birds are year-round residents, while others are summer residents (breeding), winter residents, spring and/or fall migrants, or nonbreeding visiting species that are most likely to occur in the Northwest at specific times of year. The climate and geography of the Northwest differ dramatically between the west side of the Cascade Range and the east side of this formidable spine of mountains, so in many cases the status of birds differs substantially west and east of the Cascades. So, for example, the Snowy Egret's status is an "uncommon summer resident and migrant in southeast Oregon and southern Idaho; rare but regular summer resident and migrant in eastern Washington. Rare migrant and winter visitor to western Oregon and Washington."

Such specific status details are only possible because of the defined, limited geographic scope of this book—the Northwest—and because we, the authors, are lifelong residents of the region, able to draw on decades of experience with Pacific Northwest birds. Throughout *Birds of* the *Pacific Northwest*, we have noted such differences in relative abundance and temporal status for many species to help you better discern the likelihood of finding specific birds in certain locations at specific times of year. Moreover, in some cases, we are able to provide detailed information on locations, such as in the case of the Greater Scaup: "Fairly common winter resident and migrant; large numbers winter in Puget Sound, the Victoria and Vancouver, British Columbia, areas, and the Columbia River; occasional summer visitor in the Columbia River mouth and coastal waters."

Key sites For many species described in this book, we have included a short list of key sites where these species are likely to be found. We chose these locations based on frequency of sightings, but in most cases, the bird in question can be found at many other sites in the region. Nonetheless, birders seeking to find specific birds will find these key site suggestions a valuable resource. In choosing each site,

we considered the likelihood of finding the bird in question at that location, so if you visit a key site at the proper time, you should stand a good chance of finding that species. Naturally you must still seek the specific habitat preferred by the bird you wish to see within that key site, and must also consider the bird's habits. Please note that when mentioning national wildlife refuges, we have used the abbreviation N W R.

Similar species For 16 species described herein we have added a brief description of a bird close in appearance to the species in question, primarily to describe very rare vagrants that look similar but which are not common enough to merit full species accounts. We also use this subhead to describe the Northwestern Crow (essentially identical to the American Crow); the Eurasian Skylark, which is superficially similar to the widespread Horned Lark, and comprises a small introduced population in Victoria, British Columbia; and to provide some additional help in differentiating between Golden Eagles and juvenile Bald Eagles.

BIRDS *of the*
PACIFIC NORTHWEST

Ducks, Geese, and Swans

Commonly and collectively called waterfowl (order Anseriformes), ducks, geese, and swans come in myriad varieties and sizes, ranging from the tiny Green-winged Teal, which weighs a scant 2/3 of a pound or less, to the massive Trumpeter Swan, which can weigh 30 pounds. Swans and geese are large and stately; ducks exhibit some of the most striking and colorful patterns in the avian world, making them favorites among birdwatchers. During migration and winter, most waterfowl form flocks, and the Northwest hosts massive congregations of numerous species, such as the thousands-strong flocks of Snow and Ross's Geese that annually stage in several Northwest locations. Such flocks rank among the must-see spectacles in birdwatching.

Hooded Merganser

Greater White-fronted Goose
(*Anser albifrons*)

Harkening the arrival of autumn, Greater White-fronted Geese typically begin showing in the Pacific Northwest in September after summering on their Arctic breeding range. Their distinctive call can be heard overhead at any time of day or night during migration. Oregon's Klamath Basin is a critical staging area for the entire West Coast population of this species.

LENGTH 26–31 inches. **WINGSPAN** 53–60 inches. Grayish brown overall, with buffy breast and belly marked with large irregular black splotches or bars; pinkish bill, with white around face at base of bill, forming low white forehead; orange legs; ventral area is bright white; black rump, and white tail with broad black subterminal band and thin white tip. **JUVENILE** Similar to adult but not as strongly marked, and usually lacking the dark belly mark; paler legs; minimal white around base of bill. **VOICE** High-pitched, slightly raspy, laughing *caa-AAR-aaha*, as well as similar-toned single notes; overhead flocks emit a cacophony of such intermixed calls. **BEHAVIORS** Frequently grazes in open fields; often migrates at night; migrating flocks often fly very high. **HABITAT** Large agricultural fields, pastures, shallow wetlands, estuaries. **STATUS** Fairly common migrant, uncommon winter resident. **KEY SITES** Idaho: Lower Boise River (Parma area). Oregon: Klamath Basin; Willamette Valley NWR Complex. Washington: Nisqually NWR; Ridgefield NWR. British Columbia: Saanich Peninsula and Victoria area.

Greater White-fronted Goose

Greater White-fronted Goose in flight

Greater White-fronted Goose range

Snow Goose (*Chen caerulescens*)

Thousands of Snow Geese taking to the air is one of the iconic sights in Northwest birding, and this phenomenon occurs regularly in key wintering and staging sites in the region. Snow Geese, headed for the Arctic nesting grounds, depart the Northwest by early spring; best viewing time for large flocks west of the Cascades is November through March; in the interior Northwest, highest concentrations occur from March through mid-April.

LENGTH 27–32 inches. **WINGSPAN** 53–56 inches. Pure white, with black primary flight feathers; fairly large pinkish bill has distinct black "grin patch," separating this species from the very similar Ross's Goose; head and upper neck feathers occasionally lightly stained from feeding in mud; juvenile birds show light gray wash or patches. Morph describes a color variation within a species; rarely, a dark morph Snow Goose ("Blue Goose") occurs with flocks in the Northwest. Coloring will be steel-gray overall, or a steel-gray back and collar and white breast and belly; base of neck gray, head and neck white; white-edged black wing coverts; in flight, silvery gray wings with dark primaries; juvenile is ashy gray overall. **VOICE** High, hornlike honks, and low grunting honks, with various pitches in between. **BEHAVIORS** Migrating and wintering flocks typically feed in densely packed large flocks; in flight, they form wavering V shapes. **HABITAT** Large agricultural fields and fallow, pastures, shallow wetlands, estuaries. **STATUS** Locally abundant migrant in eastern Oregon and Washington and southern Idaho; locally common winter resident in Skagit Valley, Washington; uncommon migrant and winter resident elsewhere. **KEY SITES** Idaho: American Falls Reservoir; Camas NWR; Fort Boise Wildlife Area; Roberts area (Jefferson County). Oregon: Malheur NWR; Silvies River floodplain (Burns). Washington: Skagit River Delta. British Columbia: Richmond.

Snow Goose range

Snow Geese in flight

Snow Goose

Dark morph Snow Goose ("Blue Goose")

Ross's Goose (*Chen rossii*)

Noticeably smaller than the Snow Goose, with which it often occurs, the diminutive Ross's Goose takes a migratory path that does not include Washington and British Columbia, so it is a rare sight in much of the Northwest. Oregon's Klamath Basin as well as Silvies River floodplain (Burns) host the largest concentrations. Large migratory flocks stage in these areas, primarily in early spring.

LENGTH 23–25 inches. **WINGSPAN** 47–51 inches. Pure white, with black primary flight feathers; stubby pinkish bill lacks the obvious black "grin patch" of the very similar Snow Goose. **JUVENILE** Slight gray wash over mantle and crown. Very rarely, blue morph Ross's Geese occur in the Northwest (see description of dark morph Snow Goose). **VOICE** Abrupt grunting honks and a high-pitched, rusty *kee* and *kee-kee*. **BEHAVIORS** During feeding and staging, often occurs within or adjacent to flocks of Snow Geese; typically forms wavering V shapes in flight. **HABITAT** Large agricultural fields and fallow, pasturage, shallow wetlands. **STATUS** Locally common migrant (March–April) in Oregon; uncommon in Idaho; rare in Washington and British Columbia. **KEY SITES** Idaho: Camas NWR; Roberts area (Jefferson County). Oregon: Klamath Basin; Malheur NWR; Silvies River floodplain (Burns).

Ross's Goose

Ross's Goose range

Ross's Geese in flight

Emperor Goose (*Chen canagica*)

A goose of the high Arctic that winters mostly on the Alaska Peninsula and Aleutians, the Emperor Goose is a rare winter straggler to the Northwest, with a few individuals appearing almost annually, typically in saltwater environs, but with occasional sightings in the Columbia River corridor east to Portland, and in the Willamette Valley.

LENGTH 26 inches. **WINGSPAN** 46 inches. Gray overall, with distinct scaled appearance, and diagnostic white head and nape in adult, with gray foreneck and chin; dark undertail coverts and ventral surface; large, stubby, pinkish bill; yellow-orange legs; juvenile has gray-flecked head and nape. **VOICE** High-pitched barking yelps and low, drawn-out growling grunts. **BEHAVIORS** Generally feeds on intertidal invertebrates and marine vegetation, such as eelgrass and sea lettuce. **HABITAT** Estuaries, saltmarshes, bays, agricultural fields, pastures. **STATUS** Rare winter vagrant, often within flocks of other waterfowl.

Emperor Goose

Brant (*Branta bernicla*)

A saltwater-specific, medium-sized goose, Brants nest in the high Arctic, and return to the Pacific coast as far south as Baja California for the winter and early spring. Oregon's wintering population totals only about 3000 birds, while Washington hosts some 25,000, and British Columbia about 8000.

LENGTH 22–26 inches. **WINGSPAN** 42–47 inches. Blackish brown above; blackish brown belly (lightest in Western High-Arctic breeding population that winters primarily in Puget Sound); black breast, neck, and head, with partial white neck ring; silvery white flanks; white rump and ventral area. **JUVENILE** Less strongly patterned, with gray flanks. **VOICE** Raspy, grunting honks. **BEHAVIORS** Feeds by grazing on tide-exposed eelgrass and other vegetation; very fast and direct flight, in staggered lines. **HABITAT** Estuary environments, especially eelgrass flats and shallows in bays; also salt-marshes, saltwater mud flats, coastal lakes, pasturage. **STATUS** Common winter resident (November–May) in Washington and British Columbia; locally uncommon in Oregon. **KEY SITES** Oregon: Netarts Bay and Tillamook; Yaquina Bay. Washington: Dungeness Spit and Dungeness NWR; Padilla Bay and Fidalgo Bay; Willapa Bay. British Columbia: Boundary Bay; Parksville area and Qualicum Beach.

Brant

Brant range

Brant in flight

Cackling Goose
(*Branta hutchinsii*)

The Cackling Goose, also known as the Cackling Canada Goose and colloquially as "cackler," was, until 2004, considered part of the Canada Goose species. This species now co-opts the 4 smallest former Canada Goose subspecies; all of them nest in the Arctic, but winter to varying degrees in the Northwest. *Branta hutchinsii minima*, the traditional cackler, is the smallest, but size alone (because of substantial overlap) is an ineffective field mark in separating Cackling Geese from Canada Geese. Bill size helps substantially in identifying *minima*, less so in the other cackler subspecies detailed below. All told, differentiating the various subspecies of Cackling and Canada Geese is often very difficult, but a tiny specimen not much larger than a Mallard, and with a short, stubby bill, is safely identified as a Cackling Goose.

Cackling Goose range

LENGTH 22–30 inches. **WINGSPAN** 42–48 inches. Brownish above, pale grayish tan below; black head and neck with distinctive white cheek patches that form chinstrap; black tail with wide white subterminal band. *Cackling Goose*: 3–4 pounds; dark overall; darker breast often with a purplish cast; small, stubby bill and short neck; sometimes a white neck ring; diagnostic high-pitched yelping ("cackling") call. Common in Willamette Valley, Oregon, north to Columbia River. *Taverner's Cackling Goose*:

Cackling Goose in flight

Cackling Goose

4.5–6 pounds; light brown above, with silvery gray breast; no neck ring; short, thick neck; steep forehead in contrast to nearly inseparable Lesser Canada Goose. Common in western Oregon and western Washington. *Aleutian Cackling Goose*: 4.5–5.5 pounds; broad white collar at base of neck is complete or nearly so; chinstrap usually divided by narrow black band (difficult to see); short neck, and blocky, square-ish head; fairly stubby bill; dark brown above; dark breast, often with purplish cast.

Locally common along coast. *Richardson's Cackling Goose*: 3–7 pounds; complete white chinstrap, no neck ring; pale breast and light color overall; short neck, stubby bill. Rare in Northwest. **VOICE** Distinctive honking (high-pitched yelping in *minima* subspecies). **BEHAVIORS** Migrating and wintering Cackling Geese readily mix with Canada Geese and other waterfowl. **HABITAT** Fields, pasturage, marshes, estuaries, lakes, mudflats, parks. **STATUS** Common migrant and winter resident.

Cackling Geese

Canada Goose
(*Branta canadensis*)

Comprising several subspecies, the Canada Goose is one of the most familiar birds in the Northwest, with year-round populations in or around many cities and towns as well as in many other locations in the region. From fall through spring, numbers are bolstered further by migrants and wintering flocks. The Northwest is home to the Western Canada Goose (*Branta canadensis moffitti*), the most widespread and common, even introduced in many areas; the Lesser Canada Goose (*B.c. parvipes*), which winters in much of the region; the Vancouver Canada Goose (*B.c. fulva*), which lives along the British Columbia and southeast Alaska coast, with occasional birds wandering down to Washington and Oregon; and the Dusky Canada Goose (*B.c. occidentalis*), which winters primarily in Oregon's Willamette Valley, where a complex of federal refuges were created specifically for these migratory winter residents. Identifying subspecies and separating smaller Canada Geese from Cackling Geese often requires close scrutiny, especially because some subspecies readily hybridize, and because Canada Geese hybridize with Cackling Geese.

LENGTH 25–45 inches. **WINGSPAN** 45–60 inches. Brownish to grayish tan overall, typically lighter on breast and belly; long black neck and head; distinctive white chinstrap; black tail with wide white subterminal band. Subspecies range from the very dark medium-sized Dusky Canada Goose and the light colored medium-sized Lesser Canada Goose, to the dark-colored medium- to large-sized Vancouver Canada Goose, and the light colored very large Western or Great Basin Canada Goose. **VOICE** Familiar nasal honking. **BEHAVIORS** Frequently feeds in flocks on agriculture fields, especially grain stubble, alfalfa, and grass seed, as well as golf courses, parks, playing fields, and other large lawns. **HABITAT** Marshes, estuaries, lakes, rivers, fields, farmlands, pasturage, parks. **STATUS** Abundant year-round resident, migrant, winter resident.

Canada Goose range

Canada Geese

Trumpeter Swan
(*Cygnus buccinator*)

Largest of North America's waterfowl, the Trumpeter Swan—most common in winter and migration—breeds in limited numbers at a few specific locations in Idaho and Oregon. The Oregon breeding population, just a few pairs spread between Malheur National Wildlife Refuge and Summer Lake, arose from birds captured in and transplanted from Idaho.

LENGTH 57–60 inches. **WINGSPAN** 84–96 inches. Very large; pure white, with black legs; black bill has red border on lower

Trumpeter Swan

mandible (sometimes present on Tundra Swan, but pale pink); neck longer and bill larger than in Tundra Swan, and head and bill slope profile is more angular; black base of bill envelopes eye; bobs neck and head up and down, especially when disturbed and when a flock is ready for takeoff, whereas Tundra Swan only bows head. **JUVENILE** Pale grayish brown or pale gray-white overall, darkest on head and neck; pinkish bill. **VOICE** Loud, rich, trumpetlike single- and double-note honks. **BEHAVIORS** Rather maladroit on land, with a pronounced slow waddling as compared to the sprightlier Tundra Swan. On takeoff, holds neck very briefly in a shallow S curve; strong, fast flight, typically in V shapes when aloft. **HABITAT** Large agricultural fields and fallow, pastures, shallow wetlands, estuaries. **STATUS** Locally common winter resident; uncommon migrant; locally uncommon breeder in Idaho; locally rare breeder in Oregon. **KEY SITES** Idaho: Harriman State Park and Island Park. Oregon: Malheur NWR; Sauvie Island. Washington: Chehalis River Valley; Skagit Valley. British Columbia: Comox Valley (Vancouver Island).

Trumpeter Swan range

Trumpeter Swan in flight

Tundra Swan
(*Cygnus columbianus*)

Large and stately, Tundra Swans are nearly identical to the larger and less common Trumpeter Swans. Where they occur together—Washington's Skagit River Valley, for example—side-by-side comparison allows birders to more easily notice the size difference in the species, as well as the smaller bill profile of the Tundra Swan. Most adult Tundra Swans also have yellow lore spots that can vary in size; and the Tundra Swan's black eye tends to be distinct from the black base of its bill, unlike in the Trumpeter Swan. The infallible way to differentiate the Tundra and Trumpeter Swans is by their calls.

LENGTH 48–54 inches. **WINGSPAN** 72–80 inches. Very large; all-white, with black bill and legs; at close range, adult usually shows yellow spot on lores; shorter neck and more rounded head profile than Trumpeter Swan, with smaller bill, and eyes more distinct from base of bill than in Trumpeter Swan. **JUVENILE** Pale grayish tan overall; pinkish bill. **VOICE** High-pitched quavering honks, typically one to 3 syllables. **BEHAVIORS** More agile on land than the Trumpeter Swan; on takeoff, holds neck straight the entire time; strong, fast flight, typically in V shapes as flocks. **HABITAT** Large agricultural fields and fallow, pastures, shallow wetlands, lakeshores. **STATUS** Fairly common migrant; locally common winter resident. **KEY SITES** Idaho: Coeur d'Alene basin; Hagerman Wildlife Area and Hagerman Valley; Market Lake Wildlife Area and Camas NWR. Oregon: Klamath Basin (Lower Klamath Lake Road, Stateline Road, Township Road); Sauvie Island; Willamette Valley NWR Complex. Washington: Ridgefield NWR; Samish Flats; Skagit Wildlife Area. British Columbia: Creston area; Okanagan Valley.

Tundra Swan

Tundra Swan range

Tundra Swan in flight

Wood Duck (*Aix sponsa*)

Cavity-nesting Wood Ducks rely on natural holes or woodpecker holes in trees, and have benefited from widespread installment of nesting boxes. By the early 1900s, this species had been nearly exterminated by habitat destruction and unregulated hunting, but protections, including regulations on harvest, stemmed the tide and populations rebounded.

LENGTH 18 inches. **WINGSPAN** 30 inches. Long tail; blue-green speculum with white rear border. **MALE** Combination of brilliant colors, with iridescent greenish purple head accented by white chin with 2 white stripes reaching up along the face and neck, and another running up the forehead from the bright red base of the multicolored bill and extending along the puffy, drooping crest; pale yellow flanks bordered above with brilliant black then white bars; chestnut breast with white specks separated from flanks by distinct vertical white bar then black

bar; blackish back. Eclipse male resembles female but retains white chinstrap stripes. **FEMALE** Grayish overall, flecked with white; distinct white eye ring tapering back into soft, drooping gray crest. **VOICE** Male produces a high, rising, whistled *zaaweeeet*. Female call is a high, wailing *waak waak waak* or *aweek aweek aweek*. **BEHAVIORS** Lounges and preens perched on logs and limbs protruding from or hanging over the water. Detecting an intruder, often swims quietly away by slinking through lily pads and other aquatic vegetation. Direct and fast flight when aloft; highly agile in flying through timber. **HABITAT** Secluded, timbered ponds, sloughs, swamps, lakes; small flocks can occur on more open water during migration; will feed in oak groves (acorns) and filbert (hazelnut) orchards. **STATUS** Fairly common year-round west of Cascades; uncommon interior.

Male Wood Duck

Wood Duck range

Female Wood Duck

Gadwall (*Anas strepera*)

Though fairly widespread across the
Pacific Northwest, the Gadwall is best
known as the common dabbling duck of
the northern Great Basin and Columbia
Plateau, where they usually nest on virtually every marsh, pond, and lake margin with ample cover. Despite being quite
common in their nesting range, Gadwalls
are easily overlooked because their subdued plumage competes against more
brilliantly colored ducks for the attention
of birdwatchers.

Male Gadwall

LENGTH 21 inches. **WINGSPAN** 33 inches.
MALE Rich grayish overall, with fine vermic-
ulation visible at close range; black rump;
head more brownish gray; white upper
wing patch in flight; orange legs and feet,
black bill; eclipse male resembles female.
FEMALE Mottled brown overall, somewhat
darker than the similar female Mallard;
white wing patch in flight; bill blackish on
top with dirty orange edges, thinner than
Mallard's bill. **VOICE** Flush and flight calls
are a series of husky quacks; male's calls
include a truncated quack and high-pitched
whistles. **BEHAVIORS** Courtship among Gadwalls occurs in fall and early winter, and
pair bonds are established then, well ahead
of the spring nesting season. Fast, direct
flight; launches straight up from water's surface. **HABITAT** Freshwater marshes; ponds
and lakes with emergent vegetation; flooded
fields. **STATUS** Locally common summer resident and migrant; uncommon winter visitor
west of Cascades.

Female Gadwall

Male Gadwall in flight

Gadwall range

59

American Wigeon
(*Anas americana*)

Among the most common Northwest ducks during migration and winter, especially in the valleys west of the Cascades and along the coast, American Wigeons often form large flocks and congregate at prime feeding locales with Northern Pintails, Mallards, and Green-winged Teal. They are equally at home on flooded fields and saltwater estuaries.

LENGTH 19 inches. **WINGSPAN** 32 inches. Relatively short, black-tipped, bluish bill. **MALE** Pale rosy brown breast, flanks, and upper back; speckled grayish cheeks and neck, with broad, iridescent green, crescent-shaped eye stripe extending down the nape, and distinctive ivory forehead and crown. Male with eclipse plumage resembles female. **FEMALE** Mottled grayish brown above; pale grayish rust breast, finely speckled, usually contrasts subtly but noticeably with speckled grayish head and neck. **VOICE** Male's is a high, clear series of single-note whistles, *pipe pipe pipe*; also a distinctive wavering musical whistle, *zwee-ZI-zi*. Females make deep, husky, grunting quacks, more persistent as a flush call. **BEHAVIORS** Outside of breeding season, often feeds by grazing in open fields, croplands, shortgrass prairielike habitats, and even park and golf-course lawns (their bill is ideally adapted to this feeding strategy). Fast, direct flight; launches straight up from water's surface. **HABITAT** Breeds on freshwater marshes, sloughs, ponds; migration and winter habitat includes marshes, ponds, lakes, rivers, estuaries, mudflats, tidal creeks, fields, and wet prairies. **STATUS** Common migrant and winter resident; uncommon summer resident.

Male American Wigeon

Female American Wigeon

American Wigeon range

American Wigeon flock

Eurasian Wigeon (*Anas penelope*)

Male Eurasian Wigeon

Eurasian Wigeon range

Closely related to the similar American Wigeon, the Eurasian Wigeon, as its name suggests, is native to the Old World, where its massive breeding range stretches from coast to coast across Eurasia. Its wintering range, likewise geographically expansive, includes Europe, Asia, Africa, and, to a limited extent, both coasts of North America. Eurasian Wigeons occur annually in the Northwest, and while more common along the coast and in the valleys from western Oregon to southwest British Columbia, they usually show up annually in the interior of the region as well. The female Eurasian Wigeon is so similar to the female American Wigeon that unless Eurasian Wigeon drakes are present as well, the Eurasian hens may go unnoticed by observers.

LENGTH 19 inches. **WINGSPAN** 32 inches. Relatively short, black-tipped, bluish bill. **MALE** Silvery gray back and flanks, pale rose-red breast; rufous head with cream-colored forehead and crown. **FEMALE** Very similar to female American Wigeon, but head is plain, warm brown, often almost rufous (instead of grayish) and bill lacks the black edge along the gape usually present on the American Wigeon; wing linings uniformly gray instead of white. **VOICE** Male call is a distinctive fast, emphatic, accelerating musical whistle, *zaaweeEEEEL*. Female produces deep, husky grunting calls. **BEHAVIORS** Typically occurs within migrating and wintering flocks of American Wigeon. Fast, direct flight; launches straight up from water's surface. **HABITAT** Freshwater and saltwater marshes and mudflats; fields, croplands, wet prairies, large lawns in parks and playing fields. **STATUS** Rare but regular winter visitor.

Mallard (*Anas platyrhynchos*)

The familiar ducks of city parks and ponds, not to mention virtually every other watery habitat, Mallards nest throughout the Northwest and their numbers are bolstered in late winter and fall by migrants that nest at more northern latitudes. They are as likely to be found on secluded beaver ponds as coastal estuaries, city park ponds, and large rivers—and just about anyplace else with water; those that inhabit residential areas even drop in on backyard feeders from time to time.

LENGTH 21–24 inches. **WINGSPAN** 32–36 inches. Iridescent blue speculum with white borders on both edges. **MALE** Silky iridescent green head, bright yellow bill, white neck ring, chestnut-brown breast, grayish back, silvery flanks, white belly, black rump; eclipse male resembles female but darker, with dark breast and yellowish bill. **FEMALE AND JUVENILE** Mottled brown-tan overall; head and neck finely speckled; brown crown and eye stripe; orange-brown bill. **VOICE** Male's is a grunting, low-pitched quack. Female makes a familiar hoarse quacking, usually in a series. Both sexes produce low, drawn-out flight quacks, often after more hurried flush quacking. **BEHAVIORS** Adaptable to human activity, Mallards commonly colonize city parks with ponds, lakes, or creeks; fast, fairly direct flight; launches straight up from water's surface. **HABITAT** Freshwater ponds, marshes, lakes, rivers, sloughs, ditches; saltwater estuaries; city and park ponds and lakes. **STATUS** Common year-round resident and migrant.

Male Mallard

Female Mallard

Male Mallard in flight

Mallard range

Northern Pintail (*Anas acuta*)

In breeding plumage, the drake pintail—named for its long, slender tail—ranks among the most handsome of ducks. The Northern Pintail population has continued to decline markedly over the past 50 years and they are now on the list of "Common Birds in Steep Decline" compiled by the North American Bird Conservation Initiative.

LENGTH 22–28 inches. **WINGSPAN** 33–35 inches. Long neck; long, narrow wings. **MALE** Chocolate-brown head, nape, and throat; vertical white neck stripes; blue bill with black stripe along the top; silvery gray back, white breast and belly; black rump and long, pointed tail; iridescent green speculum. Eclipse male resembles female and is separable from other eclipse ducks by the pointed tail and blue-edged bill. **FEMALE** Mottled brown and beige; pale tan neck and head, gray bill, bronze speculum. **VOICE** Male produces a slightly warbled, airy, peeping whistle; also a high, quick *zuuweeee*. Females make deep, resonating, grunting quacks and rapid clucks. **BEHAVIORS** Occurs in large migrating flocks on several expansive marsh systems—especially the

Klamath Basin and Malheur Lake region in Oregon—that serve as critical staging areas for Northern Pintails headed for nesting grounds on the Canadian prairies. Fast, direct flight; launches straight up from water's surface. **HABITAT** Freshwater marshes, lakes, sloughs; saltwater estuaries; flooded fields, large grain fields. **STATUS** Common migrant; fairly common summer resident; common winter resident west of Cascade Range.

Eclipse plumage Pintails (August)

Male Northern Pintail in flight

Male (front) and female Northern Pintails

Northern Pintail Range

Northern Shoveler
(*Anas clypeata*)

The Northern Shoveler is easily recognized by its unique, massive, spatula-shaped bill, adapted to strain water, allowing the duck to filter out tiny crustaceans, insect pupae and larvae, plant matter, and other foods. Because of the size and shape of its bill, the Northern Shoveler has earned nicknames such as "spoonbill" and "spoonie."

LENGTH 19 inches. **WINGSPAN** 30 inches. Very large spatulate (spoonlike) bill, black in male, dirty orange and often speckled or mottled with black in female. **MALE** Iridescent green head and neck, white breast, rich chestnut-brown flanks and belly, black rump and back, bright yellow eye; in flight, note conspicuous chesnut belly bordering white breast, and iridescent green speculum divided from extensive blue shoulder by white lateral stripe. Eclipse male resembles female, but with gray head, or may retain muted male pattern, but with gray head. **FEMALE** Mottled tan or dark brown overall; speckled brown or grayish tan head and neck; speculum pattern as in male, but colors more subdued. **VOICE** Female has a variety of quacking calls, including a series of persistent husky quacks as a flush call. The male makes hollow, bubbling, cluck-like grunts. **BEHAVIORS** Frequently feeds by holding the bill underwater and swimming forward, dabbling and straining invertebrates. Also tips up in very shallow water to strain food from bottom detritus. Fast, direct flight; launches straight up from water's surface. **HABITAT** Shallow freshwater marshes, ponds; vegetated ditches; shallow saltwater estuaries and tidal creeks outside of breeding season. **STATUS** Common migrant; common summer resident east of the Cascade Range; fairly common summer resident and common winter resident west of Cascades.

Male Northern Shoveler

Female Northern Shoveler

Male Northern Shoveler in flight

Northern Shoveler range

Blue-winged Teal *(Anas discors)*

Breeding on marshes and vegetated ponds of the interior Northwest, the Blue-winged Teal is the last duck to arrive on its breeding grounds and the first to depart for wintering regions, generally arriving in the Northwest between mid-April and early May and departing in early September.

Male Blue-winged Teal

LENGTH 15 inches. **WINGSPAN** 23 inches. Broad blue shoulder patches, green speculum with white leading edge. **MALE** Rich brownish tan breast and flanks with dark spots, turning to vertical bars above legs; steely gray neck and head with bold white vertical crescent on face in front of eyes; black bill. Eclipse male resembles female but may retain partial white face crescent. **FEMALE** Mottled brownish gray-tan overall, darkest on back; pale grayish brown face with white spot at the base of the bill; dark crown and eye line; dark bill. **VOICE** Male produces typically plaintive whistled *peep* calls. Female recognized by high-pitched scratchy quacking; flush call sometimes a double-note grunting *kwaak-ahh*, rapidly repeated. **BEHAVIORS** Often feeds by dabbling in very shallow water and grazing on adjacent mudflats and wet shorelines. Very fast and agile flight; can change course very rapidly and often flies in small, twisting and turning flocks; launches straight up from water's surface. **HABITAT** Freshwater ponds, marshes; mudflats; weedy ditches; prairie and steppe potholes with emergent vegetation. **STATUS** Uncommon summer resident; rare west of Cascades.

Female Blue-winged Teal

Male Blue-winged Teal in flight

Blue-winged Teal range

Cinnamon Teal (*Anas cyanoptera*)

The male Cinnamon Teal is aptly named for its rich reddish plumage, and close examination likewise reveals his striking bright red eye. The Cinnamon Teal hen, however, in her speckled brown-tan camouflage, is very difficult to distinguish from the female Blue-winged Teal, making the presence of drakes a helpful indicator to the female's identity.

LENGTH 15.5 inches. **WINGSPAN** 22 inches. Broad blue shoulder patches and green speculums with white leading edges (wing colors generally more prominent on males); rounded head, fairly large bill. **MALE** Rich cinnamon-red overall; red eye. Eclipse male resembles female. **FEMALE** Mottled brownish gray-tan overall, darkest on back; warm beige face more uniform, lacking the strongly darker crown and eye line of the female Blue-winged Teal, but the difference is subtle; dark bill is slightly larger than in female Blue-winged Teal, and lacks the white spot at the base of the bill as in female Blue-winged teal. **VOICE** Males commonly make a thin, whistled *peep*. The female call is a guttural quacking, sometimes slightly wavering. **BEHAVIORS** Like other dabblers, the Cinnamon Teal often feeds by "tipping up," reaching underwater with its head and neck while propping its rump upward to reach deeper, but teal also frequently skim the water's surface for plant material. Very fast and agile flight; can change course rapidly and often flies in small, twisting and turning flocks; launches straight up from water's surface. **HABITAT** Shallow alkaline freshwater marshes; well-vegetated freshwater sloughs, ponds, ditches. **STATUS** Summer resident; common interior; uncommon to locally common west of Cascades.

Male Cinnamon Teal

Female Cinnamon Teal

Female (left) and male Cinnamon Teal

Cinnamon Teal range

Green-winged Teal (*Anas crecca*)

The spritely and colorful little Green-winged Teal is among the tiniest of ducks, typically weighing in at a mere 12 ounces. Widespread and common in migration, these ducks are easily overlooked, owing to their penchant for feeding in concealed, well-vegetated marsh and pond edges. Outside of the breeding season they are equally at home in freshwater and brackish water. The vast majority of Green-winged Teal nest on the Canadian prairies; in the Northwest, breeding populations are widely scattered and rarely dense.

LENGTH 14 inches. **WINGSPAN** 23 inches. Speculum shiny bright green on the inner half, black on the outer half, with white rear border and light tan front border. **MALE** Rich cinnamon head and neck with broad, striking, comma-shaped, iridescent green stripes enveloping the eyes and curving down the sides of the head; small black bill; silvery gray flanks with vertical white strip at the shoulders; triangular yellow patch on each side of rump; eclipse male similar to female. **FEMALE** Mottled brown overall, white chin and belly, small black bill, gray legs. **VOICE** Male calls are commonly bright, high-pitched whistles; also a very rapid, short, wavering, ringing *cha-cha-cha*. Females produce sharp, raspy quacks.

Male (left) and female Green-winged Teal

Male Green-winged Teal

Female Green-winged Teal

Green-winged Teal range

BEHAVIORS Very fast and agile in flight, often twisting and turning; launches straight up from water's surface. **HABITAT** Freshwater ponds, lakes, sloughs, marshes; saltwater estuaries; flooded fields. **STATUS** Locally uncommon breeder; common migrant; common winter resident west of Cascades; uncommon winter resident interior.

SIMILAR SPECIES The Eurasian (Common) Teal, currently considered conspecific (of the same species) with the Green-winged Teal, is a rare vagrant to the Northwest, generally during winter and early spring. It is best distinguished by the lack of a vertical white bar on the sides of the drake's breast, a field mark seen on the drake Green-winged Teal. The Eurasian Teal has a horizontal white stripe along the lower scapulars that can be partially or sometimes fully obscured by surrounding feathers, and its white facial stripes can be more defined than on the Green-winged Teal.

Canvasback (*Aythya valisinaria*)

Among the swiftest of ducks, Canvasbacks can attain flight speeds of more than 70 miles per hour. These large, regal-looking birds nest on freshwater marshes of the inland Northwest, where the breeding-plumage males are easily identified even at distance by their conspicuous ivory-colored bodies. Like each of the duck species that follow herein, Canvasbacks feed primarily by diving and they are expert underwater swimmers.

LENGTH 21 inches. **WINGSPAN** 32 inches. In profile, distinctive long, gently curving slope from tip of lengthy black bill to peak of crown. **MALE** Deep rusty red head and neck and blackish red forehead, lustrous in good light; black breast and rump contrasting with white flanks and mantle; red eyes; in flight, white wings with light gray primaries. Eclipse male similar to breeding male but colors more muted. **FEMALE** Softly mottled tones of light grayish tan and brown overall; light brown breast, neck, and head, darker on nape and crown; whitish or pale tan chin and throat; in flight, grayish wings with white underwing linings. **VOICE** Female flush and flight call is a coarse, guttural grunting; males utter a variety of muted piping cooing calls during display. **BEHAVIORS** Male Canvasbacks use a variety of courtship and breeding displays, including a head throw in which the male lays its head backward over its body and then snaps it back forward. Very fast and direct flight; runs on water to take flight. **HABITAT** Breeds on freshwater marshes, vegetated ponds, seasonally flooded wetlands; during migration and winter, found on estuaries, bays, large rivers, lakes, marshes. **STATUS** Uncommon summer resident; common migrant; fairly common winter resident. **KEY SITES** (breeding) Idaho: Camas NWR; Grays Lake NWR. Oregon: Klamath Basin; Ladd Marsh Wildlife Area; Malheur NWR. Washington: Potholes Reservoir; Turnbull NWR; Winchester Wasteway.

Female Canvasback

Canvasback flock

Male Canvasback

Canvasback range

Redhead *(Aythya americana)*

The loud, wailing courtship call of the male Redhead is among the iconic sounds of the shallow marshes of the interior Northwest, where these handsome ducks nest in substantial numbers. Waterfowl ecologist and artist Hans Albert Hochbaum, in *The Canvasback on a Prairie Marsh* (1944), perfectly captured the idyllic din of a calm early summer morning on the marshlands: "The laughter of Franklin's Gulls and the whistle of yellowlegs cut the morning air, in the reeds a marsh wren bubbles, and drifting across the bay comes a soft, incessant, violinlike *whee-ough*, the courting note of the drake Redhead."

LENGTH 19 inches. **WINGSPAN** 30 inches. Steep forehead and rounded crown; in flight, dark gray shoulders contrast with black-tipped light gray flight feathers. **MALE** Lustrous rusty red head and neck, jet-black breast, light gray flanks and back, black rump; large soft-blue bill with black tip. Eclipse male similar to female, but with rustier head. **FEMALE** Soft unspeckled brown overall; rich brown crown and nape; pale tan cheeks and throat; blue-gray bill with black tip. **VOICE** During courtship, males make a wailing *keee-ow* or more abrupt *meow*, soft but penetrating. Females produce a hoarse quacking flush and flight call, plus various low grunts. **BEHAVIORS** Male's courtship display includes a head throw in which he lays his head back over his body and rapidly snaps it back forward

Redhead range

Male Redhead

Female Redhead

while uttering his unique *meow* call. Strong, fast, and direct flight. Runs on water to take flight; skids to a stop when landing on water. **HABITAT** Shallow freshwater marshes and seasonally flooded wetlands during breeding season; marshes, lakes, large rivers, estuaries during migration. **STATUS** Fairly common summer resident interior, rare west of Cascades; fairly common migrant; uncommon winter resident. **KEY SITES**

Idaho: Camas NWR; Hagerman Wildlife Area; Indian Creek Reservoir (Ada County). Oregon: Klamath Basin; Ladd Marsh Wildlife Area; Malheur NWR. Washington: Columbia River (Priest Rapids Lake, Wanapum Lake, Lake Pateros); Desert Wildlife Area; Phillio Lake and Turnbull NWR; Potholes Reservoir. British Columbia: Okanagan Lake; Vaseux Lake.

Male Redhead in flight

Ring-necked Duck
(*Aythya collaris*)

In breeding plumage, the male Ring-necked Duck is among the handsomest of waterfowl, but the neck ring for which it is named is the least obvious part of its bold plumage: a band of rich brown at the base of the iridescent purple neck. The ring is generally only visible in excellent direct light. Ring-necked Ducks generally nest on secluded forested ponds, so while their Northwest breeding range is fairly extensive, they are far easier to find when congregated during migration and winter.

LENGTH 17 inches. **WINGSPAN** 26 inches. Medium-sized duck with knobbed crown; in flight, gray flight feathers contrast with black shoulders. **MALE** Lustrous black breast and back; iridescent purple head and neck (appears black in poor light); silvery, finely vermiculated flanks with white patch at front edge; white belly and underwing linings;

bright blue bill with white base, white subterminal band, and black tip. Eclipse drake resembles female, but darker overall, with uniformly dark head. **FEMALE** Brownish gray overall, with brown flanks; dark gray crown and nape, light gray face; white around entire base of bill, and white eye ring; dark gray or blue-gray bill with black tip and sometimes a white subterminal band. **VOICE** Male makes short, musical, piping whistles. Female calls are hoarse, husky grunts. **BEHAVIORS** To gain momentum to dive, Ring-necked Ducks first propel themselves upward before doubling over to enter the water. Very fast and agile in flight; often drops toward water very quickly at steep, twisting angles; usually runs short distance to take flight from the water. **HABITAT** Breeds on small wooded ponds and sloughs; ponds, sloughs, lakes, rivers, reservoirs, marshes during migration and winter. **STATUS** Fairly common summer resident and migrant; common winter resident west of Cascades and along Columbia River through Oregon and Washington.

Male Ring-necked Duck

Female Ring-necked Duck

Female (left) and male Ring-necked Ducks in flight

Ring-necked Duck range

Tufted Duck (*Aythya fuligula*)

A rare vagrant to the Pacific Northwest, the Tufted Duck is a Eurasian species very similar to the scaups and the Ring-necked Duck. As with other exotic waterfowl, some individuals seen in the Pacific Northwest are perhaps escaped or released captive birds, but Tufted Ducks occur with enough regularity that most are probably true vagrants.

LENGTH 17 inches **WINGSPAN** 26 inches. Rounded head; long, prominent white wing stripes. **MALE** black breast, back, rump; bright white flanks; black head appears iridescent purple in good light; long, slender crest curves down along back of head and neck; bright yellow eyes; pale bluish bill with black tip and small white subterminal band. **FEMALE** Very similar to female Scaup, but minimal or no white around the base of the bill, short tuft. **BEHAVIORS** When found in the Northwest, Tufted Ducks are often seen with Scaups or other diving ducks. **HABITAT** Lakes, ponds, marshes, slow rivers, sloughs, estuaries, bays. **STATUS** Rare winter vagrant.

Male Tufted Duck

Female Tufted Duck

Greater Scaup (*Aythya marila*)

A large diving duck of open water, the Greater Scaup is very similar to the Lesser Scaup, with the safest field marks generally being the comparative head shapes and, though a subtle difference, the extent of white in the flight feathers. The Greater Scaup drake, in excellent light, shows an iridescent green head rather than the iridescent purple of the Lesser Scaup, but head color can be an unreliable field mark. Both species are especially common in the Pacific Northwest during the nonbreeding season.

LENGTH 18.5 inches. **WINGSPAN** 29 inches. Rounded head; bill more robust than in Lesser Scaup; in flight, dark shoulders contrast with extensive white in flight feathers, with the white stripe extending nearly to the wingtips in contrast to Lesser Scaup, where white wing stripe is less extensive. **MALE** Shiny black (or deep iridescent green) head; broad bluish bill with black tip ("nail") broader than on male Lesser Scaup; silky black breast, bright white flanks; white back with fine black barring appears light gray at distance; black rump; yellow eye; first-fall male resembles female. **FEMALE** Brownish overall, lighter on flanks; rounded dark brown head, with extensive white around the base of large, grayish bill. **VOICE** Rarely heard during nonbreeding season; rolling, tinny, growl-like grunts and softer purring grunting by female. **BEHAVIORS** During winter, Greater Scaups often congregate in large water-roosting flocks called "rafts" in open water (as do Lesser Scaups). The Columbia River near Interstate-84 (Oregon) and SR 14 (Washington) is an excellent place to see these rafts, which can contain hundreds of ducks. Fast and direct in flight, with large flocks typically twisting and shifting. When scaups surface from a dive, they pop to the surface like a cork. **HABITAT** Bays, estuaries, large rivers. **STATUS** Fairly common winter resident and migrant; large numbers winter in Puget Sound, the Victoria and Vancouver, British Columbia, areas, and the Columbia River; occasional summer visitor in the Columbia River mouth and coastal waters.

Female Greater Scaup in flight

Raft of Greater Scaup

Female (front) and male Greater Scaup

Greater Scaup range

Lesser Scaup (*Aythya affinis*)

Lesser and Greater Scaup are difficult to distinguish from one another, with the safest field marks being the comparative head shapes and, though the difference is subtle, the extent of white in the flight feathers. The Lesser Scaup drake, in excellent light, may show an iridescent purple head rather than the iridescent green of the Greater Scaup, but this head color can be unreliable. Both species are especially common in the nonbreeding season.

LENGTH 17 inches. **WINGSPAN** 26 inches. In relaxed posture, head is peaked toward the rear of the crown, and narrower and taller than in the Greater Scaup; in flight, dark shoulders contrast with white inner flight feathers and gray outer flight feathers, with the white stripe not extending as far out the wings as in the Greater Scaup; bill is straighter and less robust than in Greater Scaup. **MALE** Shiny black (or iridescent purplish) head; broad bluish bill with small black nail at the tip; silky black breast, white flanks; white back with fine black barring appears gray at distance; black rump; yellow eye; juvenile and eclipse males resemble female. **FEMALE** Brownish overall, lighter on flanks; dark brown head, with extensive white around the base of grayish bill. **VOICE** Seldom heard; during courtship, males make soft whistled *weee-eew* and single-note *whew*; females more vocal, with grunting and purring growl-like quacks. **BEHAVIORS** During migration, both Lesser and Greater Scaups usually flock in rafts on open water to roost or feed; Lesser Scaups often disperse to shallow wetlands to feed early in the morning, returning to the open-water roost late in the day. Fast and maneuverable in flight; high-speed landings with feet extended forward and skidding to a stop. **HABITAT** Lakes, reservoirs, large rivers, open marshes, estuaries, bays. **STATUS** Common winter resident; rare summer resident (Malheur NWR, Oregon).

Male Lesser Scaup

Female Lesser Scaup

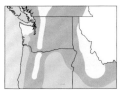

Lesser Scaup range

Male Lesser Scaup

Harlequin Duck
(*Histrionicus histrionicus*)

The striking male Harlequin Duck, with his vaudevillian plumage, departs the swift mountain stream where his hen raises her young as soon as she sits on her eggs. Because the drakes spend so little time in the montaine breeding habitat, they are much more easily found when concentrated along the coastal wintering grounds.

LENGTH 16 inches. **WINGSPAN** 25 inches. Compact, roundish body; short, stubby, grayish bill; all-dark wings. **MALE** Slate-gray overall, distinctly marked with white stripes and patches, including large white teardrop-shaped patches in front of eyes; broad white neck ring and vertical shoulder stripe both bordered by black; rich chestnut flanks. Eclipse male similar to female, but grayer, and often with remnant white patches and flecks. **FEMALE** Gray-brown overall with 2 or 3 white patches on each side of face, with the small round white spot on the face, along with the smaller bill, distinguishing it from the superficially similar female Surf Scoter. **VOICE** Males make high-pitched barking squeaks. The females produce high-pitched grunting pips, singly and in series. **BEHAVIORS** Feeds by diving, and highly maneuverable at diving and clambering about in rushing water; lounges on rocks in and along rivers and streams, in the surf zone, and on jetties; swift flight, typically low over the water. **HABITAT** Breeding: swift mountain rivers and streams, from foothills to about 4000 feet. Winter: coastal; rugged, rocky surf; bays, especially along jetties. **STATUS** Uncommon summer resident; fairly common in winter as birds are concentrated along coast. **KEY SITES** Oregon: Cape Arago State Park; Cascade Head; Yaquina Bay. Washington: Neah Bay and Cape Flattery area; San Juan Islands; Ocean Shores and Westport jetties. British Columbia: Burrard Inlet and Point Roberts (Vancouver); Victoria area.

Female Harlequin Duck

Male Harlequin Duck

Harlequin Duck range

Surf Scoter
(*Melanitta perspicillata*)

The striking male Surf Scoter is easy to identify, but the female can be mistaken for the similar female White-winged Scoter. Both species, along with the Black Scoter, are common winter residents all along the Pacific Northwest coast. The 3 species of scoter commonly form loose aggregations, drawn to the same feeding areas, and even flock together at times. Scoters typically fly in fast flocks or lines very low over the surface of the water.

Male (rear) and first-winter female Surf Scoters

LENGTH 19–21 inches. **WINGSPAN** 30–32 inches. **MALE** Black overall, with 2 conspicuous white patches, one on the forehead and the other on the nape; large multicolored bill has yellowish tip, red around the nostrils, white at the sides with a black spot on each side; black culmen extends nearly to nostrils; bright white eye. First-winter male brownish overall, with lighter face; very large bill mostly black or identical to adult male but with muted colors; formative white nape patch. **FEMALE** Dark grayish overall; large black bill; 2 small white face patches, one behind the eye and one in front of the eye at the base of the bill, are generally better defined than on female White-winged Scoter; front white patch, often vertically oval in shape, does not extend onto the culmen as it generally does in the female White-winged Scoter. **VOICE** Rarely heard and poorly documented; a variety of croaks and short squawks have been described. **BEHAVIORS** Feeds by diving in fairly shallow water for mollusks and crustaceans. Swift, direct flight, low over the water after a running takeoff; small flocks often fly in a line (also called a string). **HABITAT** Bays, coves, coastal inlets, estuaries, outer surf zone, shallow ocean waters. **STATUS** Common coastal migrant and winter resident (August–April); rare inland.

Adult female Surf Scoter

First-winter male Surf Scoter

Surf Scoter range

White-winged Scoter
(*Melanitta fusca*)

Nesting on freshwater from the Canadian prairies to the high Arctic, the White-winged Scoter winters in substantial numbers along the Pacific Northwest coast, and—as with the closely related Surf Scoter—individuals tend to show up on inland lakes or reservoirs somewhere in the region annually, often far from salt water.

LENGTH 21 inches. **WINGSPAN** 34 inches. All-dark body with distinctive white wing patches (not always visible when the bird is at rest). **MALE** Jet-black body; white teardrop-shaped eye patch, often reduced to small white speck in fall-plumage and transitional-plumage birds; bill has long, bulbous, black base and red or reddish tip; in breeding plumage, colorful bill ranges from reddish orange or bright pink-red to scarlet, edged in black, with white and yellow accents. First-winter male more grayish black, sometimes with limited white spackling on face. **FEMALE** Dark grayish black or brownish black, with 2 highly variable white face patches; forward patch at the base of the bill and extending onto the culmen is typically round (rather than vertically oval as in female Surf Scoter) and often nondescript and grayish. In both sexes and all plumages, white wing patches differentiate White-winged Scoter from other scoters. **VOICE** Raspy, low-pitched grunts. **BEHAVIORS** Feeds by diving in fairly shallow water for mollusks and crustaceans. Swift, direct flight, low over the water after a running takeoff; small flocks often fly in a line (also called a string). **HABITAT** Bays, coves, coastal inlets, estuaries, outer surf zone, shallow ocean waters. **STATUS** Fairly common coastal winter resident (August–April); rare inland.

Male White-winged Scoter

Female White-winged Scoter

White-winged Scoter
range

Black Scoter (*Melanitta nigra*)

Male Black Scoter

When observed at close range without substantial background noise from crashing waves, male Black Scoters sometimes utter a fairly continuous plaintive whistle, making them the most vocal of the scoters. Like the Surf Scoter and White-winged Scoter, these robust sea ducks congregate for winter all along the Pacific coast; in prime feeding areas, all 3 species often occur together.

LENGTH 18 inches. **WINGSPAN** 30 inches. **MALE** Velvety black overall; large black bill with bright yellow-orange culmen hump. First-year male is brownish or gray-brown overall, with light brown face and dark crown; yellow, or mottled yellow and gray-brown culmen. **FEMALE** Plain, soft brown overall, with distinctive light tan face contrasting with dark brown cap and nape; large blackish bill. **VOICE** Male's is a flat, plaintive whistle, somewhat drawn-out and sometimes wavering. Female has various low, raspy grunting calls. **BEHAVIORS** Feeds by diving, often in the outer surf zone of rocky beaches and along sea stacks and headlands, often in congregations with other scoter species. Swift and powerful flight, usually low over the water's surface; undersides of flight feathers appear silvery in good light. **HABITAT** Bays, coves, coastal inlets, estuaries, outer surf zone, shallow ocean waters. **STATUS** Fairly common coastal migrant and winter resident (September–April).

Female (top) and male Black Scoters

Black Scoter range

Long-tailed Duck
(*Clangula hyemalis*)

Breeding in the Arctic and subarctic, Long-tailed Ducks winter in substantial numbers along the southern British Columbia coast and in Puget Sound, and in small numbers along the remainder of the Pacific coast. The male, with its winter plumage even more striking than its darker breeding-season plumage, carries a rapierlike tail nearly the length of his body. Occasionally Long-tailed Ducks show up on freshwater in the interior Northwest.

LENGTH 15–18 inches. **WINGSPAN** 30 inches. **MALE NONBREEDING** Very long, pointed black tail plumes; white crown, neck, and nape, with gray face and large circular back patch on each side of neck; black breast; black mantle with pale gray edges; pale gray flanks; black rump, white belly; stubby black bill with large pink subterminal band; in flight, note adult male's broad black breast band and streaming tail. **MALE BREEDING** Black breast, neck, cheeks, and crown, with large whitish, teardrop-shaped eye patch, and a brown and black mantle. First-winter male resembles female. **FEMALE** Predominantly white below, gray or gray-brown above; white neck and face, dark crown; large dark neck spot; short, stubby, dark-colored bill. **VOICE** Resonant squawking *ayauk auk auk auk* and rising *aaa-aayit*. **BEHAVIORS** Perhaps the deepest diver among all the ducks, Long-tailed Ducks can reach depths of 200 feet. Swift, direct flight on very fast, narrow wings. **HABITAT** Saltwater bays, inlets, sounds, straits, outer surf zone. **STATUS** Fairly common winter resident (especially November–March) of north Puget Sound, Washington, and coastal British Columbia; uncommon but regular winter resident in Oregon; rare inland.

Male nonbreeding Long-tailed Duck

Female nonbreeding Long-tailed Duck

Long-tailed Duck range

Male (right) and female Long-tailed Ducks in flight

Bufflehead (*Bucephala albeola*)

The tiny Bufflehead is a common and familiar winter resident, especially along the coast, where it generally occurs in small flocks on sheltered waters. They are especially fond of small inlets, tidal creeks, and estuary sloughs, where they dive for mollusks, crustaceans, and other invertebrates. Seen in good light, the male's black head flashes an oily sheen of iridescent purple and green.

Male Bufflehead

LENGTH 13.5 inches. **WINGSPAN** 21 inches. Compact, roundish body; stubby gray bill. **MALE** Striking black-and-white pattern overall; bulbous black head (flashes iridescent green and purple in excellent light) with large, white, wedge-shaped patch from eye to crown and around nape; black back, bright white flanks, breast, and belly; broad white wing patches in flight. **FEMALE** Depending on light, dark steely gray to brownish gray crown and back; grayish face with soft white cheek patches; small white wing patches in flight. **VOICE** Generally quiet; during courtship, females use high-pitched, wavering guttural grunts, and males make squeaky squealing and chattering calls during display. **BEHAVIORS** In small groups, Buffleheads begin their frenetic courtship display while on the wintering grounds; the male bobs his head quickly, rises up on his rump while flapping his wings, and makes short display flights and landings around the female. Swift and direct flight, typically in small groups low over the water; runs quickly along the surface to take flight. **HABITAT** Bays, inlets, tidal creeks; saltwater and freshwater sloughs; marshes, ponds, lakes; slow lowland rivers. **STATUS** Common winter resident, especially west of the Cascades (September–May); rare summer resident.

Female Bufflehead

Male Bufflehead in flight

Bufflehead range

Common Goldeneye
(*Bucephala clangula*)

The 2 species of Goldeneye are quite similar, and differentiating between females can be especially problematic, typically requiring close and persistent observation. Though often touted as a basic difference between the 2 species, relative head shape is highly variable based on the bird's attitude and angle, but can still be helpful in identifying the Goldeneyes: Common Goldeneye generally has a more gently sloped forehead and peaked crown, whereas the Barrow's Goldeneye typically has a steeply sloped forehead and a flattened crown. More useful is comparative bill shape: the Common Goldeneye's

bill is strongly triangular when viewed directly broadside, while the Barrow's Goldeneye's bill has an upward-arcing "smile" on the bottom, and a more prominent nail at the tip of the bill, which gives the impression of a shallow divot just rear of the nail. In addition, note the typical difference in bill color between female Goldeneyes during winter.

LENGTH 19 inches. **WINGSPAN** 30 inches. Compact and roundish body; short neck; large, bulbous head; black bill in summer. **MALE** Iridescent deep greenish black head (often appears black), with large white spot in front of yellow eye; stubby black bill; bright white neck, breast, flanks, and belly; center of back black with extensive white to the sides; black rump; in flight, extensive

Male Common Goldeneye

Female Common Goldeneye

white on inner half of wing; juvenile male not as strongly marked or as colorful as adult male. **FEMALE** Rich chocolate-brown head, yellow eye; in winter, black bill with pale yellow near the tip; white neck, gray

First-winter male Common Goldeneye

breast, brownish gray flanks, dark gray back and wings; white wing patches. **VOICE** Generally silent. Female has laughlike chuckles, *agh-agh-agh-agh-agh*; flush and flight call is a husky, grunted quacking. Male produces reedy, buzzing rattles during courtship displays. **BEHAVIORS** Cavity nesters, Common Goldeneye require natural holes made by large woodpeckers or other means, and will also use nest boxes made for them. **HABITAT** In winter, bays, sounds, estuaries, tidewater rivers; also lakes and reservoirs; open-water marshes; slow rivers. Breeds on forested lakes, ponds, and wetlands. **STATUS** Common winter resident; rare breeder in northeast Washington and northern Idaho, north into British Columbia.

Common Goldeneye range

Male (showing wing pattern) and female Common Goldeneye

Barrow's Goldeneye
(*Bucephala islandica*)

Northwesterners who enjoying summer days hiking into small mountain lakes might be pleasantly surprised to occasionally find a female Barrow's Goldeneye and her brood of fuzzy chicks, for these handsome ducks breed in the high country. But don't expect to find the striking male Barrow's Goldeneye with his mate unless you've trudged through snowbanks to get there, because he departs the breeding territory shortly after the hen begins sitting on her eggs.

Male Barrow's Goldeneye

Female Barrow's Goldeneye

LENGTH 17–19 inches. **WINGSPAN** 30–32 inches. Compact and roundish body; short neck; large, bulbous head; black bill in summer. **MALE** Iridescent deep purple head (often appears black), with large white crescent in front of yellow eye; stubby black bill; bright white breast, flanks, and belly; black back with row of white spots (secondary feathers) on each side; black rump; in flight, black wings with white patch on trailing edge; juvenile male not as strongly marked or as colorful as adult male. **FEMALE** Rich chocolate-brown head, yellow eye; in winter, dirty orange bill, sometimes with varying amounts of black; bill has an upward arcing "smile" on the bottom, and a prominent nail at the tip, which gives the impression of a shallow divot just rear of the nail; white neck, gray breast, brownish gray flanks, dark gray back and wings; white wing patches. **VOICE** Male call is a short, reedy *kaa kaa kaa* and *ki-KAA* during courtship displays. Females offer short, grunting, wavering, crowlike caws. **BEHAVIORS** Cavity nesters, Barrow's Goldeneyes rely on holes made by large woodpeckers, or other natural cavities, including tree chimneys, in both standing and fallen timber, generally along lakeshores; they occasionally nest in riparian rock crevices, and will also use nest boxes made for them. **HABITAT** Breeds on forested mountain lakes and ponds, and high-elevation rangeland lakes and ponds; winters primarily on salt water in bays, harbors, inlets, inshore waters, and also in small densities on large lakes and rivers. **STATUS** Common winter resident of British Columbia and Puget Sound regions; uncommon elsewhere; uncommon summer resident of British Columbia, Washington, and Oregon; rare in Idaho.

Male Barrow's Goldeneye (left) and male Common Goldeneye (right) for comparison

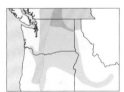

Barrow's Goldeneye range

First-winter male Barrow's Goldeneye

Hooded Merganser
(*Lophodytes cucullatus*)

Generally secretive and retiring, Hooded Mergansers are the smallest of the 3 mergansers found in North America. Cavity nesters, they depend upon existing holes (often created by woodpeckers), including nest boxes, so the typical nesting habitat is well-wooded ponds, lakes, and sloughs, often with substantial emergent vegetation.

LENGTH 17.5 inches. **WINGSPAN** 25 inches. White inner wing patch in flight; fairly long, stiff tail often cocked upward or laid flat on or in the water. **MALE** Striking white crest fully framed in black (broad white stripe trailing back from head when crest is lowered); black face, crown, throat; thin black bill; yellow eye; white breast with 2 black vertical bars; rich chocolate-brown flanks; black back with trailing white-edged black tertials; eclipse male similar to female. **FEMALE** Dark brown overall, with lighter lustrous brown head and cinnamon-toned crest; buffy breast fading to white belly; thin, dark bill with yellowish base. **VOICE** Low-pitched, raspy, grunting quack; low-pitched frog-like growl by displaying male. **BEHAVIORS** Very fast and direct flight, usually low over water, but capable of agile maneuvering. Dramatic courtship ritual begins in winter and can be observed just about anywhere "hoodies" congregate, including small farm ponds, narrow tidal creeks, borrow pits, and mountain lakes. The male, crest fully raised, stretches his neck high in the air, lays his head backward over his back, then dips forward; he snaps his bill rapidly, and emits a froglike, low-pitched croaking. Typically several or more males will display for one or more females. **HABITAT** Migration and winter: ponds, lakes, large reservoirs, marshes, estuaries, tidal creeks, sloughs, slow rivers. Breeding: small wooded ponds and lakes, wooded sloughs and backwaters, up to about 3000 feet elevation. **STATUS** Uncommon but widespread in breeding season; fairly common winter resident and migrant west of Cascades; uncommon migrant and winter resident inland.

Male Hooded Merganser

Female Hooded Merganser

Female (left) and male Hooded Mergansers

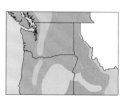

Hooded Merganser range

Common Merganser
(Mergus merganser)

Large, conspicuous, and widespread, Common Mergansers breed and raise their young along rivers of all sizes, making them a familiar bird to boaters, anglers, and other river users. The striking, green-headed male departs the breeding grounds shortly after nesting, so the crested, cinnamon-headed hens and juveniles are most familiar.

Male Common Merganser

LENGTH 23–26 inches. **WINGSPAN** 34–36 inches. Long streamlined body; long, slender red bill with hooked tip. **MALE** Lustrous white neck, breast, flanks, belly; black back, gray rump; dark iridescent green head; in flight, white shoulders and secondaries, black primaries. **FEMALE** Gray overall; cinnamon-red head with ragged crest, but more uniform crest than female Red-breasted Merganser; reddish head contrasts markedly with gray body and white breast (in female Red-breasted Merganser, brownish red head and neck blend into gray body); bright white chin (lacking in female Red-breasted). **VOICE** Hollow, reedy caws from males during courtship. Deep, chuckling grunts from females. **BEHAVIORS** Females raise the young, generally along rivers at low to middle elevations. Broods acclimate to the presence of people (anglers, rafters, boaters) and become fairly approachable. Swift and powerful flight, usually low over water; broods typically fly in lines or fairly tight flocks, following the course of the river; runs along the surface to take off. In flight, head often held below center line of body. **HABITAT** Rivers, large lakes and reservoirs, estuaries, bays. **STATUS** Common year-round resident in western and central Oregon, Washington, and British Columbia; fairly common summer resident and migrant, uncommon winter visitor eastward through interior.

Female Common Merganser

Common Merganser range

Red-breasted Merganser
(*Mergus serrator*)

A fish-eating specialist, the Red-breasted Merganser has a heavily serrated bill ideally adapted to catching small fish underwater. They are ducks of the Far North, and in the Pacific Northwest, they are most easily found on their saltwater wintering grounds, especially in Puget Sound, the Strait of Juan de Fuca, and the Strait of Georgia.

LENGTH 21–25 inches. **WINGSPAN** 30–34 inches. Long streamlined body; long slender bill; shaggy crest. **MALE** Dark iridescent green head, red eye, reddish bill, wide white neck ring; black back, brown breast, gray flanks; in flight, white inner half of upper wing has 2 black stripes and leading edge of shoulder is black. First-winter male resembles female but with male wing pattern.
FEMALE Grayish overall, with dingy grayish white breast; dirty reddish brown head not as cinnamon colored as female Common Merganser, and crest is wispier; lacks obvious bright white chin and breast of female Common Merganser, creating less contrast between brownish head and grayish body; reddish orange bill and eye; in flight, white secondaries and black-tipped white secondary coverts. **VOICE** Female makes raspy low-pitched grunts; males seldom heard.
BEHAVIORS In placid waters, Red-breasted Mergansers will swim with their heads underwater to look for fish. Swift, direct flight, often low over the water; runs along the surface to take flight. **HABITAT** Winters primarily in coastal waters, including bays, inlets, estuaries, and shallow inshore waters. **STATUS** Fairly common winter resident and migrant (September–April) in coastal waters; rare inland.

Male Red-breasted Merganser

Red-breasted Merganser range

Female Red-breasted Merganser

Ruddy Duck (*Oxyura jamaicensis*)

With its cartoonish profile, the Ruddy Duck ranks among the more entertaining waterfowl, especially for birders lucky enough to witness the male's intricate mating behavior, called a "bubbling display." The male extends his tail and neck upward and then, bowing, rapidly and audibly slaps his bill against his breast and the surface of the water, creating bubbles, and ending with a low, froglike belching sound.

LENGTH 15 inches. **WINGSPAN** 22 inches. Broad spatulate bill with slightly upturned "smile" shape; long, stiff tail, often cocked upward; compact body; short, slender wings. **MALE BREEDING** Cinnamon-red overall; bright blue bill, large white cheek patch, black crown and nape. **MALE NONBREEDING** Similar to female, but more extensive white cheek patch without brown stripe. **FEMALE** Mottled or speckled brown-tan overall, darkest on back; dark brown crown; pale cheek patch with horizontal brown line from bill to nape. **VOICE** Generally silent except for male's froglike call during mating display; females may utter a squeaky *eee-eee-eee*. **BEHAVIORS** Ruddy Ducks feed extensively on Chironomid (midge) larvae or pupae, which densely populate the shallow freshwater marshes and ponds the ducks prefer for nesting; they dive to gather mouthfuls of detritus to strain through their bills; on wintering grounds they eat a variety of invertebrates as well as plant material. Swift, direct flight after long, running takeoff. **HABITAT** Breeds on freshwater marshes, seasonal wetlands, shallow lakes, potholes; winters on estuaries, bays, large lakes and reservoirs. **STATUS** Common summer resident and migrant in interior regions; common winter resident west of Cascades.

Female Ruddy Duck

Male nonbreeding Ruddy Duck

Ruddy Duck range

Male breeding Ruddy Duck

Pheasants, Partridges, Quails, Grouse, and Turkeys

The order Galliformes includes pheasants, partridges, quails, grouse, and turkeys, which are colloquially known as upland gamebirds, a moniker reflecting their longstanding popularity with hunters. The Northwest is home not only to a variety of indigenous species of upland gamebirds, but also several introduced species. In fact, the first colorful Asian Ring-necked Pheasants released in North America were stocked in western Oregon in the early 1880s. Later, both Chukars and Gray Partridges were successfully introduced into the Northwest. Many upland gamebirds— especially quail and partridge—remain in family groups called coveys most of the year, and during winter, many families can combine into very large coveys.

Displaying male Sooty Grouse

California Quail
(*Callipepla californica*)

The familiar, handsome, and gregarious California Quail (also called the Valley Quail) is the most widespread gallinaceous bird in the region. Some small towns—Burns and Hines in southeast Oregon, for example— host winter coveys that typically number in the hundreds, creating a spectacle for birders. Interestingly, the California Quail's original native range reaches its northern extent in southern Oregon, but the species has been widely introduced elsewhere and has also expanded its range alongside human expansion, as it thrives in cultivated areas.

LENGTH 10 inches. **WINGSPAN** 14 inches. Small and plump, with short, broad wings; black forward-drooping topknot is larger on males; grayish overall with scaled nape appearing almost spotted; white-streaked flanks; markedly scaled beige belly bordering gray breast. **MALE** Striking; bold black face framed by white necklace stripe reaching up to eyes and white crown stripe wrapping around forehead, bordered above by narrow black stripe; cinnamon-brown crown; brown belly patch. **FEMALE AND JUVENILE** Similar overall to male but grayish tan face lacks black and white accents; smaller topknot. **VOICE** Numerous calls for a variety of situations; emphatic *chi-CA-go* call is most familiar; also single and multiple *waaa* calls; mournful *waaauu*; alarm call is a sharp *wit wit wit wit wit*. **BEHAVIORS** Frequently runs rather than flies to escape; strong, low, direct, short flights on noisy wings. Coveys typically flush a few at a time in quick succession. Forms large coveys in winter and visits backyard feeders in rural, suburban, and some urban areas. **HABITAT** Dry shrubland, brushy draws, shrubby grasslands;

farms and ranch lands; fallow and scrublands; riparian corridors; small towns and suburbs; city parks with appropriate cover. **STATUS** Common year-round resident.

Male California Quail

Female California Quail

California Quail range

Gambel's Quail
(*Callipepla gambelii*)

Similar in appearance to the widespread California Quail, the Gamble's Quail is native to the American desert Southwest, but was introduced near Salmon, Idaho, in 1917 and this remains the only place in the Pacific Northwest where these handsome little birds are found.

LENGTH 10 inches. **WINGSPAN** 14 inches. Small and plump, with short, broad wings; black forward-drooping topknot is larger on males; both sexes grayish overall with lightly scaled napes and beige checked chestnut flanks. **MALE** Exquisitely patterned; bold black face framed by white necklace stripe reaching up to eyes and white crown stripe wrapping around forehead, bordered above by narrow black stripe; head, rich chestnut; light buff belly with large central black spot. **FEMALE** Grayish tan face and head, light buff belly. **VOICE** Loud persistent *ha-HAA-ha-ha*; short, wailing, repeated *waa* or *uwaa*; sharp *wick-wick-wick*; various cackling and chirping calls. **BEHAVIORS** Frequently runs rather than flies to escape; strong, low, direct, short flights on noisy wings. Coveys typically flush a few at a time in quick succession. **HABITAT** Brushy draws and grasslands in and near farmlands; scrublands; riparian corridors. **STATUS** Locally uncommon resident near Salmon, Idaho. **KEY SITES** Idaho: Old Lemhi Road (north of and paralleling State Route 28) southeastward from Salmon; Tendoy area.

Male Gambel's Quail

Gambel's Quail range

Female Gambel's Quail

Mountain Quail *(Oreortyx picta)*

An elusive, handsome quail typically found in the most rugged mountain terrain, Mountain Quail, which seldom fly, are more often heard than seen, although driving or hiking gravel forest roads in appropriate habitat often results in brief sightings, particularly in mid-morning and late afternoon. Look for covey footprints on dusty mountain roads and trails. Historically more widespread, these birds have been reintroduced to various parts of their former range east of the Cascade Mountains with limited success.

LENGTH 11 inches. **WINGSPAN** 15 inches. **SEXES SIMILAR** Small and plump, with short, broad wings (though seldom flies); straight black topknot, erect or laid backward; grayish nape, grayish brown back and wings; chestnut flanks with broad white and cream bars; brown throat framed on each side by white stripe running up to the eyes. **VOICE** Typical call is a high, loud, ringing, exotic *eee-ark* or *re-ork*, sometimes as a single barking syllable, *aark*; other calls include rolling *chrrrrr*, woodpeckerlike emphatic *cha-cha-cha-cha-cha-cha-cha* or *chi-chi-chi-chi-cheea*, a high chirped trill; also soft clucks and coos at very close range. **BEHAVIORS** Prefers to run rather than fly to escape; explosive, low, direct, short flights on noisy wings. Coveys typically flush one or a few at a time in quick succession, the birds quickly diving into heavy cover. Forages under dense cover, often on steep slopes; during dry, late summer weather, needs water daily (which can be obtained from morning dew); migrates downslope for winter. **HABITAT** Steep, rocky slopes near forest edges with substantial brushy cover; 10- to 30-year-old clear-cuts and burns; gravel forest roads cutting across appropriate habitat. **STATUS** Year-round resident. Common southern Oregon, becoming less common northward, rare east of Cascades; reintroduction projects completed and ongoing at various locales in southeast Oregon and west-central Idaho. **KEY SITES** Oregon: Rogue River National Forest; Umpqua National Forest. Washington: Tahuya Peninsula.

Mountain Quail range

Mountain Quail

Chukar (*Alectoris chukar*)

The Chukar, named for its distinctive call, was introduced from Eurasia beginning in the 1930s and has subsequently become one of the most popular gamebirds in the Northwest. These hardy birds require hard-to-find water during the dry summers, so various agencies have installed "guzzlers"—water-catchment devices—across the driest parts of the birds' range.

LENGTH 13 inches. **WINGSPAN** 18.5 inches. **SEXES SIMILAR** Plump and chickenlike, with broad, rounded wings; gray-buff back and wing coverts, gray breast and crown; bold black and cream-barred flanks; rusty outer tail feathers visible in flight; white throat fully bordered by black necklace extending up neck and around forehead; red bill and legs. **VOICE** Common call is a loud, clucking then squeaky *chuk chuk chuk chukah-chukah-chukah*, increasing in intensity; whistled contact call is similar but higher pitched; the flush and flight call is a raspy, screeching *cheeuk cheeuk cheeuk*. **BEHAVIORS** Powerful runners, often sprinting uphill to escape threats, and can easily climb near-vertical slopes and scree slides; approached from above, coveys flush downhill. Strong, swift, direct flight, typically escaping downhill and then veering left or right, flapping swiftly and then gliding for considerable distances. Coveys typically flush as a unit, often cackling as they fly. During dry summers, chukars visit water sites daily. Readily feeds in grain fields adjacent to steep-slope habitat; often roosts and nests in rimrocks and steep rockslides. Winter coveys can exceed 100 birds. **HABITAT** Steep, arid, rocky slopes, especially river and creek canyons with extensive cheatgrass and stands of sagebrush or other cover; riparian zones during dry summer weather. Nests on cooler north-oriented slopes; winter range is high (to at least 7000 feet) southern exposures blown free of snow, including sagebrush flats above steep south-facing slopes; or below snowline on south-facing slopes. **STATUS** Year-round resident. Fairly common in southeast and north-central Oregon; uncommon elsewhere. Annual abundance depends heavily on spring nesting conditions, summer water availability, and winter severity.

Chukar range

Chukar

Gray Partridge (*Perdix perdix*)

The introduced Gray Partridge is a secretive covey bird primarily found in agricultural lands east of the Cascade Mountains. It is difficult to find without foot access to prime habitat, and most sightings occur along country roads leading through grain-producing regions within the Gray Partridge's range. The first birds released in North America came from Hungary, hence the popular name Hungarian partridge, or hun, which is more commonly used in the Northwest than the name Gray Partridge.

LENGTH 11 inches. **WINGSPAN** 18 inches. **SEXES SIMILAR** Plump and chickenlike, larger than a quail, with broad, rounded wings. Grayish brown overall, finely speckled, with narrow brown bars on the rump; orange-rust face and throat; vertical chestnut bars on flanks; dark maroon belly patch often missing on females; rusty orange outer tail feathers. **VOICE** Flush call a rapid, high-pitched squawking; male's crowing "rusty gate" display call most common in spring. **BEHAVIORS** Fast runners, they prefer running to flying when threatened; covey rises are explosive; strong and direct; coveys typically flush as a unit, usually with flush calls; most flights are fairly short, but birds subjected to hunting pressure may fly considerable distances; well-adapted to snowy winters, they sleep in depressions in the snow. **HABITAT** Grain, alfalfa, and clover fields and field edges, shortgrass prairie, Conservative Reserve Program (CRP) lands, sparse sagebrush steppe bordering farmlands; rolling hills with cheatgrass and sparse sagebrush or other shrub component. **STATUS** Uncommon year-round resident. **KEY SITES** Idaho: Arbon Valley; Avimor area (Ada County); Moscow area. Oregon: grain-producing regions of Gilliam, Morrow, and Sherman counties; Wallowa Valley near Enterprise (Golf Course Road, Leap Lane, School Flat Road). Washington: Okanogan Highlands; SR 129 south of Asotin; Waterville Plateau. British Columbia: White Lake area (Okanagan Falls).

Gray Partridges

Gray Partridge in flight

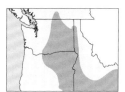

Gray Partridge range

Ring-necked Pheasant
(*Phasianus colchicus*)

America's best-known upland game-bird is actually an import, native to Asia. These striking birds have been widely introduced around the world, and the first North American stocking occurred in the early 1880s in Oregon's Willamette Valley.

LENGTH 21–35 inches. **WINGSPAN** 28–32 inches. **MALE** Unmistakable, with exotic plumage; iridescent green-blue head with bright red face, grayish crown, bold white neck ring; rich chestnut breast; golden flanks, each feather tipped with black; boldly patterned back, grayish rump; long bronze tail with black bars. **FEMALE** Camouflaging pattern of brown-black markings over soft tan background; long pointed tail. **VOICE** Male's display call a loud, echoing, raspy *cawk-cawk*; flush call an agitated, raspy cackled *ki-aak ki-aak ki-aak*. **BEHAVIORS** Explosive, powerful flush on loud, rapidly beating wings, males typically with rapid loud cackling; then alternating flapping and gliding. Pheasants can be attracted to backyard feeding stations offering corn scratch or similar poultry food; can adapt to suburban settings with sufficient cover and feed, where they become extremely stealthy and essentially silent. **HABITAT** Farmlands with row crops, especially grain and corn, with sufficient brushy cover; marsh edges and brushy riparian corridors; brushy woodland edges. **STATUS** Locally uncommon year-round resident, more common east of the Cascades.

Female Ring-necked Pheasant

Male Ring-necked Pheasant

Ring-necked Pheasant range

Ruffed Grouse (*Bonasa umbellus*)

Sometimes tame and approachable, Ruffed Grouse avoid detection through stealth and camouflage, and are often most easily found by walking or driving gravel forest roads through appropriate habitat or following tiny streams through aspen groves. They are well-adapted to long, snowy winters, feeding heavily on buds of willow, aspen, alder, and other plants. The male's courtship display includes fanning its tail, expanding the black ruffs on its neck, and producing a characteristic rapid thumping with its wings.

Gray-phase Ruffed Grouse

LENGTH 15–18 inches. WINGSPAN 19–22 inches. SEXES SIMILAR Plump and chickenlike; handsomely mottled above, lighter below; barred fanlike tail with a wide, black subterminal band; black ruffs on neck; thin crest. Two color phases: red-phase (rich rust-red) birds are found west of the Cascades; gray-phase birds predominate in the interior. Parts of the Cascade Range have both color phases as well as intermediate colors. VOICE Female uses a variety of calls with broods, including a rolling whiny alarm call, soft scratchy clucks, and low cooing; on rare occasions a Ruffed Grouse will turn on a human intruder and deliver what can only be described as a tongue-lashing of various scolding calls. BEHAVIORS Fast, explosive flushes followed by rapid escape flight on noisy wings. HABITAT Mixed woodlands near springs, seeps, streams, rivers in mountains and foothills; aspen groves and alder groves; mixed-age forests with strong young tree component. Strongly riparian in drier parts of range. STATUS Uncommon year-round resident.

Red-phase Ruffed Grouse

Ruffed Grouse range

Sooty Grouse
(*Dendragapus fuliginosus*)

Closely related to and very similar to the Dusky Grouse, Sooty Grouse are found primarily from the Cascades (both slopes) westward, whereas Dusky Grouse are found east of the Cascades. Their ranges in Oregon may overlap in the Ochoco Mountains; Idaho only has Dusky Grouse. But from north-central Washington northward, the range of the 2 species meets just east of the Cascade crest, and hybrids occur within and near the overlap zone. Hens are very difficult to distinguish, but males can be identified by their tails (dark gray with a light gray terminal band in Sooty Grouse, blackish without the gray band in Dusky Grouse).

LENGTH 16–22 inches. WINGSPAN 22–26 inches. Plump and chickenlike, with broad, rounded wings, but typically seen on ground. MALE Dark gray overall; long deep gray tail with light gray terminal band; yellow comb and air sac most evident during courtship display. FEMALE Richly mottled gray, brown, and tan, with white-ticked flank feathers. JUVENILE Similar to female but shorter tail. VOICE Male produces a series of loud, haunting, hollow hoots during courtship display; hens and broods communicate with soft clucks and coos; the occasional alarm call is a rapid, laughlike *chuk-chuk-chuk-chuk*; soft wails by chicks and juveniles. BEHAVIORS In late spring courtship display, the male fans its tail, puffs up its air sacs and breast, and emits low-pitched hoots; displaying males can be quite approachable; they often display from trees or stumps. In winter, Sooty Grouse become largely arboreal (a tree-dwelling species), feeding primarily on evergreen needles in dense, mature coniferous forest. Powerful and noisy flight, but fairly slow except when birds launch downhill from trees. HABITAT Mature coniferous forests and forest edges from near sea level to timberline, with strong component of true fir and Douglas fir; steep ridgelines with mature forest edges and diverse understory are preferred; also open, parklike conifer woodlands at or near timberline; hens with broods in particular may occupy mixed woodlands and regrowth areas; adult males prefer mature conifer stands. STATUS Uncommon year-round resident.

Female Sooty Grouse

Male Sooty Grouse

Sooty Grouse range

Dusky Grouse
(*Dendragapus obscurus*)

Closely related to and very similar to the Sooty Grouse, the Dusky Grouse is found primarily east of the Cascade Range, whereas Sooty Grouse is found from the east slope of the Cascades westward. Its range in Oregon may overlap in the Ochoco Mountains; Idaho only has Dusky Grouse. But from north-central Washington northward into British Columbia, the range of the 2 species meets just east of the Cascade crest, and hybrids occur within and near the overlap zone. Hens are very difficult to distinguish, but males can be identified by their tails—dark gray with a light gray terminal band in Sooty Grouse, blackish without the gray band in Dusky Grouse.

LENGTH 16–23 inches. **WINGSPAN** 22–26 inches. Plump and chickenlike, with broad, rounded wings, but typically seen on ground. **MALE** Dark gray overall; long black tail with very thin silvery tip; yellowish comb and red air sac most evident during courtship display. **FEMALE** Richly mottled gray, brown, and tan, with white-ticked flank feathers. **JUVENILE** Similar to female

but shorter tail. **VOICE** Male produces a series of quiet, very low-pitched, hollow hoots during courtship display; hens and broods communicate with soft clucks, coos, and wails. **BEHAVIORS** In late spring courtship display, male fans its tail, puffs up its air sacs and breast, and emits quiet, low-pitched hoots; loud single hoots often given near female; displaying males can be quite approachable; they usually display on the ground. In winter, Dusky Grouse becomes largely arboreal, feeding primarily on evergreen needles in mature coniferous forest. Powerful and noisy flight, but fairly slow. **HABITAT** Mature coniferous forest edges; high-elevation sagebrush and shrub steppe near timber stands; open mountain meadows; steep ridgelines with forest edges; parklike conifer woodlands at or near timberline. Hens with broods in particular may occupy mixed woodlands, aspen groves, and regrowth areas. **STATUS** Locally common year-round resident.

Dusky Grouse range

Displaying male Dusky Grouse

Female Dusky Grouse

Spruce Grouse
(*Falcipennis canadensis*)

Though generally secretive and retiring, Spruce Grouse are often so approachable and seemingly tame when encountered by humans that these handsome birds long ago earned the nickname of fool hen. This sobriquet, however, belies one of their primary defense mechanisms, which is to rely on excellent camouflage to avoid detection.

LENGTH 15–17 inches. **WINGSPAN** 21–23 inches. Plump and chickenlike, with broad, rounded wings. **MALE** Striking pattern of black, gray, and white; jet-black throat, with black extending down the breast but ticked at the sides and belly with white bars; grayish flanks with white-tipped feathers; gray rump, crown, and nape; jet-black tail with white-tipped gray tail coverts; bright red combs during courtship display. **FEMALE** Mottled brown, rusty tan, and gray above, brown-on-beige or white barring pattern on breast and belly; white-ticked mottled flanks; brown terminal band on tail. **VOICE** Generally silent, but low, soft, cooing clucks, especially by hens with broods. **BEHAVIORS** Loud, explosive flush; generally avoids flying, with most flights into or out of

trees. In courtship display, male, strutting about with bright red combs, puffs up his plumage, fans his tail and tail coverts, and, at the end of short flights, claps his wings together above his back, usually twice, making sharp, echoing, gunlike reports. **HABITAT** High-elevation fir and spruce forests and successional lodgepole pine forests. **STATUS** Uncommon year-round resident, except rare Wallowa Mountains, Oregon.

Female Spruce Grouse

Male Spruce Grouse

Spruce Grouse range

White-tailed Ptarmigan
(*Lagopus leucura*)

Without use of hiking boots, don't expect to see a White-tailed Ptarmigan. These small grouse, which turn pure white in winter, inhabit rocky mountain slopes and alpine tundra above timberline, so birders wishing to see one have little choice but to trek into the bird's oft-inhospitable habitat.

White-tailed Ptarmigan, summer plumage

LENGTH 12 inches. **WINGSPAN** 22 inches. Plump and chickenlike; pure white during winter; substantial variation in summer and fall plumage. **MALE** (summer) Rich, finely vermiculated shades of gray, tan, and black above with varying degrees of white; black spots across breast and upper flanks; white belly and leggings. **FEMALE** (summer) Richly mottled gray-black above, often with brown tones; black-on-white chevron pattern on breast and flanks; white belly and leggings. **VOICE** Rapid trilling chatter, high-pitched clucking, high-pitched screech. **BEHAVIORS** Secretive but often approachable; relies on excellent camouflage to avoid detection; strong and direct flight; capable fliers even though rarely flushed. **HABITAT** Rocky alpine tundra above timberline, including scree slopes. **STATUS** Year-round resident. Uncommon and very local in British Columbia and Washington. Very rare winter vagrant in northern Boundary County, Idaho. **KEY SITES** Washington: Pasayten Wilderness; Ptarmigan Ridge (Mount Baker area). British Columbia: Cathedral Lakes area, Cathedral Provincial Park.

White-tailed Ptarmigan, fall plumage

White-tailed Ptarmigan range

103

Greater Sage Grouse
(*Centrocercus urophasianus*)

The Greater Sage Grouse, perhaps the iconic bird of the northern Great Basin, was historically much more widespread, but populations are highly susceptible to changing land-use patterns, in part because these largest of the North American grouse have low reproductive rates compared to other gallinaceous gamebirds, and are longer lived. The early spring courtship display of the male Greater Sage Grouse is a must-see spectacle for Northwest birders.

LENGTH 22–30 inches. **WINGSPAN** 33–38 inches. Very large and chickenlike; long, broad wings; long, pointed tale; mottled brown, black, and white overall, with black belly; large, heavy bill. **MALE** Substantially larger than female; dark chin and black throat divided by white necklace; bright white breast. **FEMALE AND JUVENILE** Mottled plumage; buffy throat, speckled neck. **VOICE** Rapid clucking flush call at times; low-pitched murmuring clucks and coos used as contact calls among females and brood members. **BEHAVIORS** On communal courtship display areas called leks, typically used over many years annually in the spring, males perform a spectacular early morning courtship display. They fan their long-feathered tails, strutting around, puffing up their white ruffs, and inflating the yellow air sacks on their chests to produce hollow, watery popping sounds. Flight is strong and direct on loud, fairly slow wingbeats; Greater Sage Grouse often fly considerable distances. **HABITAT** Open sagebrush steppe at middle to high elevations; alfalfa and grain fields adjacent to typical sagebrush habitat; hens with broods will use riparian areas with good cover, including aspen and willow groves; post-breeding males disperse to high-elevation range

(over 9000 feet in places); winter movement downslope or to areas windswept of snow. **STATUS** Uncommon year-round resident. **KEY SITES** Idaho: widespread across southern half of state. Oregon: Hart Mountain National Antelope Refuge; Steens Mountain. Washington: Waterville Plateau (Douglas County).

Male Greater Sage Grouse

Female Greater Sage Grouse

Displaying male Greater Sage Grouse

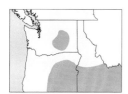

Greater Sage Grouse
range

Sharp-tailed Grouse
(*Tympanuchus phasianus*)

In both Oregon and Washington, land-use changes—conversion of native habitats to farmlands and ranch lands—virtually extirpated once-widespread and common Sharp-tailed Grouse. Today, Washington continues ongoing efforts to save and bolster 5 remaining geographically isolated populations, while in Oregon, efforts to reestablish a population in Wallowa County have met with only minimal success. In the Northwest, populations are strongest in eastern Idaho.

LENGTH 16–19 inches. **WINGSPAN** 20–24 inches. **SEXES SIMILAR** Plump and chicken-like, with fairly long, rounded wings, and long primary flight feathers; richly mottled in white, brown, and tan; breast appears scaled, transitioning to white belly flecked with rows of dark chevrons; tail short and pointed with white outer edges; breeding male has yellow comb and purple neck patch exposed during courtship dance. **VOICE** Flush and flight call a low rapid clucking; display calls include low, squeaky cooing, hooting, whining squeals, sometimes in

rapid succession. **BEHAVIORS** Flight is strong and direct; coveys often flush as a unit, typically with flush calls; escape flights can cover considerable distances. Like Greater Sage Grouse, uses traditional leks for spring courtship; males puff up their purple air sacs, droop and spread their wings, and bow with tail cocked up at 90 degrees. **HABITAT** Shrub steppe, meadow steppe, grain farm borders, CRP lands and riparian shrublands and woodlands within such areas; dense shrublands and trees required for winter cover. **STATUS** Year-round resident. Locally uncommon in Idaho; rare elsewhere. **KEY SITES** Idaho: Curlew National Grasslands and Arbon Valley; Hixon Sharptail Preserve (north of Weiser); Sand Creek Wildlife Area and Red Road (Fremont County). Washington: Scotch Creek Wildlife Area and Conconully Highway (Okanogan County); Swanson Lakes Wildlife Area.

Displaying male Sharp-tailed Grouse

Adult Sharp-tailed Grouse

Sharp-tailed Grouse range

Wild Turkey (*Meleagris gallopavo*)

Though not native to the Northwest, the Wild Turkey has largely thrived here after being introduced as a game species. So successful has this massive bird been that in some parts of its range the Wild Turkey is a familiar backyard species in suburban and rural areas.

LENGTH 45–47 inches (male), 34–36 inches (female). **WEIGHT** 15–24 pounds (male), 8–12 pounds (female). **SEXES SIMILAR** (except for size) Huge, largely terrestrial bird, with bronze plumage, featherless head and neck, long fanlike tail with black subterminal band and light tip; black-and-white-barred flight feathers; hens less iridescent and smaller. **VOICE** Males make familiar rolling gobble; both sexes use various clucking, purring, and grating yelping calls, commonly *chop chop chop-chop* and *tut-tut tut-tut*. High-pitched *zwit-zwit-zwit* from juveniles. **BEHAVIORS** Male's courtship display is iconic: gobbling, fully fanned tail, puffed-up chest, feathers fluffed up, wings drooping, and considerable strutting and posing. Night roosts can include high tree branches from which the birds depart in early morning gliding plunges, flapping furiously to slow before reaching the ground. **HABITAT** Highly adaptable; oak woodlands, ponderosa pine forests, open mixed woodlands, savannahlike grasslands, brushy riparian areas, farmlands, suburban areas (especially in southwest Oregon). **STATUS** Common year-round resident.

Wild Turkey

Displaying male Wild Turkey

Wild Turkey range

Loons and Grebes

Interesting aquatic birds, and accomplished divers, loons (Gaviidae family) and grebes (Podicipedidae family) are known for their elaborate courtship rituals. Western and Clark's Grebes, for example, perform a ritualized dance in which both sexes use head bobs and head throws, present billfulls of aquatic weeds to one another, and ultimately lift their bodies vertically above the water, crane their necks, flap their wingtips, and literally run along the water's surface in unison. Other grebe and loon species exhibit their own displays, and all 5 breed within the Pacific Northwest. Their larger cousins, the loons, are primarily Far North breeders, though the best-known of them—the Common Loon—breeds at a few locations within the region. Common Loons are renowned for their haunting calls.

Western Grebes

Red-throated Loon
(*Gavia stellate*)

Red-throated Loons nest in the Arctic, but many winter along the Northwest coast. Because they arrive on the wintering grounds as early as late summer and depart as late as midspring, some birds occur annually with partial or even fully developed breeding plumage, but most are in nonbreeding plumage.

LENGTH 25 inches. **WINGSPAN** 36 inches. Red eye; sharp bill with outer end of lower mandible angling upward (unlike Pacific Loon); bill frequently held angled slightly upward. **ADULT NONBREEDING** (October–April) Bright white face, throat, breast, and belly; white on face extends above eyes; dark gray back with dense rows of white spots; gray nape with white specks; gray crown. **ADULT BREEDING** (May–September) Full breeding plumage unmistakable, with pearl-gray head, rust-red throat, silvery nape and breast. In transitional plumage, partial red throat. **JUVENILE** (August–December) Similar to nonbreeding plumage, but dusky gray neck and pale bill; brownish gray speckled

back; first-spring adult has narrow reddish streaks on each side of throat. **VOICE** Generally silent in winter; flight call is a rapid, hollow, guttural *ack-ack-ack-ack*. **BEHAVIORS** Unlike other loons, can take off from water with only short running start, often just 20 to 30 yards; **HABITAT** Winters on salt water, especially bays, harbors, sheltered coves; rare inland. **STATUS** Common migrant and winter visitor (August–April); rare summer resident. **KEY SITES** Oregon: Tillamook Bay; Yaquina Bay. Washington: Grays Harbor; widespread in Puget Sound and Strait of Juan de Fuca. British Columbia: Clover Point (Victoria); Iona Island.

Adult breeding Red-throated Loon

Adult nonbreeding Red-throated Loon

Red-throated Loon range

Pacific Loon (*Gavia pacifica*)

Widely distributed along the Pacific coast during winter and migration, the Pacific Loon is a Far North breeder and the most numerous loon in North America. It nests on freshwater on the Arctic and near-Arctic tundra and taiga, then spends the rest of the year on salt water. By the time Pacific Loons reach the Northwest, most have molted at least partially into winter plumage, but many birds linger long enough in spring to transition into their striking breeding plumage.

LENGTH 25 inches. **WINGSPAN** 36 inches. Straight, sharp bill held horizontal (unlike Red-throated Loon, which typically holds bill angled upward). **ADULT NONBREEDING**

Adult nonbreeding Pacific Loon

(September–March) Dark grayish brown above, white below, often with narrow, partial dark chinstrap; dark crown, with grayish brown extending down to eye level; sides of neck are dark (nonbreeding Red-throated Loon has much more extensive white on face and neck). **ADULT BREEDING** (April–August) Light pearl-gray crown and nape; black throat flanked on sides of neck by thin, vertical, alternating black-and-white stripes; blackish back patterned with rows of square white spots; white breast and belly. **JUVENILE** (August–February) Similar to nonbreeding adult, but paler above, with faint speckling on back. **VOICE** Generally silent outside of breeding grounds. **BEHAVIORS** During migration, especially in spring, may form flocks of hundreds and sometimes thousands of birds over the open ocean; generally prefers open ocean, including near shore, more than other loons. **HABITAT** Primarily coastal; bays, harbors, coves, inlets, offshore. **STATUS** Common coastal migrant and winter resident (August–May); rare inland, but fairly regular lower Columbia River from Portland area westward. **KEY SITES** Oregon: Fort Stevens State Park and Columbia River mouth; Tillamook Bay; Yaquina Bay. Washington: Grays Harbor; widespread in Puget Sound and Strait of Juan de Fuca. British Columbia: Cattle Point and Clover Point (Victoria); widespread in Strait of Georgia.

Adult breeding Pacific Loon

Pacific Loon range

Common Loon (*Gavia immer*)

Much larger than Red-throated and Pacific Loons, the Common Loon is the famous yodeler of lakes in the north woods of Canada and the northernmost fringe of the United States. These widespread loons only breed in a few locales in the Northwest south of Canada, so opportunities to hear their eerie calls are limited; nonetheless, they are widespread and generally easy to find during migration and winter. Upon arrival in the Northwest in late summer and early fall, most Common Loons still wear a transitional plumage, sometimes very near breeding plumage, and birds lingering into spring usually wear the full breeding plumage by sometime in April.

LENGTH 33 inches. **WINGSPAN** 50 inches. Heavy bill with slightly curved culmen. **ADULT BREEDING** (April–August) Black head and neck, with broad necklace and breast patterned with very fine alternating black and white lines; all-black bill; black back boldly patterned with rows of square white spots. **ADULT NONBREEDING** (September–March) Grayish bill, with dark culmen; crown, nape, and sides of neck dark brown, with faint whitish partial neck band; brownish back; white throat, breast, belly. **JUVENILE** (August–January) Similar to nonbreeding adult, but lighter on back, with pale edges on feathers yielding a scaled appearance. **VOICE** Loud, wailing yodels, primarily on breeding grounds. **BEHAVIORS** Requires very long running start to takeoff from water. **HABITAT** Bays; ocean; large lakes and reservoirs; large, slow rivers. Breeds on freshwater lakes. **STATUS** Rare and local summer resident, except fairly common summer resident in British Columbia; common migrant and winter resident.

SIMILAR SPECIES The Yellow-billed Loon (*Gavia adamsii*), an Arctic species, is a rare winter visitor to coastal British Columbia and Washington, very rare to coastal Oregon. It is very similar to the Common Loon, except the entire bill is pale yellow, and the lower mandible's outer half angles upward, while the culmen is straight, not curved. Overall the bill appears upturned, an effect heightened by the bird's habit of tilting its head slightly upward. In winter plumage, the Yellow-billed Loon is paler then the Common Loon.

Adult breeding Common Loon

Adult nonbreeding Common Loon

Common Loon range

Pied-billed Grebe
(*Podilymbus podiceps*)

Though common, the little Pied-billed Grebe is unobtrusive, especially in the breeding season, when it is often more easily detected by its loud wailing calls than by sight. Widespread in the Northwest, they are most easily seen in fall and winter, when they occur nearly anywhere with open water.

LENGTH 13 inches. **ADULT BREEDING** (April–July) Distinctive heavy, pale bill with black band; black chin and forehead; dark grayish brown mantle, gray head and neck, brownish flanks; brownish gray breast; white undertail coverts. **ADULT NONBREEDING** (August–March) Brownish overall, darkest on back and crown; thick unmarked pale bill. **JUVENILE** (July–December) Similar to nonbreeding adult, but with pale tan stripes on face. **CHICK** Small and fuzzy, with vermiculated dark brown-tan plumage overall, especially on head and neck. **VOICE** Song is a loud, guttural, rapid *kuk-kuk-kuk-cheup-cheup-cheup* or some variation of such notes, such as a slower mix of high wailing *wuuk* notes and grunting notes. Calls include a rapid, tinny *cha-cha-cha-cha-cha chaa chaa chaa*. **BEHAVIORS** Often slinks quietly way from potential intruders by swimming with only head and neck above water or dives quietly and resurfaces farther away. **HABITAT** Breeds on freshwater marshes, swamps, sloughs, ponds, lake margins with emergent vegetation; winters on both freshwater and salt water. **STATUS** Common year-round resident; retreats from higher elevations after breeding season.

Adult Pied-billed Grebe

Juvenile Pied-billed Grebe

Pied-billed Grebe range

Horned Grebe *(Podiceps auritus)*

While breeding pairs of colorful Horned Grebes are very rare in the Pacific Northwest outside of British Columbia, these robust little grebes are common and widespread in the region during winter, especially on coastal waters and large, slow rivers. Late-departing spring migrants in May bear the handsome breeding plumage, but early arriving fall migrants in late summer are usually in transitional plumage.

LENGTH 14 inches. Short, daggerlike bill with white tip, and both mandibles gently curved (lower mandible angles upward in Eared Grebe); red eyes. **ADULT NONBREED-ING** (September–April) Dark grayish above; white below, with throat and white face extending up to eye and back to edges of nape; dark cap, sharply delineated from white face, unlike in nonbreeding Eared Grebe; head more rounded rather than peaked as in Eared Grebe. First-fall juveniles resemble adult, but throat and face washed in pale, dusky gray. **ADULT BREEDING** (May–July) Rusty red flanks and neck; bulbous, glossy black head with large, straw-colored to yellow ear tufts. **VOICE** Generally silent outside of breeding season. **BEHAVIORS** Often gathers in small groups in prime feeding locations; dives frequently when feeding. **HABITAT** Breeds on shallow, vegetated wetlands, ponds, lakes; winters extensively on bays, harbors, coves, open sea, as well as large inland lakes and reservoirs. **STATUS** Common winter resident, fairly common migrant; very rare summer resident, except uncommon summer resident of south-central British Columbia. **KEY SITES** Idaho: C.J. Strike Reservoir; Deer Flat NWR. Oregon: Broughton Beach and Marine Park (Portland); Fort Stevens State Park; Tillamook Bay; Yaquina Bay. Washington: Clarkston area (Snake River); McNary NWR and Sacajawea State Park; widespread in Puget Sound and Strait of Juan de Fuca. British Columbia: Kokanee Creek Provincial Park and Nelson area; Okanagan Valley; Richmond and Vancouver waters; Victoria area.

Adult nonbreeding Horned Grebe

Adult breeding Horned Grebe

First-winter Horned Grebe

Horned Grebe range

Eared Grebe (*Podiceps nigricollis*)

In its sharp breeding plumage, the Eared Grebe is a resplendent member of the alkaline marshland community of the interior Northwest, but its simple gray-and-white nonbreeding plumage is so nondescript that birdwatchers are well rewarded for making a trip to this small grebe's nesting locations to see it at its colorful best in late spring and early summer.

LENGTH 13 inches. Often a rear-heavy appearance with body highly domed; lower mandible angles upward at its tip. **ADULT BREEDING** (April–August) Black crown with ear tufts of sinuous yellow plumes fanning out from behind bright red eyes; black crest, when raised, creates sharply peaked crown; black neck; blackish mantle; cinnamon-red flanks. **ADULT NON-BREEDING** (September–March) Dark gray above, whitish below, with grayish neck and pale gray flanks; whitish breast; dark gray to dusky gray cheeks, unlike clean white cheeks of adult Horned Grebe; white throat, with white extending back below cheeks to sides of nape; head somewhat peaked, with a steep forehead rather than rounded as in Horned Grebe. **JUVENILE** Similar to adult nonbreeding, but eye is pale instead of red. **VOICE** On breeding grounds, song is a high, squeaking *owoy-ICKa* or *ohh-WICK*. **BEHAVIORS** Colonial nesters, Eared Grebes build platform nests of floating vegetation anchored to emergent marsh plants. **HABITAT** Breeds on vegetated ponds, lakes, wetlands, especially shallow alkaline waters; winters primarily on salt water. **STATUS** Locally common summer resident; fairly common migrant; fairly common winter resident along coast and Puget Trough, less common in British Columbia; rare winter resident inland. **KEY SITES** Idaho: Bear Lake NWR (breeding); C.J. Strike Reservoir; Market Lake Wildlife Area (breeding). Oregon: Upper Klamath Lake (breeding); Malheur NWR (breeding); Tillamook Bay; Yaquina Bay. Washington: Lenore Lake and Soap Lake; Turnbull NWR (breeding); widespread in Puget Sound and Strait of Juan de Fuca. British Columbia: Okanagan Lake; Robert Lake (breeding); White Rock Pier area (Vancouver).

Adult breeding Eared Grebe

Adult nonbreeding Eared Grebe

Eared Grebe range

Red-necked Grebe
(Podiceps grisegena)

Because they nest on isolated wetlands, lakes, and ponds, Red-necked Grebes are difficult to find during the breeding season. But for the balance of the year, they retreat to larger, more open waters, making them easier to observe. Estimates place the Northern Rockies breeding population at perhaps 500 pairs, most of them in British Columbia, but with several locations in both Idaho and Washington hosting nesting pairs. In decades past, a few pairs have nested near Rocky Point on Upper Klamath Lake, Oregon.

LENGTH 19–21 inches. **WINGSPAN** 32 inches. Long, daggerlike, yellowish bill, with slightly decurved culmen. **ADULT BREEDING** (April–August) White cheeks, chin, and throat; reddish neck; black crown; dark brownish gray above, whitish below. **ADULT NONBREEDING** (September–March) Grayish brown above, often with a tinge of pale rust on neck; grayish flanks; white breast and belly; dark crown contrasts with pale cheeks and pale ear patch. **JUVENILE** (August–March) Similar to nonbreeding, but lacks distinct face pattern, with pale cheek. **VOICE** Typical song is a rapid, squealing rattle followed by loud, wailing, grunts of *ar-ar-ar-whaaa-whaaa-whaaa*. Calls include a variety of grunts, squeals, and quacks. **BEHAVIORS** Aggressively chases other waterbirds away from nest area, often from underwater. Generally solitary in winter. **HABITAT** Breeds on shallow, vegetated wetlands, including lake margins, with sufficient open water for foraging; winters extensively on bays, harbors, coves, open sea, as well as large inland lakes and reservoirs. **STATUS** Fairly common summer resident of British Columbia, northeast

Red-necked Grebe range

Adult breeding Red-necked Grebe

Washington, and northern Idaho; uncommon summer resident in central and eastern Idaho. Common winter resident and migrant of Northwest coast, and fairly common winter resident and migrant of large interior lakes within summer range; uncommon winter resident and migrant in southern Idaho, eastern and central Oregon, and southeast and central Washington. KEY SITES Idaho: Coeur d'Alene Lake; Hayden Lake (breeding); Henrys Lake (breeding); Lake Pend Oreille. Oregon: Siuslaw River mouth; Tillamook Bay; Yaquina Bay. Washington: Little Pend Oreille NWR (breeding); Spectacle Lake (breeding); widespread in Puget Sound. British Columbia: Duck Lake and Creston area (breeding); Swan Lake (breeding).

Adult nonbreeding Red-necked Grebe

First-winter Red-necked Grebe

Western Grebe
(Aechmophorus occidentalis)

The Western Grebe, along with closely related Clark's Grebe, enact perhaps the most elaborate and showy courtship display of all Northwest birds, a performance that ranks among the must-see spectacles in the region. The 2 very similar grebes were considered conspecific until 1985, when they were divided into 2 species, and because their Northwest ranges are similar, close scrutiny is needed to differentiate them, especially in winter.

LENGTH 25 inches. Long neck, large head; daggerlike, yellowish olive bill; black above, white below, with bright white belly, breast, throat, chin, face; in breeding plumage, red eye surrounded by dark gray (usually white in Clark's Grebe). In nonbreeding plumage, color around eye is less diagnostic, often dusky (like Clark's Grebe), so bill color and flank color are best identifying characteristics. **VOICE** On breeding grounds, call is a high, reedy, *kreeer-keep*. **BEHAVIORS** Often forms flocks, sometimes numbering hundreds of birds, on wintering grounds; at

rest, both Western and Clark's Grebes often lay their long necks down rearward so their heads tuck up against their backs. **HABITAT** Breeds on freshwater marshes and wetlands with emergent vegetation; winters on coastal waters as well as large lakes and reservoirs, and large, slow rivers. **STATUS** Common summer resident; abundant winter resident and migrant. **KEY SITES** Idaho: C.J. Strike Reservoir; Deer Flat NWR; Lake Cascade; Lake Pend Oreille; Market Lake Wildlife Area; Mud Lake Wildlife Area. Oregon: Upper Klamath Lake; Malheur NWR; widespread on coast. Washington: Banks Lake; Moses Lake and Potholes Reservoir; Priest Rapids Lake and Wanapum Lake; Tri-Cities (Columbia and Snake Rivers); widespread on coast. British Columbia: Okanagan Valley.

Western Grebe range

Western Grebe

Clark's Grebe
(*Aechmophorus clarkia*)

The Clark's Grebe, like its very similar and one-time conspecific cousin, the Western Grebe, breeds on shallow wetlands of the interior Northwest, where it performs a remarkable must-see courtship display in late spring. These 2 largest of North American grebes share somewhat similar ranges in the Northwest, so close scrutiny is needed to properly identify them.

LENGTH 25 inches. Long neck, large head; daggerlike, yellow to orange-yellow bill; black above, white below, with bright white belly, breast, throat, chin, and face; in breeding plumage, red eye surrounded by white (dark gray in Western Grebe). In nonbreeding plumage, color around eye is less diagnostic, often very similar to Western Grebe, so bill color, as well as flanks that are visibly paler and whiter than on Western Grebe, are the best identifying characteristic. **VOICE** On breeding grounds, call is a high, clear, whistled *keeee* or *keeeeip*. **BEHAVIORS** At rest, both Western and Clark's Grebes often lay their long necks down rearward so their heads tuck up against their backs. Chicks of both species frequently hitch a ride on their parents' backs. **HABITAT** Breeds on freshwater marshes and wetlands with emergent vegetation; winters on coastal waters, large lakes and reservoirs, and large, slow rivers. **STATUS** Locally common summer resident (mid-April–September); uncommon migrant and winter resident. **KEY SITES** Idaho: Bear Lake NWR; Deer Flat NWR; Market Lake Wildlife Area. Oregon: Upper Klamath Lake; Malheur NWR; The Narrows (Malheur Lake). Washington: Moses Lake and Potholes Reservoir.

The Western Grebe (front) and Clark's Grebe (rear) are very similar. The Clark's Grebe has white around the eye, a bright yellow-orange bill (dusky yellow in the Western Grebe) and substantial white along the flanks.

Clark's Grebe

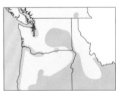

Clark's Grebe range

Albatrosses, Shearwaters, Fulmars, and Storm-Petrels

The Pacific Ocean off the coast of the Northwest, from nearshore waters to far offshore, is home to a variety of pelagic birds of the order Procellariiformes: albatrosses, shearwaters, fulmars and storm-petrels. Pelagic is a word meaning of or relating to the open ocean, and indeed these birds are ideally suited to life at sea, including having the ability to spend many hours in flight. Many of these birds cover vast distances in migration and in transit, only making landfall to nest, typically on islands and islets far out to sea. Only rarely are these pelagic birds seen from the shores of the Pacific Northwest, typically in the midst of or following violent oceanic storms, so birders seeking them often charter offshore boating trips, which are organized frequently by Audubon Society chapters and other groups in the region.

Sooty Shearwater

Black-footed Albatross
(*Phoebastria nigripes*)

The only all-dark albatross (other than the much-larger juvenile Short-tailed Albatross, which is very rare) in the Pacific Ocean off the Northwest Coast, and the only common albatross, the Black-footed Albatross is much larger than the all-dark species of shearwaters. They nest in the Hawaiian Islands, but spend the non-breeding season at sea, only very rarely coming within sight of land along the West Coast.

LENGTH 28 inches. **WINGSPAN** 82 inches. Very large, dark gray seabird with very long, slender wings; face around base of large, dark bill is silvery white (absent on juvenile); some adults have white undertail coverts. **VOICE** When flocking to feed, sometimes utters low hornlike groan and high wailing squeals; otherwise silent. **BEHAVIORS** Takes prey on the wing from on or near the surface; follows fishing boats for scraps, and readily zeroes in on chum set out to attract fish or birds. **HABITAT** Pelagic. **STATUS** Fairly common summer visitor (May–October).

Black-footed Albatross

Black-footed Albatross

Black-footed Albatross
range

Black-footed Albatross

Laysan Albatross
(*Phoebastria immutabilis*)

Like the closely related Black-footed Albatross, the Laysan Albatross nests on the islands of the Hawaiian Archipelago and a few other Pacific atolls and islands. Though widespread in the Pacific, it is seldom seen off the Northwest coast.

LENGTH 31 inches. **WINGSPAN** 80 inches. Very large; white below, with white head and neck; varying amount of dark in the otherwise white underwings, especially in juvenile; dark gray-brown mantle and dorsal wing surfaces; white rump, short black tail; very large pinkish bill; small, dark eye patch. **VOICE** Usually silent, but may squawk and squeal when flocking to feed. **BEHAVIORS** Takes prey on the wing from on or near the surface; follows fishing boats for scraps. **HABITAT** Pelagic. **STATUS** Rare year-round visitor; mostly December–March and August–October.

Laysan Albatross

Laysan Albatross

Laysan Albatross range

Laysan Albatross

Northern Fulmar
(*Fulmarus glacialis*)

Breeding in the Arctic, but dispersing throughout the North Pacific for the balance of the year, the Northern Fulmar is notable for its polymorphism: plumages range from a silvery light morph to a deep gray-brown dark morph, with every tone in between.

LENGTH 18 inches. **WINGSPAN** 42 inches. Fairly long, slender wings; stubby, thick yellowish bill; small, dark eye patch; thick neck. Very light morph is nearly all white, but usually with gray edging in the flight feathers and tail; light morph is gull-like, with a pale gray mantle and dorsal wing surfaces, and otherwise white overall; intermediate morph is ashy gray overall, slightly darker on mantle and dorsal wing surfaces; dark morph is uniformly sooty gray. Light and intermediate morphs generally have dark wingtips with a light central spot ("window"). **VOICE** Usually silent. **BEHAVIORS** Direct flight with quick wingbeats, often gliding on stiff wings. **HABITAT** Pelagic; sightings from shore are uncommon but regular, especially after powerful storms. **STATUS** Fairly common winter visitor (August–April); rare summer visitor.

Northern Fulmar

Northern Fulmar

Northern Fulmar

Northern Fulmar range

Sooty Shearwater
(*Ardenna griseus*)

The common Sooty Shearwater and the typically rare Short-tailed Shearwater are nearly identical. Key field marks when the bird is at rest on the water include length of primary flight feathers, head shape, bill length, and the shade of the throat relative to the breast and face. Flight patterns between the 2 species differ as well. Flocks of Sooty Shearwaters can number in the hundreds, sometimes in the thousands.

LENGTH 18 inches. **WINGSPAN** 40 inches. Dark gray brown to brownish overall; silvery white wing linings, especially on median primary coverts, contrasting with dark flight feathers; crown less rounded than in Short-tailed Shearwater, with sloping forehead and bill about same length as length of head. When the bird is at rest on the water, the wingtips extend to the end of the tail (or slightly beyond). **VOICE** Usually silent at sea, except high-pitched squealing during feeding frenzies. **BEHAVIORS** Flight pattern is usually strong, direct, and repetitive, with a series of flaps followed by short glides, whereas flight pattern of Short-tailed Shearwater is more frenetic, with more gliding, abrupt directional adjustments, and sometimes a side-to-side rocking motion while flapping. **HABITAT** Pelagic, but fairly often seen from shore. **STATUS** Common April–October, rare November–March.

Sooty Shearwater

Sooty Shearwater range

Sooty Shearwater

Short-tailed Shearwater
(*Ardenna tenuirostris*)

Short-tailed and Sooty Shearwaters are nearly identical. Key field marks when the bird is at rest on the water include length of primary flight feathers, head shape, bill length, and the shade of the throat relative to the breast and face. Flight patterns between the 2 species differ as well, and the Short-tailed Shearwater has more plumage variation, with some individuals nearly as uniformly dark as the Sooty Shearwater.

LENGTH 16 inches. **WINGSPAN** 38 inches. Dark head, nape, and mantle contrasting with light brown throat and breast; underwings usually uniformly pale or nearly so; rounded head with steep forehead and short, thin bill (shorter than the length of the head). When the bird is at rest on the water, the wingtips extend well past the end of the tail. **VOICE** Usually silent at sea. **BEHAVIORS** In contrast to the Sooty Shearwater's normally strong, direct, repetitive flight pattern, Short-tailed Shearwater is more frenetic, with more gliding, abrupt directional adjustments, and sometimes a side-to-side rocking motion while flapping. If present in the area, easily drawn to chum set out to attract birds on pelagic trips. **HABITAT** Pelagic. **STATUS** Uncommon fall and winter visitor (October–April).

Short-tailed Shearwater
range

Short-tailed Shearwater

Flesh-footed Shearwater
(*Ardenna carneipes*)

A very rare but fairly regular autumn transient off the Northwest coast, the Flesh-footed Shearwater is difficult to distinguish from the Sooty Shearwater and Short-tailed Shearwater, 2 species that are more common.

LENGTH 18 inches. **WINGSPAN** 40 inches. Uniformly dark chocolate brown, sometimes appearing almost black, with heavy, dark-tipped pinkish bill, and pale legs and feet. **VOICE** Usually silent at sea. **BEHAVIORS** Powerful flight on stiff wingbeats, often gliding; sometimes seen with far-more-common Sooty Shearwaters or other pelagic species. **HABITAT** Pelagic. **STATUS** Rare transient September–October; very rare May–August.

Flesh-footed Shearwater

Flesh-footed Shearwater range

Flesh-footed Shearwater

Pink-footed Shearwater
(*Ardenna creatopus*)

The most common white-bellied Shear-water off the Northwest Coast, the Pink-footed Shearwater actually nests on islands off far-away Chile, but outside of the breeding season, spreads out widely over the Pacific.

LENGTH 19 inches. **WINGSPAN** 43 inches. Dark brown mantle and dorsal wing surfaces; dusky brown rump, tail, neck, and head; white belly, throat, and underwing linings; pale grayish flanks; dark-tipped pinkish bill; pinkish legs and feet. **VOICE** Usually silent at sea. **BEHAVIORS** Flies on stiff wings, often gliding, and often tilting so that one wing dips down toward the water's surface. Usually occurs alone or in small groups, often mixed with Sooty Shearwaters, but large flocks (sometimes hundreds of birds) occur off the Northwest coast. **HABITAT** Pelagic, but often seen near shore and sometimes from shore. **STATUS** Fairly common summer and fall visitor (May–October).

Pink-footed Shearwater

Pink-footed Shearwater range

Pink-footed Shearwater

Buller's Shearwater
(*Ardenna bulleri*)

Owing to its unique and striking plumage pattern, the Buller's Shearwater is one of the easiest pelagic species to identify. These graceful birds occur in small flocks, but are often seen with other pelagic species. Like other shearwaters, they often glide into a tilt, with the lower wing nearly touching the water's surface.

LENGTH 16 inches. **WINGSPAN** 40 inches. Gray mantle separated from silvery gray secondary coverts by wide, diagonal, brown stripes on dorsal wing surfaces, and with dark gray wingtips—the entire dorsal surface pattern forms a broad M shape; black crown, nape, and tail; bright white breast, belly, flanks; white underwings completely outlined in dark gray and black. **VOICE** Usually silent at sea. **BEHAVIORS** Graceful in flight, frequently gliding; they sometimes feed by dipping bills into water while fluttering and gliding agilely into the wind. **HABITAT** Pelagic. **STATUS** Fairly common fall visitor (July–November).

SIMILAR SPECIES Manx Shearwater (*Puffinus puffinus*), a rare summer visitor (May–September), with most Northwest sightings off the Washington coast, is smaller than Buller's Shearwater (much smaller than Pink-footed Shearwater); clean white below, including flanks and throat, with white undertail coverts extending to end of short tail; also slender bill, and a small white crescent extending up from throat to behind eye.

Buller's Shearwater

Buller's Shearwater range

Buller's Shearwater

Fork-tailed Storm-Petrel
(Oceanodroma furcate)

The dainty little Fork-tailed Storm-Petrel, the only all-pale-colored Storm-Petrel, nests in burrows on Pacific Northwest islands—but is seldom seen at its breeding colonies because it only departs and returns from feeding and foraging trips at sea under cover of darkness. It is most easily seen on offshore birding trips.

LENGTH 8 inches. **WINGSPAN** 18 inches. Light steel-gray above, with silvery carpal bar (a stripe along the leading edge of the inner wing) contrasting with dark gray diagonal shoulder stripes in flight; pale gray below, with dark wing linings; light gray face with dark gray forehead and black eye patch; forked tail. **VOICE** Silent at sea. **BEHAVIORS** Storm-petrels often feed by fluttering along just above the ocean's surface and dipping down agilely to pluck small morsels from the water. **HABITAT** Pelagic, but nests on nearshore islands and islets; occasionally seen in bays and harbors. **STATUS** Locally common breeder in Washington and British Columbia, very rare breeder in Oregon; uncommon spring–fall (May–November) visitor, primarily offshore.

Fork-tailed Storm-Petrel range

Fork-tailed Storm-Petrel

Leach's Storm-Petrel
(*Oceanodroma leucorhoa*)

The tiny, wide-ranging, oceanic Leach's Storm-Petrel nests in burrows on islands but is rarely seen on its breeding grounds because it uses the cover of darkness to embark on pelagic foraging trips. It is the only common dark-colored Storm-Petrel in Northwest waters.

LENGTH 8 inches. **WINGSPAN** 18 inches. Dark brown overall, with black flight feathers and pale brown carpal bars; white rump; forked tail. **VOICE** Silent at sea. **BEHAVIORS** Often snatches prey from the ocean surface by hovering gracefully just over the water, sometimes with legs dangling and feet pattering on surface; also frequently feeds while sitting buoyantly on the surface. **HABITAT** Pelagic. **STATUS** Common summer resident and migrant (mostly April–August) well offshore.

Leach's Storm-Petrel

Leach's Storm-Petrel range

Leach's Storm-Petrel

Cormorants

Sleek and shiny in the water, cormorants are fish-eating, expert divers that sometimes incur the wrath of people concerned about the plight of salmon and steelhead in the Pacific Northwest. In reality, though, only the Double-crested Cormorant, equally at home in freshwater and salt water, is arguably culpable as a major predator of juvenile salmonids, but of course the birds can hardly be faulted for taking advantage of readily available food supplies. The other 2 species—the Pelagic Cormorant and the Brandt's Cormorant—are entirely marine.

Double-crested Cormorant

Brandt's Cormorant
(Phalacrocorax penicillatus)

Entirely marine, unlike the similar-sized Double-crested Cormorant, the Brandt's Cormorant is commonly seen feeding in bays and lounging on jetties and breakwaters, especially outside of the late spring and early summer breeding season.

LENGTH 34 inches. **WINGSPAN** 49 inches. Large and stocky with long neck; all black, appearing oily iridescent or iridescent green (or purple in breeding plumage) in good light; pale throat pouch in all seasons. Breeding adult has blue throat pouch bordered below by a pale straw-yellow; wispy white plumes on the upper neck and back. First-year birds all brown, darkest on back. **VOICE** Very low, hoarse grunts and growls generally heard only at close range or at breeding colonies. **BEHAVIORS** Birds traveling long distances, such as between night roosts and feeding areas, often fly in long straggling lines; flight is direct and fairly slow; laborious running takeoff from water. **HABITAT** Bays, outer surf zone, shallow seas; breeds on sea stacks and islands. **STATUS** Fairly common year-round resident.

Nonbreeding Brandt's Cormorant

Brandt's Cormorant in flight

Brandt's Cormorant range

Breeding Brandt's Cormorant

Double-crested Cormorant
(*Phalacrocorax auritus*)

The most widespread of the 3 Northwest cormorants, the Double-crested Cormorant is the only one that frequents freshwater as well as salt water. It is commonly seen on roosts, often with its wings spread open, apparently to dry its feathers.

LENGTH 33 inches. **WINGSPAN** 53 inches. All black, with conspicuous yellowish throat pouch year-round; breeding adult has white "ear" tufts. Juvenile is dark brown above, pale gray-tan below, lightest on throat and breast. **VOICE** Low, hoarse grunts and deep froglike croaks. **BEHAVIORS** When not fishing, typically perches on rocks, piers, trees, snags, and logs. Often swims low in the water with little more than head and neck above the surface. Flight is direct and fairly slow, on long, pointed wings; neck often slightly kinked in flight; laborious running takeoff from water. **HABITAT** Lakes, reservoirs, large rivers, swamps, bays, estuaries, tidal creeks and sloughs, nearshore ocean. **STATUS** Common year-round resident west of Cascades and Columbia River; fairly common summer resident and migrant in the interior.

Breeding Double-crested Cormorant

Nonbreeding Double-crested Cormorant

Double-crested
Cormorant range

Pelagic Cormorant
(Phalacrocorax pelagicus)

Unlike the other Northwest cormorants, the Pelagic Cormorant is not a large-colony nester, although loose aggregations of nesting birds occupy prime nest sites, typically on steep, rocky slopes or cliffs on sea stacks and headlands. Smallest of the cormorants in the North Pacific, the Pelagic Cormorant has a noticeably slighter build than the Double-crested and Brandt's Cormorants. This is especially obvious when Pelagic Cormorants and other species are seen in close proximity.

LENGTH 28 inches. **WINGSPAN** 39 inches. All black, oily iridescent in good light; long, snaky neck; breeding adult has white patches at rear end of flanks, distinctly visible in flight, as well as a small red throat pouch and inconspicuous slender white plumes on sides of neck; first-year birds uniformly dark charcoal brown. **VOICE** Low-pitched grunts and croaks. **BEHAVIORS** Despite its name, the Pelagic Cormorant nests and fishes near shore and frequently perches on rocks, jetties, buoys, and other objects for extended periods; direct, fairly slow flight, though swift when covering long distances; laborious running take-off from water. **HABITAT** Bays, outer surf zone, shallow seas; breeds on sea stacks and steep island faces. **STATUS** Fairly common year-round resident.

Pelagic Cormorant range

Breeding Pelagic Cormorant

Nonbreeding Pelagic Cormorant

Pelagic Cormorant in flight

Pelicans, Herons, and Ibises

Three families of birds compose the order Pelecaniformes: Pelicans (Pelecanidae family); herons, night-herons, egrets, bitterns (Ardeidea family); and ibis (Threskiornithidae). Herons and their allies are long-legged, long-necked predatory birds with daggerlike bills ideal for capturing and killing a wide variety of prey, from fish to small mammals. The White-faced Ibis, the Northwest's lone representative of this family, is unique in the region, with its lustrous plumage and long, curved bill. Two species of pelican live in the Northwest, the American White Pelican, an inland freshwater species, and the Brown Pelican, a marine species.

Great Blue Heron

American White Pelican
(*Pelicanus erythrorhynchos*)

The unmistakable American White Pelican is among the largest birds in North America. These huge white birds with massive pouched bills breed on wetlands of the interior Northwest, where they feed heavily on fish. During the migration seasons, observant birders may occasionally spy the glint of wings at extreme altitudes, as flocks of pelicans sometimes pass over or soar at nearly the extent of human vision.

Breeding American White Pelican

LENGTH 60–65 inches. **WINGSPAN** 100–110 inches. Massive white bird with huge bill and black flight feathers. **BREEDING** Orange bill and legs, with an orange finlike ridge on the upper mandible during early breeding season. **NONBREEDING** Yellowish bill and legs. **VOICE** Adults generally silent; chicks and juveniles on nesting colonies produce loud, raspy begging calls. **BEHAVIORS** Feeds by herding and scooping fish; several to numerous individuals often cooperatively herd fish, often into shallower water against shorelines, for easier capture. Flies on slow, powerful wingbeats; often glides; often soars when aloft. **HABITAT** Shallow marshlands, large lakes and reservoirs. **STATUS** Locally common summer resident; fairly common migrant in interior and lower Columbia River; uncommon migrant west of Cascades. **KEY SITES** Idaho: American Falls Reservoir; Deer Flat NWR; Hagerman Valley and Hagerman Wildlife Area; Harriman State Park. Oregon: Ladd Marsh Wildlife Area; Malheur NWR; Sauvie Island. Washington: Columbia River near Vantage; Columbia NWR and Potholes Reservoir; McNary NWR. British Columbia: Creston-area lakes.

American White Pelican in flight

American White Pelican range

Brown Pelican
(*Pelicanus occidentalis*)

Though it nests only as far north as Southern California, the Brown Pelican extends its range northward along the Pacific coast after breeding season, with many individuals remaining into midautumn, and a few staying into December or January, as far north as southern British Columbia.

LENGTH 46–52 inches. **WINGSPAN** 79–84 inches. Huge bill; grayish brown above, brownish below; white neck and pale yellow crown in nonbreeding plumage; chestnut-brown neck, yellow crown, reddish bill pouch in prime breeding plumage (into midsummer). **JUVENILE** Uniformly brown above, dirty white below. **VOICE** Mostly silent. **BEHAVIORS** Typically feeds primarily on fish, plunge-diving headfirst from the air into the water, wings folded at the wrists. Schools of prey fish may draw swarms of diving pelicans and other birds, sometimes within the surf zone. Strong, direct flight with frequent gliding low over the water, often traveling in lines. **HABITAT** Coastal, including bays, shallow inshore waters, outer surf zone. **STATUS** Fairly common summer and fall visitor; rare winter resident. **KEY SITES** Oregon: Chetco Point and Harris Beach; Columbia River mouth; Seaside; Yaquina Bay. Washington: Columbia River mouth; Grays Harbor; Ocean Shores; Willapa Bay.

Brown Pelican range

Adult breeding Brown Pelican

136

Brown Pelican in flight

Adult nonbreeding Brown Pelican

Juvenile Brown Pelican

American Bittern
(*Botaurus lentiginosus*)

Perfectly camouflaged to blend in with its marshland habitat, the stealthy, secretive American Bittern is a widespread, medium-sized heron that is probably more often heard than seen: its odd but distinctive low, resonant call emanates from the marshes during breeding season, but is ventriloquistic and very difficult to pin down to a location. Moreover, bitterns primarily call nocturnally and around sunrise and sunset.

LENGTH 28 inches. **WINGSPAN** 42 inches. Chunky, stout-necked heron; warm tannish brown overall; mottled yellowish tan and brown above, with long dark malar stripes; buffy white below with bold brown streaks running down throat, breast, and flanks; yellowish bill, yellow legs, bright yellow eye with back pupil. Juvenile lacks dark malar stripes. In flight, wings more pointed than rounded as in the Black-crowned Night-Heron, and dark flight feathers contrast with lighter shoulders and coverts. **VOICE** Call is a low, hollow, liquid *gulp-aloink*, somewhat like a dripping faucet, preceded by bill clicks (not always audible). Flush call is a low, hoarse croak or series of croaks. **BEHAVIORS** Generally feeds by ambush, remaining largely motionless for long periods of time; to avoid threats, typically freezes with neck stretched and bill pointing upward, blending in with the surrounding marsh vegetation. **HABITAT** Primarily dense hardstem freshwater marshes; also estuary marshes during winter. **STATUS** Uncommon summer resident (May–September); rare winter resident.

American Bittern

American Bittern

American Bittern in flight

American Bittern range

Least Bittern (*Ixobrychus exilis*)

Difficult to observe because they live in dense marshes, particularly extensive stands of cattail and bulrush, Least Bitterns are more easily heard than seen. They are not particularly shy, but they are well camouflaged and live in extensive seas of heavy marsh cover. Although widespread in the eastern half of the United States, this tiniest of North American herons has only spotty distribution in the West, where its northern breeding range limit is the extensive marshland of Oregon's Klamath Basin.

Least Bittern

Least Bittern range

LENGTH 12.5 inches. **WINGSPAN** 17 inches. **MALE** Tiny heron; rich, buffy tan overall; black back and crown; alternating buffy stripes running down pale throat and breast; yellow bill and legs; yellow feet with extremely long toes. **FEMALE** Similar to male, but paler overall, with brownish instead of black back and crown. **JUVENILE** Similar to female, but typically mottled above. **VOICE** Song is a series of low, cooing croaks, typically early in the morning; calls include a series of raspy *klick* notes, and a squawking *chek-chek-chek* alarm call. **BEHAVIORS** Often feeds by ambush, remaining motionless from favorite perches near the water's surface; escapes by skulking away through dense cover, or by flying weakly, often with legs dangling like a rail; also freezes in place to avoid detection. Flight is generally weak and of short duration; note contrast between dark flight feathers and brown shoulders. **HABITAT** Hardstem marshes dominated by cattail and/or bulrush. **STATUS** Rare and very local summer resident. **KEY SITES** Oregon: Klamath Marsh NWR; Upper Klamath Lake.

Great Blue Heron
(*Ardea herodias*)

Largest and commonest of the herons in the Northwest, the Great Blue Heron is equally at home in freshwater and brackish water. It feeds by stalking, often barely moving as it tracks the movements of potential prey, such as fish and frogs, until the moment of the kill, when it jabs with almost imperceptible speed to spear the unlucky creature that ventures too close. While strongly associated with wetlands, Great Blue Herons commonly hunt small rodents in fields.

LENGTH 44–48 inches. **WINGSPAN** 68–74 inches. **SEXES SIMILAR** Very long legs and long neck, often held in an S curve; long, daggerlike bill; grayish overall, with pale foreneck; black shoulders and black crown plumes most pronounced in breeding season; breeding season birds also have long gray plumes on the back, and long ashy gray plumes on breast and throat; dark blue-gray flight feathers. Juvenile lacks plumes and black shoulders. **VOICE** Call is a deep, froglike croaking. **BEHAVIORS** Nests in colonies called rookeries (colonial nesting sites of herons, egrets, and ibises), building huge nests in trees. In flight, holds neck in a compact S curve (unlike Sandhill Cranes, which fly with neck extended). **HABITAT** Freshwater and saltwater wetlands, including marshes, estuaries, sloughs, ponds, rivers, shallow lakes, and flooded and dry fields. **STATUS** Common year-round resident.

Adult Great Blue Heron

Great Blue Heron in flight

Juvenile Great Blue Heron

Great Blue Heron range

Great Egret (*Ardea alba*)

One of 2 pure-white heron species that regularly occur in the Northwest, the Great Egret is larger than the Snowy Egret and more widely distributed in the region. Standing out starkly against their surroundings, these 3-foot-tall birds hunt marshes and fields for fish, amphibians, and rodents. In prime hunting areas, especially in western Oregon during winter, sometimes several dozen Great Egrets will gather in the same place.

LENGTH 38 inches. **WINGSPAN** 50 inches. Pure white overall, with long black legs and black feet (Snowy Egret has yellow feet); long, daggerlike yellow bill (Snowy Egret has black bill). In breeding season, back is adorned with long white plumes. **VOICE** Call is a deep, hoarse, froglike croak. **BEHAVIORS** Hunts by stealth, stalking very slowly and often freezing in position for considerable lengths of time; snatches prey with lightning-fast jabs of bill. **HABITAT** Freshwater and saltwater marshes; wet and dry fields; lake and pond edges. **STATUS** Uncommon summer resident, fairly common winter resident and migrant in Oregon; common summer resident, uncommon winter resident and migrant in Washington; rare summer resident, uncommon migrant in Idaho. **KEY SITES** Idaho: Deer Flat NWR; Eagle Island State Park (Eagle). Oregon: Coos Bay; Fern Ridge Reservoir; Klamath Basin; Malheur NWR; Sauvie Island; Willamette Valley NWR Complex; Yaquina Bay. Washington: McNary NWR; Nisqually NWR; Potholes Reservoir area; Ridgefield NWR.

Great Egret

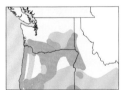

Great Egret range

Great Egret

Snowy Egret *(Egretta thula)*

Smaller and less robust than the Great Egret, the beautiful Snowy Egret breeds in the interior Northwest, especially in the marshes of southeast Oregon and southern Idaho. Post-breeding dispersal can take Snowy Egrets well outside their breeding range, and they regularly show up west of the Cascades, especially along the Oregon coast in places such as the Coos Bay area and Bandon Marsh.

LENGTH 24 inches. **WINGSPAN** 40 inches. Pure white overall, with long black legs and yellow or yellow-orange feet (Great Egret has black feet); long, thin, daggerlike black bill (Great Egret has yellow bill). In breeding season, back and breast adorned with long white plumes. **VOICE** Generally silent away from nesting areas; calls are hoarse croaks. **BEHAVIORS** Generally more active in hunting prey than larger, more robust Great Egret; often runs ahead quickly in pursuit of fish, amphibians, and other prey. **HABITAT** Breeds on freshwater marshes; migrants and winter visitors occur on freshwater and saltwater marshes. **STATUS** Uncommon summer resident and migrant in southeast Oregon and southern Idaho; rare but regular summer resident and migrant in eastern Washington. Rare migrant and winter visitor to western Oregon and Washington. **KEY SITES** Idaho: Bear Lake NWR; Fort Boise Wildlife Area; Market Lake Wildlife Area; Roberts Sewage Ponds and Roberts Slough (Jefferson County). Oregon: Klamath Basin; Malheur NWR.

Snowy Egret range

Snowy Egret

Cattle Egret (*Bubulcus ibis*)

Native to parts of Asia, Europe, and Africa, the Cattle Egret has made a massive range expansion since first arriving in South America in the 19th century. By the 1950s, these stocky little white egrets had arrived in North America and subsequently spread rapidly across much of the United States. The Cattle Egret is common in the South, has spotty distribution in the West, is highly migratory, and can show up just about anywhere with appropriate habitat, although it is rare in the Northwest.

LENGTH 21 inches. **WINGSPAN** 36 inches. Small, short-legged, stocky, white heron; nonbreeding adult (late summer through early spring) has dark legs and yellow bill; breeding-season adult (midspring through midsummer) has yellowish legs; yellow-orange bill; tannish orange plumes on crown, breast, and back. Juvenile has dark bill. **VOICE** Generally silent away from breeding grounds. **BEHAVIORS** Often feeds on insects in fields and open marshlands. **HABITAT** Open grasslands, fields, golf courses, pastures, shallow marshes and marsh edges, open lakeshores. **STATUS** Rare vagrant. Most sightings in southeast Oregon and southern Idaho occur in spring and summer; elsewhere, most sightings are from late autumn to winter.

Cattle Egret

Green Heron (*Butorides virescens*)

The small, rather exotic-looking Green Heron is a secretive denizen of a variety of wetland habitats in the Northwest. It is an astute angler and will bait fish with everything from bread crumbs to insects by dropping these lures on the water, then spearing any fish that comes in to feed. This angling behavior is best known from the Southern states, especially in locales where the birds have become accustomed to people, and so will tolerate close observation—and filming with cameras and phones—but it testifies to the ingenuity of these colorful little herons.

LENGTH 17 inches. **WINGSPAN** 26 inches. Deep chestnut-colored neck (less extensive on first-summer adults); pale tannish white throat and breast with bold chestnut streaks; fairly long neck is frequently tucked so tightly against the body that the bird appears almost neckless; oily, blackish green crown can be raised to a shaggy crest; grayish green back and wing coverts; yellowish orange feet; bright yellow irises; black bill with yellow underside. **JUVENILE** Brownish gray above, with buff-tipped wing coverts; brown streaks down whitish breast; dark crown; pale yellow legs. **VOICE** Varied repertoire of squawks, grunts, and other noises; common calls include a sharp, abrupt, slightly raspy *cheeow*, and a rapidly repeating *kuk-kuk-kuk-kuk*. **BEHAVIORS** Feeds by ambush, remaining motionless on a favorite perch just above the water, and also actively forages; very agile in heavy cover, including overhanging tree limbs and branch tangles. In suburban and urban settings, such as parks with good habitat, Green Herons seem to acclimate to people and become a bit less secretive and more easily observed. **HABITAT** Swamps and marshes, especially with wooded fringes; wooded sloughs and ponds; canopied creeks; riparian corridors with tree and shrub cover, from small streams to large rivers. **STATUS** Uncommon to locally common summer resident and migrant; rare winter resident. Very rare east of Cascades, except locally uncommon in Klamath Basin, Oregon.

Adult Green Heron

Juvenile Green Heron

Green Heron range

Black-crowned Night-Heron
(Nycticorax nycticorax)

A striking small heron most commonly found in the drier regions of the interior Northwest, the Black-crowned Night-Heron is largely crepuscular (active at dawn and dusk) and nocturnal, although it is often found foraging during daylight hours in the nesting season, when growing chicks demand constant feeding. When not active, it tends to roost in trees immediately adjacent to and within its wetlands habitat, and it will use the same roost for years on end.

LENGTH 25 inches. **WINGSPAN** 45 inches. Pearly gray overall, with white face and glossy black back; black crown with long, slender white plumes trailing over back; yellow legs; heavy black bill (sometimes yellowish with black tip and black upper mandible); red eyes; long, stout neck is rarely extended. Second-year birds are duskier overall with brownish rather than black backs. **JUVENILE** Brownish above, heavily patterned with white spots on shoulders and wing coverts; large white spots on back fade by first spring; pale tan below with brown streaks on breast, neck, and throat; speckled crown molting to brownish crown in first spring. Juvenile is easily confused with American Bittern, especially in poor light, but American Bittern lacks white dorsal spots. **VOICE** Guttural *woolk woolk woolk* and raspy *aaack*, mostly at rookeries and roosting colonies; flight call is a somewhat crowlike *walk*. **BEHAVIORS** Feeds by remaining motionless on a perch just above the water, and by stalking very slowly through marshes, mudflats,

Adult Black-crowned Night-Heron

Juvenile Black-crowned Night-Heron

ditches, and road edges. Black-crowned Night-Herons are colony nesters and their colonies sometimes comingle with other herons and egrets. **HABITAT** Wide variety of wetland habitats; breeds in marsh environments with appropriate trees or cattail stands for nests; colony nesting sites and day-roost sites are used repeatedly over the years. **STATUS** Locally uncommon summer resident; uncommon year-round resident and migrant. **KEY SITES** Idaho: Camas NWR; Deer Flat NWR and Wilson Spring Ponds (Nampa); Hagerman Wildlife Area; Lake Walcott State Park and Minidoka NWR; Market Lake Wildlife Area and Roberts Slough (Jefferson County). Oregon: Klamath Basin; Ladd Marsh Wildlife Area; Malheur NWR; Umatilla NWR and McNary Dam ponds. Washington: Columbia NWR and Potholes Reservoir; McNary NWR; Sprague Lake.

Black-crowned Night-Heron in flight

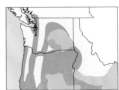

Black-crowned
Night-Heron range

White-faced Ibis (*Plegadis chihi*)

Often seen in large flocks during migration, White-faced Ibises are unlikely to be confused with any other Northwest bird. Their dark, glossy plumage sheens in good light, but in low light these wetlands birds can appear all black, especially in flight. Often foraging in grassy marshlands, pastures, and fields, large flocks can be so active and densely packed that from a distance they almost resemble a colony of black ants.

LENGTH 22 inches. **WINGSPAN** 36 inches. Long decurved bill, long legs, glossy plumage. **BREEDING** Deep glossy maroon from belly and breast to head, neck, and upper back; dark, glossy, oily green (sometimes appears purplish) wing coverts and lower back; bright red legs; reddish face framed by narrow white border. **NONBREEDING** Paler overall than breeding adult, with subdued speckling on head and neck. First-summer adults and juveniles similar to nonbreeding adult. **VOICE** Common calls include a nasal, ducklike *woink-woink*, and a deep, reedy *awk-awk-awk*. **BEHAVIORS** Commonly flies in wavering lines and V formations. **HABITAT** Marshes, fields, pastures, and lake edges. **STATUS** Summer resident and migrant (April–September). Common in southeast Oregon and southeast Idaho;

Juvenile White-faced Ibis (late July)

White-faced Ibis range

White-faced Ibis in flight

uncommon in southwest Idaho and western Oregon; rare in Columbia Basin, Washington, and central Idaho. **KEY SITES** Idaho: Bear Lake NWR; Camas NWR and Market Lake Wildlife Area; Duck Valley Indian Reservation; Grays Lake NWR. Oregon: Lower Klamath NWR; Malheur NWR and Burns area; Summer Lake Wildlife Area; Warner Valley; Wood River Wetland.

SIMILAR SPECIES The Glossy Ibis is a rare vagrant from the east, very similar to the White-faced Ibis; note dark face, dark iris, and darker legs.

Adult breeding White-faced Ibis

Hawks, Eagles, Falcons, Vultures, and Osprey

Patrolling skies throughout the Northwest, from ocean shores to alpine slopes, from barren desert to lush temperate forest, birds of prey—raptors—come in many varieties: the Turkey Vulture soaring aloft on summer days; Bald Eagle, iconic national symbol, and Golden Eagle, emblematic of the region's wildest places; half a dozen species of large, broad-winged *Buteo* hawks, including the ever-familiar Red-tailed Hawk; accipiters (Northern Goshawk, Cooper's Hawk, Sharp-shinned Hawk), the incredibly agile bird killers of the forests; 5 species of sleek falcons, including the diminutive and colorful American Kestrel and the flight speed champion Peregrine Falcon; Osprey, the fishing specialist; and the elegant and graceful White-tailed Kite and Northern Harrier. One kind or another of these predatory birds fill every imaginable niche.

Ferruginous Hawk

Turkey Vulture (*Cathartes aura*)

Circling overhead, wings held slightly upturned in a shallow V and often rocking left and right, the Turkey Vulture is the carrion-feeding specialist of the Northwest, using its highly developed sense of smell and superb eyesight to locate food. It expertly rides the thermals, often soaring for great lengths of time without need of flapping.

LENGTH 26–28 inches. **WINGSPAN** 67–70 inches. Grayish brown overall, but typically appears more black and gray; in flight, gray flight feathers contrast with darker wing linings; adult's head is red, juvenile's head is gray; head inconspicuous when bird is flying. **VOICE** Generally silent. **BEHAVIORS** Often roosts communally in large trees, especially dead trees, and manmade towers of various types; such roosts can host many dozens of vultures; sometimes perches with wings open, apparently to cool off or warm up. **HABITAT** Widespread, but generally in open regions, including deserts, valleys, ranch lands, road corridors. **STATUS** Common summer resident and migrant (February–October).

Turkey Vultures

Turkey Vulture

Turkey Vulture in flight

Turkey Vulture range

Osprey (*Pandion haliaetus*)

Colloquially known as fish hawks, Ospreys feed almost exclusively on the finned denizens of freshwater and salt water, plunge-diving with talons stretched far forward to snatch their prey in claws specially adapted to hold slippery fish. They build huge nests on elevated structures and have readily adapted to using utility poles and cell phone towers for nesting platforms, as well as artificial platforms placed for them in many Northwest locations.

LENGTH 22–25 inches. **WINGSPAN** 60–70 inches. **SEXES SIMILAR** Brown above; snow-white below (female has faint brown neck band); long wings often held with a crook, with dark wrists, secondaries, and wingtips; white face with dark masklike eye stripe; bright yellow eyes. **JUVENILE** Similar but somewhat speckled above. **VOICE** Call is a shrill, piercing whistle, often in series, typically *chee-earp*. **BEHAVIORS** Circles over water (often hovering) when hunting, often from considerable heights, then stoops (folds its wings in and dives from above) toward water, extending talons sharply forward to snatch shallow-swimming fish. Often seen flying overhead carrying fish headfirst. Bald Eagles, taking advantage of Ospreys' hard work, readily harass Ospreys into dropping their catch. **HABITAT** Near fishbearing lakes, reservoirs, large rivers, bays, estuaries; from sea level to mountains. **STATUS** Fairly common summer resident (March–September).

Osprey

Ospreys in flight

Osprey range

White-tailed Kite (*Elanus leucurus*)

Densities of this sleek and beautiful raptor fluctuate with the seasons, as post-breeding-season birds apparently disperse northward from California to bolster what appears to be a modest year-round population in the Northwest, making fall and winter the likeliest times to see them. These graceful birds were once nearly extirpated, and while they remain uncommon range-wide, their appearance in the Northwest is relatively recent, with annual sightings beginning in the 1970s.

LENGTH 13–15 inches. **WINGSPAN** 38–40 inches. **SEXES SIMILAR** White below, pearl-gray above, with black shoulders; white tail; white face, pale gray crown; pointed wings with gray primaries. **JUVENILE** Similar, but with a pale rust wash over mantle and breast. **VOICE** Calls include high chirping notes and *cheer* notes; a variable raspy, croaking *keraaack*. **BEHAVIORS** Hunts rodents over open country, flying gracefully and stopping often to hover at heights of up to nearly 100 feet (superficially similar male Northern Harrier tends to course low over the ground while hunting rather than frequently hovering). **HABITAT** Open wetlands, marshes, prairies, farmlands, grasslands. **STATUS** Uncommon fall and winter visitor, rare summer resident in Oregon; rare winter visitor and very rare summer resident of Washington. **KEY SITES** Oregon: Denman Wildlife Area and vicinity; Fern Ridge Reservoir and Wildlife Area; Finley NWR; Tillamook Bay wetlands.

White-tailed Kite in flight

White-tailed Kite

White-tailed Kite range

Northern Harrier (*Circus cyaneus*)

Exhibiting substantial sexual dimorphism, the Northern Harrier is a rodent hunter of open spaces and was once called "marsh hawk" due to its preference for wetland habitats. Not only do the male and female differ markedly in plumage, they also differ substantially in size; females average some 50 percent heavier than males (which are sometimes mistaken for White-tailed Kites).

LENGTH 18–20 inches. **WINGSPAN** 40–46 inches. Slender build; long, narrow wings often held upturned; long, barred tail and distinctive white rump; owl-like facial disk (most noticeable on female). **MALE** Light gray above, white below, with black wingtips; bright yellow legs and eyes. **FEMALE** Brown above; buff below with faint streaks; brown face; strongly barred wings and tail. **JUVENILE** Similar to adult female. **VOICE** Generally silent. **BEHAVIORS** Very buoyant in flight, almost appearing in slow motion at times. Courses low over hunting grounds, with frequent directional changes, often gliding and rocking side to side, with wings angled slightly upward. In migration, surprisingly, Northern Harriers often fly very high. **HABITAT** Marshes, farmlands, grasslands, estuary edges, shrub steppe. **STATUS** Fairly common year-round resident, migrant, and winter visitor; depending on season, the Northwest has a mix of resident, wintering, and migratory populations.

Male Northern Harrier

Female Northern Harrier

Northern Harrier range

Male Northern Harrier in flight

Female Northern Harrier in flight

Juvenile Northern Harrier in flight

Female Northern Harrier in flight

Bald Eagle
(*Haliaeetus leucocephalus*)

Nationwide, the iconic Bald Eagle has made a remarkable recovery since its listing as an endangered species by the federal government in 1967 and is now one of only a handful of species whose populations have rebounded sufficiently to be removed from the endangered list. The Northwest hosts significant numbers of these massive raptors, with Oregon and Washington both ranking in the top 5 states for Bald Eagle numbers, and British Columbia home to some 20,000 eagles—nearly a third of the continent's entire population.

LENGTH 29–36 inches. **WINGSPAN** 70–92 inches. **SEXES SIMILAR** Adult (5-plus years of age) is distinctive, with all-white head and tail, and dark brown body. **JUVENILE AND SUBADULT** Dark brown overall with varying degrees of white in wing linings, on coverts and back, and in tail; 4th-year birds show dingy white head and tail. **VOICE** Calls include high shrieks, barks, and wails. **BEHAVIORS** Bald Eagles adapt different foraging strategies based on relative prey abundance and location; in coastal waters

Adult Bald Eagle

Bald Eagle range

Second-year Bald Eagle

155

Juvenile Bald Eagle in flight

Third-year Bald Eagle in flight

and elsewhere they prey heavily on other birds, especially waterfowl and seabirds. On large lakes and reservoirs, live and dead fish become an important part of their diet. In ranching areas, they feed on dead livestock and afterbirth of livestock (such as sheep in Oregon's Willamette Valley). On Northwest rivers, post-spawn dead and dying salmon make up an important food source, drawing Bald Eagles in substantial numbers. **HABITAT** Winter: bays, coastal wetlands, river deltas, rivers with salmon runs, ranch lands. Breeding season: forested lakes and reservoirs; rivers with sufficient tree stands for nesting; coastal forests. **STATUS** Uncommon summer resident; locally common winter visitor and migrant. **KEY SITES** (winter congregations) Idaho: Boise River Valley; Hagerman Valley; Lake Coeur d'Alene (Wolf Lodge Bay, Higgins Point). Oregon: Columbia River Gorge; Klamath Basin. Washington: Columbia River Gorge; Lower Klickitat River; Skagit River and Skagit Delta. British Columbia: Boundary Bay; Harrison River; Squamish River; Vancouver-Richmond waterfront.

Adult Bald Eagle in flight

Golden Eagle (*Aquila chrysaetos*)

Symbolic of remote, wild, rugged lands, the majestic Golden Eagle is an imposing and deadly predator, capable of taking prey as large as mountain goat kids, pronghorn fawns, and juvenile coyotes—they've even been known to kill other raptors, including Peregrine Falcons. Often seen soaring high overhead far from human habitation, these massive raptors are even more impressive up close, when the sun reflects off the fiery brown nape for which they are named.

LENGTH 30–36 inches. **WINGSPAN** 78–88 inches. **SEXES SIMILAR** Chocolate-brown overall; golden nape, especially showy in good sunlight; large, dark, hooked bill with yellow cere; feathered legs and large yellow feet. In flight, long, broad wings and promi-

nent separation of primaries; lightly banded tail with broad dark terminal band; flight feathers palest at base and dark on tips; buffy undertail coverts and wing coverts. **JUVENILE** Similar to adult, but base of tail is white, contrasting with broad, dark terminal band; usually distinct white patches at base of inner primary flight feathers. **VOICE** Call is a high yelp. **BEHAVIORS** Feeds on mammals and birds, as well as carrion (especially in winter); mated pairs sometime hunt cooperatively, an especially effective strategy when hunting gamebirds, such as Chukar or sage grouse. **HABITAT** Open country with canyons and steep, rocky slopes, primarily in mountainous and other remote regions, including sagebrush steppe, deserts, prairies, high plains, rugged mountains. **STATUS** Uncommon year-round resident; rare west of Cascades.

Adult Golden Eagle

Adult Golden Eagle in flight

SIMILAR SPECIES Juvenile Bald Eagles are often misidentified as Golden Eagles. The first-year Bald Eagle has an all-dark breast like a Golden Eagle, but when seen in flight from below, the juvenile Bald Eagle has white "arm pits" and white in the wing linings, whereas the juvenile and subadult Golden Eagles have white patches at the base of the flight feathers toward the outer third of the wings. At close range, note that the Golden Eagle's feet are feathered all the way to the toes, whereas the Bald Eagle has bare lower legs (tarsi).

Golden Eagle range

Juvenile Golden Eagle in flight

Red-shouldered Hawk
(Buteo lineatus)

A relatively new arrival to the Northwest, having expanded its range northward from California, the colorful Red-shouldered Hawk is now fairly common in western Oregon, especially during the nonbreeding season. Smaller than the common Red-tailed Hawk, the Red-shouldered Hawk rarely strays far from the cover of at least a smattering of trees, and rarely perches at the tops of tall trees as Red-tailed Hawks often do.

Adult Red-shouldered Hawk

LENGTH 17–20 inches. **WINGSPAN** 38–42 inches. **SEXES SIMILAR** Reddish orange below (barring on breast and belly at close range); reddish orange wing linings; distinctive black-and-white-barred wing coverts and flight feathers; black-and-white-barred tail (often fanned in flight); rufous-gray mottled nape; when seen overhead or in flight from above, shows translucent,

First-winter juvenile Red-shouldered Hawk

light-colored "windows" near its wing-tips. **JUVENILE** Similar overall to adult, but off-white below, patterned heavily with reddish brown spots and streaks; brown-streaked head. **VOICE** Call is a high, ringing, repeated *keeyeeer* or *key-earup*; also brief, sharp, wailing whistles. **BEHAVIORS** Along roadways, Red-shouldered Hawks often perch on telephone wires, while Red-tailed Hawks seldom do, preferring telephone poles and fence posts, providing an initial clue to the possible identity of a wire-perching large hawk. Red-shouldered Hawks often perch quietly on tree branches below the canopy, or low snags, watching for prey below. **HABITAT** Semi-open areas with sufficient deciduous trees and snags

for perching; river bottoms, wetlands, farmlands, suburban margins, lake margins, fairly open deciduous woodlands. **STATUS** Common winter visitor and migrant and uncommon summer resident in southwest Oregon; locally fairly common winter resident of Willamette Valley, Klamath Basin, and coastal Oregon; rare winter resident and uncommon during post-breeding dispersal (July–October) in south-central and southeastern Oregon; rare winter resident of western Washington. **KEY SITES** Oregon: Beaver Creek Natural Area and Beaver Creek Road (Lincoln County); Denman Wildlife Area and White City area; Emigrant Lake; Fern Ridge Wildlife Area; Finley NWR. Washington: Ridgefield NWR.

Red-shouldered Hawk range

Adult Red-shouldered Hawk in flight

Broad-winged Hawk
(*Buteo platypterus*)

The Broad-winged Hawk is largely an eastern North American species, but its breeding range stretches northwesterly across the Canadian plains through Alberta and into northeastern British Columbia. Consequently, while their migration route is primarily east of the Rocky Mountains, a few of these attractive hawks occur in the Northwest, mostly during September and October.

LENGTH 14–16 inches. **WINGSPAN** 33–36 inches. **SEXES SIMILAR** Small *Buteo* species, generally seen in flight; wide wings often taper to a point at the tips; undersides of wings are white but outlined in black at the trailing edges and tips; broad black and white bars on squared tail; rufous-brown upper breast and throat; dark brown bars patterning breast, belly, and flanks; dark brown above. **JUVENILE** Speckled breast, varying from heavily patterned to just a few streaks along both sides of breast and neck; tail has black subterminal band and usually a series of faint narrow bands above; dark trailing edge of wings may be incomplete. **DARK PHASE** Very rare in Northwest; dark brown overall; white flight feathers with black tips contrast strongly with dark brown wing linings; adult and juvenile tail patterns same as in light morph. **VOICE** Rarely heard in the Northwest; call is a very high, even whistle. **BEHAVIORS** Wings often held slightly crooked, with wrists pointing forward and wingtips pointed. **HABITAT** In the Northwest, migrants are usually found flying over mountainous areas. **STATUS** Vagrant; rare fall migrant (September–October), very rare spring migrant (April–June).

Adult Broad-winged Hawk in flight

Juvenile Broad-winged Hawk in flight

Red-tailed Hawk
(*Buteo jamaicensis*)

The most common and widespread hawk in the Northwest, the aptly named Red-tailed Hawk is a habitat generalist, able to thrive in just about any open or semi-open area from the coast to the Rockies. Continentwide, Red-tailed Hawks come in myriad plumage variations, but the Northwest is dominated by light and dark morphs of the Western Red-tailed Hawk subspecies (*Buteo jamaicensis calurus*); the light morph varies in shade and breast and belly pattern. Uncommon in the region, the dark-colored Harlan's Red-tailed Hawk, or simply Harlan's Hawk (*B. j. harlani*), differs substantially in appearance from the typical Western Red-tailed Hawk.

LENGTH 18–23 inches. **WINGSPAN** 46–52 inches. **SEXES SIMILAR** Large, robust *Buteo* species; in all adult plumages in the Northwest, tail is rusty red; in light morph adult and juvenile, inner leading edge of wing is dark (unlike all other *Buteo* species in the Northwest); in adult plumages, pale flight feathers have dark tips, creating dark wing edges; dark head. **LIGHT MORPH ADULT** Rich light buff below, with variable density of darker streaks forming a band across the belly, or occasionally across the breast; dark brown above. Light morph juvenile white to pale yellowish tan below, with dark streaks forming distinct band across the belly; rear wing edges show thin, dark edge; pale patch on outer wings visible from above and below; pale gray to whitish tail with thin bars; white-edged scapulars form ragged V on back. **DARK MORPH ADULT** Deep rufous to dark brown overall; seen from below in flight, pale flight feathers contrast strongly with dark wing linings and wing edges. **DARK MORPH JUVENILE** Similar to adult,

Adult Red-tailed Hawk in flight

Juvenile Red-tailed Hawk

Adult Red-tailed Hawk

but grayish tail with narrow bands; rear wing edges not as distinctly dark. **HARLAN'S HAWK** Sooty black overall with variable light speckling on breast; throat ranges from white to speckled to black; tail is light gray rather than rufous; from below, pale flight feathers contrast with dark wing linings; rare light morph Harlan's Hawk is typically white below with limited dark speckling and pale (instead of rusty) tail. **VOICE** Call is a loud, raspy, shriek, *peeyaaaaw* or *peey-eeer*. **BEHAVIORS** Usually hunts by perching on utility poles, fence posts, trees, and many other objects, including hay bales, farm equipment, elevated irrigation systems, stone outcrops, road signs and billboards, even freeway overpasses; from such vantage points, the hawks plunge-dive onto prey. **HABITAT** Valleys, prairies, estuaries, open woodlands, shrub steppe, sagebrush-juniper steppe, ranch lands and farmlands, road corridors, residential areas, mountain slopes, clear-cuts and other forest openings. **STATUS** Common year-round resident and migrant.

Dark morph Red-tailed Hawk in flight

Dark morph juvenile Red-tailed Hawk

Red-tailed Hawk range

Dark morph Red-tailed Hawk

163

Swainson's Hawk
(*Buteo swainsoni*)

Often seen perched on telephone poles or soaring, wings slightly upturned somewhat like a vulture, the Swainson's Hawk is a bird of open country and one of the longest-distance migrants among raptors. The entire North American population spends the nonbreeding season in South America, largely in Argentina. Like the other Northwest *Buteo* hawks, the Swainson's Hawk occurs in both light and dark morphs, but also various color phases in between, with a rufous-colored intermediate morph; the light morph is most common.

LENGTH 19–21 inches. **WINGSPAN** 50–53 inches. **SEXES SIMILAR** In flight, narrow wings typically appear more pointed than other *Buteo* hawks, and are usually tilted upward; partially feathered legs (fully feathered in Ferruginous Hawk). **LIGHT MORPH ADULT** Whitish below, with distinctive broad, dark brown bib and white throat and chin, with white usually extending up around base of bill; dark gray-brown above; grayish brown face; in flight, whitish wing linings contrast with dark flight feathers; light tail with faint barring. Juvenile is similar, but speckled below, and often with a largely light-colored head. **DARK MORPH ADULT** Dark brown overall, usually with pale undertail coverts; reddish brown wing linings; gray flight feathers with black tips; prominent dark subterminal band on tail. Juvenile is similar but speckled; flight feathers are uniform in shade. **VOICE** Calls include a high, even screech, and high, piping shrieks. **BEHAVIORS** Adept on their feet, Swainson's Hawks sometimes chase down prey on the ground, walking and running; they eat large insects, such as grasshoppers, but generally rely on small mammals,

Adult light morph Swainson's Hawk in flight

Adult light morph Swainson's Hawk

164

reptiles, and even small birds during the summer nesting season in the Northwest, where they often perch on telephone poles, fence posts, and elevated irrigation pipes. **HABITAT** Prairies, shrub steppe, open farmlands and ranch lands. **STATUS** Summer resident (April–September). Fairly common in southern Idaho, southeast Oregon, northeast Oregon (Interstate-84 corridor), and Columbia Plateau; uncommon elsewhere in Washington. **KEY SITES** Idaho: City of Rocks National Reserve; Moscow area; Snake River Birds of Prey National Conservation Area; Treasure Valley. Oregon: Burns area; Chewaucan River Valley (Paisley to Valley Falls); Highway 20 Brothers to Burns; Malheur NWR. Washington: Columbia NWR and Desert Wildlife Area; Highway 12 Touchet to Walla Walla Valley.

Adult intermediate morph Swainson's Hawk

Adult dark morph Swainson's Hawk

Juvenile Swainson's Hawk

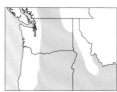

Swainson's Hawk range

Ferruginous Hawk (*Buteo regalis*)

Largest and palest of the *Buteo* hawks found in the Northwest, the Ferruginous Hawk resides in wide-open country, shunning mountainous regions and forested areas. Like the other large hawks, it occurs in both a light morph and a dark morph, but dark-plumage Ferruginous Hawks are quite uncommon, estimated at no more than 10 percent of the population.

LENGTH 23–26 inches. **WINGSPAN** 52–58 inches. **SEXES SIMILAR** Long, pointed wings; white to whitish tail and white flight feathers (with dark tips) in all plumages; yellow legs; large bill with yellow cere. **LIGHT MORPH ADULT** Snow-white belly, breast, and throat; very light wing linings only minimally contrast with white flight feathers; dark cinnamon leggings form a dark V against the belly when bird is seen in flight from below; light gray-brown above, with bright, rusty lesser secondary coverts; brown-and-white-speckled nape; pale gray face; in flight seen from above, rusty shoulders (lesser secondary coverts) contrast with

Adult Ferruginous Hawk in flight

Adult Ferruginous Hawk

Adult Ferruginous Hawk

grayish flight feathers. Juvenile similar; tail sometimes with pale bands, and pale leggings. **DARK MORPH ADULT** All dark brown when perched, with pale tail and flight feathers if visible; in flight from below, whitish tail and white flight feathers contrast with dark body and wing linings; white crescent near wrists. Juvenile is similar but with pale bands through tail. **VOICE** Call is a dry, wailing shriek, *weeaaaaaw*. **BEHAVIORS** Hunts by soaring, often quite high, and also by perching, especially on telephone poles. **HABITAT** Open grasslands, shrub steppe, sagebrush plains, desert, juniper-woodland edges. **STATUS** Uncommon summer resident (April–September); rare winter resident. **KEY SITES** Idaho: hills south of Emmett; Interstate-84, Interstate-86 to Utah border; Snake River Birds of Prey National Conservation Area. Oregon: Christmas Valley; Highway 20 Brothers to Burns; Highway 205 Burns to Malheur NWR; Fort Rock State Park. Washington: Highway 28 Wilson Creek to Odessa.

Dark morph Ferrugionus Hawk

Dark morph Ferrugionus Hawk in flight

Ferruginous Hawk range

Juvenile Ferruginous Hawk

Rough-legged Hawk
(*Buteo lagopus*)

As Swainson's Hawks and Ferruginous Hawks depart the interior Northwest in late summer and early autumn for wintering grounds farther south, Rough-legged Hawks from the Far North begin taking their place, becoming—along with ubiquitous Red-tailed Hawks—the predominant open-country *Buteo* species throughout the winter. Even more so than the other *Buteo* hawks of the Northwest, Rough-legged Hawks exhibit wide plumage variation.

LENGTH 19–20 inches. **WINGSPAN** 52–54 inches. **SEXES SIMILAR** White tail with broad dark terminal band is diagnostic in all color morphs; typical-plumage adults show dark belly contrasting with light breast, throat, and head, and black edges to light flight feathers, as well as black wrist patch on underside of wings. Lightest morph birds usually appear snowy white below, lightly speckled on belly, but retain dark wrist marks and wing edges; some light morph adults, however, can be heavily speckled.

Dark morph birds can range to all dark brown (almost black), but flight feathers (sometimes just primaries) have white bases and dark tips. **VOICE** Generally silent, but occasionally a high, wailing alarm call. **BEHAVIORS** Often hovers while hunting, unlike Red-tailed Hawk (which often occupies same regions in winter). **HABITAT** Open country; ranch lands, prairies, sagebrush steppe, large grain fields, open and frozen wetlands. **STATUS** Winter resident (November–March); fairly common east of Cascades; uncommon west of Cascades.

Light morph Rough-legged Hawk in flight

Light morph female Rough-legged Hawk

Rough-legged Hawk range

Dark morph Rough-legged Hawk in flight

Light morph male, heavily marked Rough-legged Hawk in flight

Sharp-shinned Hawk
(*Accipiter striatus*)

The sudden—usually lightning-fast—arrival of a Sharp-shinned Hawk at a backyard bird-feeding station, and the oft-resultant kill of a songbird, comes as a jolting surprise to onlookers. These small woodlands raptors, specialists at preying on smaller birds, have learned to take advantage of the windfall provided by the proliferation of bird-feeding birdwatchers, especially during winter, when "sharpies" often roam neighborhoods, routinely looking for prey in the form of songbirds at feeding stations.

LENGTH 10–13.5 inches. **WINGSPAN** 21–25 inches. **SEXES SIMILAR** Females substantially larger than males; short, rounded wings; heavily black-and-white-barred flight feathers visible when seen from below in all plumages; long tail, gray (adult) or gray-brown (juvenile), with alternating black bands and thin, white terminal band; when all tail feathers are visible and the bird is not soaring with fanned tail, the tail is noticeably squared at the corners (Cooper's Hawk has rounded tail); small head. **ADULT** Dark steel-gray above; white below, densely patterned with orange bars; dark gray hood (Cooper's Hawk has distinct dark cap contrasting with gray neck); thin, bright yellow legs; red eyes. **JUVENILE** Brownish gray above; white below, patterned with bold rufous streaks; yellow eyes. **IN FLIGHT** Wings pushed forward, making the head look tucked in between the wings, whereas the Cooper's Hawk flies with straight wings and head extended (crosslike shape). **VOICE** Call, not often heard, is a high,

Juvenile Sharp-shinned Hawk taking off

Adult Sharp-shinned Hawk

rapid *kill-kill-kill-kill-kill*. **BEHAVIORS** Often perches quietly in cover, waiting for an opportunity to make sudden, short flights after small songbirds; sometimes flies behind cover (including rooftops and other manmade structures) to approach prey, appearing suddenly for a high-speed attack. **HABITAT** Breeds in dense, fairly young coniferous and mixed woodlands; uses a wider variety of habitats in migration and winter, including urban and suburban areas; found from sea level to high in mountains. **STATUS** Uncommon year-round resident, summer resident, and migrant. Sharp-shinned Hawks are migratory, but many remain in the Northwest all year, with winter populations probably bolstered by migrants arriving from farther north.

Sharp-shinned Hawk range

Juvenile Sharp-shinned Hawk in flight

Cooper's Hawk
(*Accipiter cooperii*)

Like the very similar but smaller Sharp-shinned Hawk, the Cooper's Hawk—also called chicken hawk—is a bird-eating specialist of Northwest woodlands, adept at flying rapidly through dense forest, expertly darting through heavy cover. Also like Sharp-shinned Hawks, Cooper's Hawks will use bird feeders to their advantage, launching surprise attacks in residential areas, especially in winter. And finally—like the Sharp-shinned Hawk—Cooper's Hawks exhibit fairly dramatic sexual size differences; females are substantially larger than males. This fact makes size alone unreliable in distinguishing between the 2 species because a large female Sharp-shinned Hawk may be the same size as a small male Cooper's Hawk.

Juvenile Cooper's Hawk

Adult Cooper's Hawk

172

LENGTH 14–18 inches. **WINGSPAN** 26–34 inches. **SEXES SIMILAR** Females substantially larger than males; short, rounded wings; heavily black-and-white-barred flight feathers visible when seen from below in all plumages; long tail, gray (adult) or gray-brown (juvenile), with alternating black bands and broad white terminal band; when all tail feathers are visible and the bird is not soaring with fanned tail, the tail is noticeably rounded at the corners (Sharp-shinned Hawk has a squared tail); large head, with feathers on rear of crown often slightly raised to form a modest crest (Sharp-shinned has small head). **ADULT** Dark steel-gray above; white below, densely patterned with orange bars; dark gray crown forms a distinct cap contrasting with gray neck (crown of Sharp-shinned forms a hood); thick, bright yellow legs; red eyes. **JUVENILE** Brownish gray above; white below, patterned with thin, dark streaks on the breast, fading away to a white belly; yellow eyes. **IN FLIGHT** Flies with straight wings and head extended (crosslike shape), while Sharp-shinned Hawk flies with wings pushed forward, making the head look tucked in between the wings. **VOICE** Calls include a rapid *kak-kak-kak-kak-kak*; a slower repeated *kuck kuck kuck*, and single *kuck* notes; also raspy barking calls. **BEHAVIORS** Hunts by perching in concealed locations and attacking medium-sized birds with sudden, swift flight; also hunts on the wing, navigating through forest foliage, and from aloft, diving on prey. **HABITAT** Breeds in conifer forests, mixed forests, juniper stands, deciduous stands (including aspen and oak), riparian woodlands; disperses to a wider variety of habitats in migration and winter, including open areas and residential neighborhoods; found from sea level to high in mountains. **STATUS** Uncommon year-round resident, summer resident, and migrant, except fairly common southwest British Columbia (including southern Vancouver Island) and northwest Washington; largely migrates south from northern part of range for winter, including much of southern British Columbia and northern Washington; year-round populations (probably bolstered by migrants from the north in winter) are densest west of the Cascades.

Cooper's Hawk range

Juvenile Cooper's Hawk in flight

Northern Goshawk
(*Accipiter gentilis*)

Largest of the 3 North American accipiters, the Northern Goshawk is an impressively bold and deadly predator of forested mountains and foothills. Preying on both small mammals and small to large birds—ranging from songbirds to upland gamebirds—Northern Goshawks hunt by surprise, and are highly adept at high-speed attack flights and pursuit flights through trees and understory. Their aggressive attacks and pursuit speeds make them popular with falconers.

LENGTH 21–26 inches. **WINGSPAN** 39–45 inches. **SEXES SIMILAR** Stout and deep chested; gray above (female gray-brown); silvery below, with fine, dark gray barring (coarser on female); black cap, dark eye line, and white eyebrow; red eyes; long, broad, rounded gray tail with inconspicuous black bars. In flight from below, heavily barred flight feathers visible in all plumages; more pointed wings than Cooper's Hawk. **JUVENILE** Mottled brown-tan above; light buffy below, heavily patterned with bold, dark streaks; streaked head, usually with pronounced light eyebrow; brown tail with uneven dark bars; in flight, dorsal wing surface shows a row of pale

Juvenile Northern Goshawk

spots forming a stripe across the greater coverts. **VOICE** Call is a high, ringing, raspy *kike-kike-kike-kike-kike*, often carrying on for 24 notes or so. **BEHAVIORS** Perches for fairly brief periods to scan for prey, then changes perches to continue the hunt; also hunts on the wing, quietly maneuvering through the trees; preys heavily on both mammals (squirrels, rabbits, hares) and birds (grouse, jays, robins, woodpeckers); flight is strong and steady. **HABITAT** Mature coniferous forest, especially on slopes; also mature aspen stands and mature cottonwood stands in Northwest interior; disperses across a wider range of habitats in winter, including open juniper woodlands and isolated deciduous groves. **STATUS** Uncommon to rare year-round resident of interior, rare year-round resident west of Cascades; may withdraw from higher elevations in winter.

Juvenile Northern Goshawk in flight

Northern Goshawk range

Adult Northern Goshawk

175

American Kestrel
(*Falco sparverius*)

North America's tiniest and most colorful falcon, and among the most sexually dimorphic raptors (males and females are different in plumage color and/or pattern), the streamlined little American Kestrel is a familiar bird of relatively open country. There it often perches on utility wires, sometimes while eating its prey: mice and other small mammals, reptiles, large insects, and small songbirds. Males are more colorful and smaller than females. They nest in cavities and crevices, including holes made by Northern Flickers.

LENGTH 9–11 inches. **WINGSPAN** 21–23 inches. Narrow, pointed wings; long tail; rufous-colored back with darker barring (female) or blotches (male); 2 black face stripes contrasting sharply with white throat and cheeks; pale buff breast with darker spots or streaks. **MALE** Blue-gray shoulders and crown; rusty red tail with black subterminal band and white tip. **FEMALE** Rufous tail with thin dark bars; rufous-brown dorsal wing surface; pale tan-gray crown. **JUVENILES** Similar to adults, but paler. **VOICE** Call is a high, shrill, rapid *kill-kill-kill-kill-kill*. **BEHAVIORS** Often hovers expertly while hunting; when perched, frequently pumps its tail up and down. **HABITAT** Open to semi-open country; shrub steppe, grasslands, farms, ranch lands, open juniper woodlands, old forest burns, grassy fields in or near residential areas, road margins. **STATUS** Fairly common year-round resident.

Female American Kestrel

American Kestrel range

Male American Kestrel

Merlin (*Falco columbarius*)

An amazingly acrobatic bird hunter, the Merlin, North America's second-smallest falcon, often preys on shorebirds, such as Dunlins, but no bird up to its own size is safe when the "pigeon hawk" is on the prowl. Three North American subspecies differ by relative shade, and all occur in the Northwest: the Prairie Merlin (*Falco columbarius richardsonii*) is the palest, but is rare in the Northwest; the medium-shade Taiga Merlin is the nominate subspecies, and the Black Merlin (*F. c. suckleyi*) is the darkest.

LENGTH 10–12 inches. **WINGSPAN** 22–26 inches. Small, stocky, and powerfully built, with pointed wings; long tail with broad, dark subterminal band and narrow light gray bars; faint mustache stripe; thin white eyebrow stripe (often absent on Black Merlin); seen from below in flight, Taiga and Prairie subspecies show gray flight feathers heavily patterned with rows of white spots; Black subspecies shows very dark flight feathers without the striking rows of white spots; bright yellow legs. **MALE** Gray to blackish gray above; dark cap; pale buff below, variably patterned with dark, broad streaks and bars, very heavy in the Black subspecies so that breast and belly appears dark gray with white spots and streaks. **FEMALE** Overall pattern like adult male, but dark brown above instead of gray. **VOICE** Calls include a rapid series of high *pip* notes, *pipipipipipipipipi*, often rising then falling off in pitch and speed; also a very high, pleading *puureeeep-peep-peep-peep*. **BEHAVIORS** Often courses low and fast in hunting forays; chases down flying birds with bursts of speed and rapid directional changes. **HABITAT** Open and semi-open areas, especially estuaries, wetlands, and margins of large rivers. **STATUS** Uncommon winter visitor and migrant (September–May); rare summer resident of southwest British Columbia and Washington's Puget Trough, Olympic Peninsula, and Okanagan Highlands; very rare summer resident of Idaho and Oregon.

Black Merlin

Male Taiga Merlin

Prairie Merlin

Female Taiga Merlin in flight

Merlin range

Peregrine Falcon
(*Falco peregrinus*)

The iconic Peregrine Falcon enjoys the distinction of being one of the very few species to have been removed from the federal endangered species list. By 1970, only 39 pairs were known to exist in the entire lower 48 states, but thanks to timely and tireless intervention from many agencies and organizations, Peregrine Falcons have rebounded significantly, and the bird was removed from the endangered list to great fanfare in 1999. Today these speedy aerial hunters have even adapted to urban life, nesting and roosting on tall city buildings and bridges, and feeding largely on feral rock pigeons. Portland, Seattle, and Boise have urban populations.

LENGTH 15–20 inches. **WINGSPAN** 35–45 inches. **SEXES SIMILAR** Long tail and long, pointed wings; at rest, wingtips reach to end of tail; dark gray above, with distinct dark gray moustache stripe yielding a "helmeted" appearance; white breast and throat; white lower breast, belly, and flanks uniformly barred with black; underwings and underside of flight feathers uniformly patterned with narrow black and white bars; yellow cere and eye ring. **JUVENILE** similar to adult, but dark brown above rather than gray, and streaked rather than barred breast and belly; grayish cere and eye ring. **VOICE** Calls are typically high, harsh, scratchy yelps. **BEHAVIORS** Hunts birds up to the size of large ducks; pursuit speed can reach 70 miles per hour, and dive speed has been clocked as high as 200 miles per hour. On Pacific Northwest coast, Peregrine Falcons prey heavily on alcids (Pigeon Guillemot, Common Murre, murrelets, auklets) and other seabirds, as well as such seemingly unlikely quarry as Black Swifts; inland, diet includes a wide array of birds, from upland gamebirds and shorebirds to waterfowl and passerines (a name describing all perching birds of the order Passeriformes). **HABITAT** Coastal cliffs, sea stacks, and nearby estuaries; mountains, canyons, large rivers, large cities, wetlands. **STATUS** Uncommon year-round resident.

Peregrine Falcon range

Peregrine Falcon in flight

Peregrine Falcon in flight

Peregrine Falcon

Prairie Falcon (*Falco mexicanus*)

The vast open spaces of the Northwest's arid interior are the realm of the Prairie Falcon, a swift, powerful predator with a distinct taste for ground squirrels, as well as smaller birds, other small mammals, and reptiles. The more remote the setting, the more Prairie Falcons seem to thrive, making them symbolic of the Northwest's remote desertlike regions. The hiking birder who unknowingly approaches an active Prairie Falcon nesting site will be greeted by very noisy, agitated parents that circle overhead, shrieking incessantly.

LENGTH 15–18 inches. **WINGSPAN** 36–42 inches. **SEXES SIMILAR** Soft mottled brown above, with light brown nape and crown; white cheeks framed by brown moustache stripes in front and brown ear stripes; white throat and undertail coverts; white breast and belly with brown streaks; yellow legs; yellow cere; in flight from below, dark axillaries (armpits) are distinctive; when perched, wingtips extend nearly to the end of tail. **JUVENILE** Similar to adult but more heavily streaked below. **VOICE** Calls include high, piercing, raspy screeches and yelps; also an agitated *kek-kek-kek-kek* when disturbed near nest. **BEHAVIORS** Hunts by several strategies; often perches atop a utility pole, rock formation, or juniper, and watches for prey, which it then swoops rapidly down to kill; also forages on the wing, flying rapidly within a few yards of the ground. Feeds heavily on ground squirrels when available, but also preys on common ground-loving passerines, such as Horned Larks and Western Meadowlarks, especially in winter. **HABITAT** Open shrub steppe, grasslands, ranch lands, desert; nests on rimrock, canyon walls, and rock monoliths. **STATUS** Uncommon to rare year-round resident, except fairly common in southern Idaho and southeast and central Oregon; rare winter visitor west of Cascades. **KEY SITES** Idaho: City of Rocks National Reserve; Indian Creek Reservoir (Ada County); Snake River Birds of Prey National Conservation Area. Oregon: Fort Rock State Park; Guano Rim; Malheur NWR; Owyhee Reservoir and Leslie Gulch. Washington: Waterville Plateau; Yakima Canyon.

Prairie Falcon in flight

Prairie Falcon

Prairie Falcon range

Gyrfalcon (*Falco rusticolus*)

Large, powerful, regal-looking falcons of the Far North, Gyrfalcons expand their range during winter, with many birds wandering southward toward and into the northern tier of states. These robust falcons can show up nearly anywhere in the Northwest, and when discovered by birdwatchers, they always stir tremendous interest. In their circumpolar range, Gyrfalcons range from nearly pure white to very dark gray, but the typical medium-gray morph is commonest in the Pacific Northwest.

LENGTH 21–24 inches. **WINGSPAN** 44–50 inches. **SEXES SIMILAR** Large, heavy-bodied falcon with Accipterlike appearance; long, finely barred tail; relatively short, broad wings for a falcon; at rest, wingtips reach only to middle of the tail; pale gray to dark gray above (sometimes gray-brown), usually with lighter gray scalloping; gray nape, crown, and face; gray mustache stripes framing white throat; white below with dark spots and streaks; dark bars on white flanks; yellow eye ring, cere, and legs. **JUVENILE** Similar overall to adult, but more heavily speckled below, and lacks yellow eye ring and cere. **VOICE** Calls include a slow, loud wailing, and various *chuk* notes, but generally silent. **BEHAVIORS** Often perches on power poles and fence posts; hunts birds as large as geese. **HABITAT** Coastal lowlands, open farmlands, open wetlands, prairies, shrub steppe. **STATUS** Rare winter visitor (November–March); very rare to Idaho, Oregon, and southern Washington.

Gyrfalcon

Gyrfalcon in flight

Gyrfalcon range

Rails and Cranes

Rails are among the most secretive birds in the Pacific Northwest, so much so that the smallest species, the diminutive Yellow Rail, is practically never seen, its presence only being detected by its distinctive nocturnal call in the marshes of Klamath County, Oregon. The other 2 species of rail in the Northwest—the Virginia Rail and the Sora—are more easily seen for birders who know where and when to look, and they are also far more vocal than the Yellow Rail, their calls tending toward the exotic. Close kin to the rails, the American Coot is gregarious and easily found, often taking up residence on urban ponds and lakes. Related, but hardly similar in appearance, the Sandhill Crane is one of the largest birds in the region, standing nearly 4 feet tall. Majestic in flight, Sandhill Cranes are one of only 2 North American crane species, the other being the endangered Whooping Crane, whose population totals fewer than 400 birds.

Virginia Rail

Virginia Rail (*Rallus limicola*)

Thanks to its proclivity for foraging on mudflats and shallow pools at the edges of reed stands, the Virginia Rail is the easiest of the Northwest's rails to actually see—although easiest is relative in this case, because all rails are secretive and retiring. Unobtrusively scouting out likely openings in marshland habitat is the best way to catch a glimpse of the Virginia Rail; also keep an eye peeled for the bird's footprints: triangular impressions from 3 very long, slender toes.

LENGTH 9.5 inches. Brownish overall, with black-and-white barred flanks and gray face; light cinnamon-brown above with dark brown mottling; unmarked cinnamon-brown below, palest on belly; long, slightly decurved bill with dirty orange lower mandible and mostly black upper mandible. **JUVENILE** Paler and more grayish overall, with varying degrees of mottling on the breast. **VOICE** Calls include an odd descending and slightly acceler-

ating grunt-squeak combination; a fast, raspy *kidick kidick kidick*; a high, rapid *Kiki-kiiir* or *twiiirit-kikikiiir*; a squeaky, rusty hinge–style *chaap*; and various others.
BEHAVIORS When flushed, flies weakly for a short distance with legs dangling and plunges back into cover. **HABITAT** Breeds in freshwater marshes and swamps as well as brackish marshes with emergent vegetation (especially cattail and bulrush) and mudflat or shallow-pool openings, including heavily vegetated sloughs, creeks, and pond margins; in migration, also uses saltmarshes and occasionally wet or flooded crop fields.
STATUS Fairly common summer resident and migrant (late April–September); fairly common winter resident west of Cascades, rare winter resident interior.

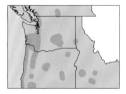

Virginia Rail range

Virginia Rail

Sora (*Porzana carolina*)

All rails tend to be secretive, some more so than others; the Sora generally sticks to heavy marsh cover and is more often heard than seen, but it sometimes wanders onto mudflats along the edges of emergent vegetation to feed, giving birders a chance for sightings. Often these handsome little rails walk and wade with their short, pointed tail cocked straight upward.

LENGTH 9 inches. Richly mottled brown-black mantle with white highlights; dark crown, black face, gray cheeks, neck, and breast; robust, bright yellow bill; yellow-green legs with very long toes; black-and-white-barred flanks. **JUVENILE** Less colorful overall, with pale gray-brown or tannish breast and neck; dingy grayish yellow bill. **VOICE** Common call is a ringing, descending whinney, slowing slightly at the end, *pi-pi-pi-pi-pi-pi-pi-pii-pii-pii*; also a resonant *pur-REE* and a squeaky *kweek*. **BEHAVIORS** Though not often seen in flight, the Sora sometimes flushes from nearly underfoot, then flies weakly away a short distance with legs dangling, only to disappear into heavy cover not to be reflushed. **HABITAT** Breeds in freshwater marshes and swamps, including heavily vegetated sloughs, creeks, and pond margins; in migration, also uses saltmarshes and occasionally wet or flooded crop fields. **STATUS** Fairly common summer resident, uncommon migrant; very rare winter resident west of Cascades.

Adult Sora

Juvenile Sora (August)

Sora range

Yellow Rail
(*Coturnicops noveboracensis*)

So secretive is the tiny Yellow Rail, that Oregon's geographically disjunct population was thought to have been extirpated until June, 1982, when 2 calling birds were reported in the upper Wood River Valley in south-central Oregon. A few years later, a systematic study carried out over several years revealed that indeed Yellow Rails still nested in Oregon; the total population is thought to comprise fewer than 300 pairs.

LENGTH 7.25 inches. **SEXES SIMILAR** Tiny and tawny colored overall, with dark back; pale buff breast transitioning to brighter yellowish neck; white throat; yellowish face with dark eye line and cap; blackish back and flanks with thin white terminal bands on black feathers yielding crosshatched appearance, and prominent yellow streaks on back; thick, short, yellow or olive-yellow bill; in flight, white trailing edges on inner wings. Juvenile is similar overall, but less colorful, with speckled nape and neck. **VOICE** Nocturnal song is a redundant, incessantly repeated *tic tic tic-tic-tic*, often described as sounding like small stones being struck together; primarily heard in May and June. **BEHAVIORS** Highly secretive, almost never seen, and almost always identified by its nocturnal song. **HABITAT** Sedge-dominated shallow freshwater wetlands and meadows in Oregon's Klamath Basin. **STATUS** Rare and very local summer resident. **KEY SITES** Oregon: Klamath Marsh NWR; Wood River Valley.

Yellow Rail

Yellow Rail range

American Coot (*Fulica americana*)

Easily adapting to manmade ponds and wetlands, American Coots are familiar residents even in urban settings, where large park waters provide habitat for them. Their nickname, "mudhen," likely derives from their chickenlike habit of bobbing their head to and fro while swimming.

LENGTH 15 inches. Dark slate-gray overall, generally more blackish on the neck and head; broad white bill with red frontal shield (at close range); red eyes. **JUVENILE** Similar but overall medium gray. **VOICE** Calls include various cackles, grunts, and reedy clucks. **BEHAVIORS** Coots often feed onshore, clustering near the water's edge to forage; they also rest on islets and shorelines, as well as logs and hummocks of vegetation in the water. A coot will often escape by simply swimming into vegetation, but is a capable flyer after a laborious running takeoff. **HABITAT** Nests on freshwater marshes, swamps, ponds, shallow lakes; in winter, expands range to shallow estuaries, large reservoirs, slow rivers (especially the Columbia River). **STATUS** Common year-round resident and migrant, retreating to open water in winter.

American Coot

American Coots

American Coot range

Sandhill Crane
(*Antigone canadensis*)

The echoing rattles of Sandhill Cranes flying in formation high overhead is a harbinger of spring in the Northwest. These stately cranes begin arriving from their wintering grounds in the Southwest, in both small groups and impressively large flocks, in February. The large flocks are generally transient, on their way to Canadian breeding grounds, whereas cranes that breed in the Pacific Northwest generally arrive in small groups. Two races of Sandhill Crane, the Lesser and the Greater, occur in the Northwest, differing primarily in size.

LENGTH 40–47 inches. **WINGSPAN** 71–78 inches. Gray overall, sometimes stained with rust from iron content in mud; long neck and legs; long, daggerlike bill; red crown and forehead. **JUVENILE** Similar overall to adult, but with rust-colored blotches

and lacks red crown. **VOICE** Call is a loud, rattling bugle, carrying for substantial distances, and given both in migratory flight and on breeding grounds. **BEHAVIORS** In its elaborate courtship dance, the cranes bow and leap into the air with wings spread. **HABITAT** Breeds on moist mountain prairies and wet meadows, including vegetated meadowlike lake margins; migrants stage on shallow wetlands, fields, and meadows. **STATUS** Fairly common summer resident and locally abundant migrant (February–October). **KEY SITES** Idaho: Bear Lake NWR; Fort Boise Wildlife Area; Grays Lake NWR. Oregon: Klamath Basin; Malheur NWR; Sauvie Island. Washington: Columbia NWR; Ridgefield NWR and Woodland Bottoms. British Columbia: Vaseux Lake and White Lake area.

Sandhill Crane range

Sandhill Crane

Sandhill Cranes in flight

Sandpipers, Plovers, Curlews, and Their Allies

In some cases, shorebirds are among the most challenging birds to identify because of the similarity between certain species—examples include the peeps, a colloquial name for the smallest members of the sandpiper genus *Calidris*, the 2 species of dowitcher, and the 2 species of golden-plover. But the order Charadriiformes also includes many distinctive, unmistakable, and beautiful shorebirds, such as the American Avocet, Black-necked Stilt, Marbled Godwit, and the familiar and boisterous Killdeer. As the name shorebirds suggests, most members of this order are found near water, especially the shallow margins of marshes, flooded fields, beaches, and estuaries. Many species breed in the Far North, only passing through the Northwest on their spring and fall migrations (when they sometimes occur in huge flocks), but a variety of intriguing shorebirds breed in the region.

A huge flock of Dunlins

Black-necked Stilt
(*Himantopus mexicanus*)

A striking bird of marshes, flooded fields, and lakeshores, distinctive with its black-and-white pattern, the Black-necked Stilt is often seen with American Avocets because the 2 species share the same habitat. Noisy and gregarious, stilts feed in fresh and alkaline water, wading and probing the shallows for brine shrimp, brine flies, fish, aquatic beetles, and insect larvae and nymphs.

LENGTH 15 inches. **WINGSPAN** 28 inches. **SEXES SIMILAR** Slender, with long neck and extremely long legs; black above, bright white below, with prominent white forehead and eye patch; long red legs; long, thin black bill. Juvenile is somewhat paler, with grayish or pinkish gray legs. **VOICE** Call is a sharp *kik-kik-kik*. When nest or young are disturbed, call becomes a more pleading *keef-keef*. **BEHAVIORS** Strong, direct flight, with long legs trailing behind. Will feign injury to lure intruders away from the nest or chicks. **HABITAT** Shallow standing water of lakes and ponds, including alkali lakes; also flooded and irrigated fields. **STATUS** Locally common summer resident; occasional visitor and rare breeder west of Cascades. **KEY SITES** Idaho: Duck Valley Indian Reservation; Kuna Sewage Ponds. Oregon: Burns area and Malheur NWR; Summer Lake Wildlife Area. Washington: Desert Wildlife Area; Tyson Ponds (Walla Walla).

Black-necked Stilt range

Black-necked Stilt

American Avocet
(*Recurvirostra americana*)

The American Avocet is a striking, long-legged shorebird most abundant on open, shallow wetlands, alkaline lakes, and other wet habitats with minimal emergent vegetation. The rust-colored head and neck of the breeding-plumage bird is distinctive, but some avocets arrive in the Northwest early enough in spring that they are still wearing their winter plumage, in which the head and neck are pale gray.

LENGTH 17–18 inches. **WINGSPAN** 32 inches. Long neck and legs; distinctive long, thin bill is slightly upturned; black and white body pattern, with long, slender, black-tipped wings; pale rust to orange-rust neck and head in breeding plumage (spring and summer), replaced by light grayish neck and head in fall and winter. Juvenile plumage more muted than that of adult. **VOICE** Call is a high, ringing *wheep-wheep-wheep*, the notes sometimes well-spaced, but quicker and more urgent as an alarm call. **BEHAVIORS** With their distinctive upturned bills, American Avocets forage actively with heads down and bills in the water, often sweeping side to side; they may even submerse their entire head to snatch prey, such as brine fly larvae and Chironomids. Strong, direct flight showing distinctive flashy black-and-white plumage pattern. **HABITAT** Very shallow, unvegetated or sparsely vegetated lakes, wet playas, lakeshores, ponds, wetlands; open mudflats. **STATUS** Locally common summer resident (mid-April–mid-September); rare west of the Cascades **KEY SITES** Idaho: American Falls Reservoir; Deer Flat NWR. Oregon: Burns area and Malheur NWR; Summer Lake Wildlife Area. Washington: Columbia NWR and Potholes Reservoir.

Breeding American Avocet

Nonbreeding American Avocet

American Avocet in flight

American Avocet range

Black Oystercatcher
(*Haematopus bachmani*)

As its name implies, this large, stocky shorebird feeds heavily on mollusks, though typically mussels, clams, limpets, chitons, and snails—not oysters. It uses its powerful bill to open bivalves, which abound in the rocky intertidal zone where it forages. Scanning the Northwest's rocky coastal shorelines with binoculars frequently reveals these common birds, and they tend to be fairly approachable for birders who look for walkable beaches fronting intertidal zones with rock escarpments.

LENGTH 17–18 inches. **WINGSPAN** 32 inches. **SEXES SIMILAR** All-black plumage; tall and stocky, with long, pinkish legs and a long, heavy red bill. **JUVENILE** Slightly scaled brown-black on back; blackish outer half of bill. **VOICE** Common call is a loud, yelping, repeated *weeep* or *weeup*, sometimes wavering; also a loud, plaintive, whistled, accelerating *pip-pip pip-pip pip-pip*. **BEHAVIORS** Oystercatchers nest on rocky islands and islets where the monogamous pair defends a territory and builds a nest of little more than a shallow scrape in gravel, shell fragments, or vegetation. Flight is direct and fairly slow on stiff wingbeats. **HABITAT** Rocky beaches, sea stacks, islands; shell beds; jetties. **STATUS** Common year-round resident. **KEY SITES** Oregon: Barview jetty (Tillamook Bay); Chetco Point (Brookings); Coquille River jetties and Kronenberg County Park (Bandon); Yaquina Head. Washington: Ediz Hook (Port Angeles); Grays Harbor jetties; LaPush area jetties and beaches; Neah Bay; Point Hudson (Port Townsend); Ruby Beach and Kalaloch; Salt Creek County Park (Clallam County); San Juan Islands (widespread). British Columbia: Cattle Point and Clover Point (Victoria); Klootchman Park and Stanley Park (Vancouver); Pipers Lagoon and Neck Point Park (Nanaimo); Sunshine Coast.

Black Oystercatcher range

Black Oystercatcher

Black-bellied Plover
(*Pluvialis squatarola*)

Far more common in the Northwest than the 2 golden-plovers, which are similar, the Black-bellied Plover is chiefly a spring and fall migrant through the region, with peak numbers from July through October, but a few nonbreeding adults stay during summer. With such a lengthy migration period, the species occurs year-round. Fall migrants wear the nonbreeding plumage or worn breeding plumage in transition to nonbreeding plumage, but birds seen in spring are often in their striking breeding plumage.

LENGTH 11.25 inches. **WINGSPAN** 29 inches. Diagnostic black "armpit" in flight in all plumages; short, fairly heavy bill. **ADULT NONBREEDING** (August–April) Ranges from lightly marked drab gray to strongly mottled above; lighter below, sometimes with gray blotches, with white belly; pale eyebrow stripe. **ADULT BREEDING** (April–September) Black belly, breast, throat, and face; light crown; white undertail coverts; mottled black and white back and wing coverts. **JUVENILE** (July–December) Similar to nonbreeding adult, but more boldly patterned above; white breast has faint streaks; distinct white eyebrow stripe. **VOICE** Calls include a high, drawn-out, whistled *puureeee* and *peeoreee*. **BEHAVIORS** Typically occurs in small groups; swift flight, on long, sharply pointed wings. **HABITAT** Coastal and freshwater mudflats; beaches, wet fields, pastures. **STATUS** Common migrant of coastal Washington and British Columbia; fairly common migrant in coastal Oregon; uncommon migrant western valleys, except regular at Fern Ridge Reservoir, Oregon; uncommon migrant interior. Occasional summer resident. **KEY SITES** Oregon: Coos Bay area; Fern Ridge Reservoir; Siuslaw River estuary. Washington: Dungeness NWR area; Grays Harbor; Skagit Wildlife Area. British Columbia: Boundary Bay area; Oak Bay (Victoria).

Adult breeding Black-bellied Plover

Adult nonbreeding Black-bellied Plover

Adult nonbreeding (top left) and breeding (top right) Black-bellied Plovers in flight

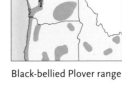

Black-bellied Plover range

Juvenile Black-bellied Plover

American Golden-Plover
(*Pluvialis dominica*)

Because they are typically in juvenile or at least partial nonbreeding plumage by the time scant numbers of them reach the Pacific Northwest from Far North breeding grounds, American Golden-Plovers are difficult to differentiate from Pacific Golden-Plovers and Black-bellied Plovers. The adult Pacific Golden-Plover in nonbreeding plumage has distinctive yellowish-edged dorsal feathers, but the juvenile tends to be paler, and thus easily confused with the American Golden-Plover. A key field mark to separating the golden-plovers is the number of primary flight feather tips extending beyond the longest tertials when the bird is at rest: 2 or 3 on the Pacific Golden-Plover, 4 or 5 on the American Golden-Plover—but making such distinctions in the field can be challenging without scrutiny from close range. American Golden-Plovers often appear more compact and short-legged than Pacific Golden-Plovers. A few golden-plovers arrive each fall still wearing their breeding plumage, though it is typically worn and in transition to nonbreeding plumage; likewise, some northward spring migrants arrive in partial and sometimes full breeding plumage.

LENGTH 10.5 inches. **WINGSPAN** 26 inches. **ADULT BREEDING** (March–September) Black below, including undertail coverts, with broad white stripes along sides of neck extending up to frame top edge of black face; rich, mottled black, golden tan, and white above. **ADULT NONBREEDING** (September–April) Mottled or speckled black, gray, and white above, often with a slight tannish wash; late summer birds may retain full or partial black breast, throat,

and flanks, or black chevrons or blotches on breast; dark cap and white eyebrow stripe; 4 or 5 wingtips extending beyond tertial feathers, and wingtips extending past tail; plain pale gray underwings in flight. **JUVENILE** (July–November) Similar to nonbreeding adult, but more boldly mottled above. **VOICE** Flight call is a high, whistled *chureep* or *cheep*. **BEHAVIORS** Typically occurs singularly or in small groups, often associated with Black-bellied Plovers or other shorebirds. Swift flight on long, pointed wings. **HABITAT** Coastal mudflats and vegetated estuaries; sand beaches; wet fields, pastures. **STATUS** Rare late summer and fall migrant (late August–October). **KEY SITES** The most consistent locations are the Northwest's prime shorebird locations, especially Boundary Bay at Vancouver, British Columbia, and Grays Harbor, Washington.

Adult breeding American Golden-Plover

American Golden-Plover range

Juvenile American Golden-Plover

Adult nonbreeding American Golden-Plover
molting to breeding plumage (April)

Pacific Golden-Plover
(*Pluvialis fulva*)

The Pacific Golden-Plover and the American Golden-Plover—especially in nonbreeding plumage typical of their arrival in the Northwest between August and October—are extremely difficult to distinguish. The adult Pacific Golden-Plover in nonbreeding plumage has distinctive yellowish-edged dorsal feathers, but the juvenile tends to be paler, and thus easily confused with the American Golden-Plover. A key field mark is the number of primary flight feather tips extending beyond the longest tertials when the bird is at rest: 2 or 3 on the Pacific Golden-Plover, 4 or 5 on the American Golden-Plover—but making such distinctions in the field can be challenging without scrutiny from close range. Pacific Golden-Plovers often appear leaner and longer legged than American Golden-Plovers. A few golden-plovers arrive each fall still wearing their breeding plumage, though typically worn and in transition to nonbreeding plumage.

Breeding Pacific Golden-Plover

Nonbreeding Pacific Golden-Plover

LENGTH 10 inches. **WINGSPAN** 24 inches. **ADULT BREEDING** (March–September) Nearly identical to American Golden-Plover, but with variable white flanks dividing black belly and breast from mottled mantle, and white undertail coverts, often with black markings; white neck stripes narrower than in American. **ADULT NONBREEDING** (September–April) Mottled or speckled black, gray, and white above, typically with a more yellow-tan wash than American Golden-Plover, and often substantial yellow edging on the dorsal feathers and face; dark cap and yellowish to white eyebrow stripe; 2 or 3 wingtips extending beyond tertial feathers, and wingtips even with or very slightly projecting past tail; plain pale gray underwings in flight; late summer birds may retain full or partial black breast, throat, and flanks, or black chevrons or blotches on breast. **JUVENILE** Similar to nonbreeding adult, but often paler, with little or no yellow on dorsal feathers, and only faint yellow wash on face; can be very similar in shade to nonbreeding-plumage American Golden-Plover. **VOICE** Flight call a high, whistled *chureep* or *cheep*. **BEHAVIORS** Typically occurs singularly or in small groups, often associated with Black-bellied Plovers or other shorebirds. Swift flight on long, pointed wings. **HABITAT** Coastal mudflats and vegetated estuaries; sand beaches; wet fields, pastures. **STATUS** Rare late summer and fall migrant (almost all sightings are late August–October). **KEY SITES** The most consistent locations are the Northwest's prime shorebird locations, especially Boundary Bay at Vancouver, British Columbia, and Grays Harbor, Washington.

Juvenile Pacific Golden-Plover

Pacific Golden-Plover range

Snowy Plover (*Charadrius nivosus*)

This tiny shorebird breeds and winters in small numbers along the Northwest coast, and breeds on a few alkali and saltpan playas in southeast Oregon; they are rarely seen in flight, and blend in well with their habitat, making them difficult to locate. On Oregon's massive Lake Abert (increasingly more playa than actual lake), a spotting scope is essentially a necessity to find these sparrow-sized plovers. And on the Oregon coast, seasonal beach access and activity restriction are in place to protect nesting Snowy Plovers (watch for signage and boundary markers). The Snowy Plover is listed as threatened under the federal Endangered Species Act and by the state of Oregon; the state of Washington lists them as endangered. For conservation information, see www.westernsnowyplover.org.

LENGTH 6 inches. **SEXES SIMILAR** Front-heavy appearance; pale grayish brown above, white below, with a partial black breast band; black ear patch and forehead; dark gray legs; thin black bill. Juveniles lack black markings. **VOICE** Calls include a husky *quirr* and a whistled *toor-weet*. **BEHAVIORS** Rarely flies, preferring to run; swift direct flight, with stiff, shallow wingbeats; hunts insects, its primary food, on foot, running and darting about; also eats small crustaceans. When disturbed, female walks or runs from the nest, flattens out on the ground, and spreads her wings in an effort to draw the intruder away. **HABITAT** Coastal beaches and dunes, and desert playas; often found on uppermost edges of beaches. **STATUS** Rare and very local summer resident; some coastal birds winter from southern Washington through Oregon. **KEY SITES** Oregon: Oregon Dunes Overlook; Siltcoos Beach and Siltcoos River Estuary; Lake Abert; Summer Lake Wildlife Area. Washington: Grayland Beach State Park and Midway Beach; Leadbetter Point (Pacific County).

Snowy Plover

Snowy Plovers

Snowy Plover range

Semipalmated Plover
(*Charadrius semipalmatus*)

Mostly occurring in the Northwest as a spring and late summer migrant, the Semipalmated Plover could be described as a miniature version of the far more common Killdeer, to which it bears a fairly strong family resemblance, except that both adult and juvenile Killdeers have 2 dark breast bands instead of one (Killdeer chicks have a single band). Semipalmated Plovers tend to show up individually and in small groups in appropriate habitat throughout the Northwest, especially from August through October, on the return journey from their Far North breeding grounds.

LENGTH 7.5 inches. **WINGSPAN** 18 inches. Dark sooty brown above, white below, with single black breast band; bright white collar and throat; black face mask with white forehead patch; broad black band at end of white-edged tail; orange legs; pale orange bill with black tip. **JUVENILE** Paler, more grayish brown accents instead of jet black; paler bill. **VOICE** Typical call is a somewhat raspy *chureeup chureeup* or *cheeup cheeup*; flush and flight call a raspy, persistent *cheeup-cheeup-cheeup-cheeup*. **BEHAVIORS** Generally forages alone or in small loose groups, but sometimes flocks with other shorebirds; quick, direct flight, sometimes twisting and turning. **HABITAT** Saltwater and freshwater mudflats and beaches, broad shorelines of shallow lakes and drawn-down reservoirs. **STATUS** Uncommon fall migrant (August–October), rare to uncommon spring migrant (April–May). Rare winter resident and visitor along coast. **KEY SITES** Idaho: Deer Flat NWR. Oregon: Fern Ridge Reservoir; Fort Stevens State Park. Washington: Dungeness NWR; Grays Harbor; Midway Beach and Grayland Beach (Pacific County); Ocean Shores and Westport; Ebey's Landing National Historic Reserve (Whidbey Island); Walla Walla River delta. British Columbia: Boundary Bay and Iona Island; Robert Lake (Kelowna); Whiffin Spit Park (Sooke).

Breeding Semipalmated Plovers

Semipalmated Plover
range

Nonbreeding Semipalmated Plover

Killdeer *(Charadrius vociferus)*

Named for its distinctive call, the Killdeer is among the most widespread and familiar shorebirds in the Northwest, able to adapt to a variety of habitats. It is well-known for its broken-wing act in which a parent feigns an injury, drooping a wing and fanning its tail, and walks away from the nest or young, hoping to lure intruders to follow.

LENGTH 10.5 inches. **WINGSPAN** 24 inches. Fairly tall and long legged, and long tailed for a shorebird. Brownish above, with rust-red rump; white below with 2 distinctive black breast bands; white collar; white and black forehead; white eyebrow stripe. **VOICE** Calls include a loud, piercing *cheeep cheeep*; high, piping *teedee teedee* or *chi-cheep chi-cheep* (or befitting the name, *kill-deer kill-deer*); and a fast, rolling, trilled *teeteeteeteeteetee*. **BEHAVIORS** Relying on camouflage, Killdeer nest on open ground, typically among rocks or gravel, including driveways, empty suburban lots, and pathways. Flocks of several dozen—sometimes hundreds—of birds can accumulate in prime feeding areas. Swift and direct flight on fairly stiff wings, often vocalizing, especially upon flushing. **HABITAT** Open fields, mudflats, golf courses, lakeshores, broad river banks, beaches, parking lots, and roadsides; from coastal estuaries to mountain valleys to desert playas. **STATUS** Common year-round resident with local movement out of higher elevations for winter.

Killdeer range

Killdeer

Spotted Sandpiper
(*Actitis macularius*)

A widespread, nonflocking sandpiper of riparian habitats, the Spotted Sandpiper is found on small mountain streams and large lowland rivers and just about every waterway in between. They tend to be boisterous and fairly bold, often allowing close approach, and they frequently perch on river rocks, distinctively bobbing up and down. So too, Spotted Sandpipers sometimes remain quiet and motionless until almost underfoot and then flutter away to the surprise of an onlooker walking along the stream.

LENGTH 7.5 inches. White breast and belly speckled with black spots; rich grayish brown or brown back with subdued dark chevrons; brown crown and eye stripe divided by white eyebrow stripe; pale legs, orange bill. Juveniles and nonbreeding adults are paler gray-brown above than breeding adult, and lack dark bars; white below, with partial brown collar patches. **VOICE** Song is a high, musical, whistling *cherEEE-cherEEE* transitioning to a rolling *purpuree-purpuree-purpuree-purpuree*. Calls include a repeated *pee-peet, pee-peet*, and a single sharp *whit*. **BEHAVIORS** Constantly bobs its rear-end up and down (often called "teetering"). Flight is fairly slow, on stiff wings, alternating quickly flapping and briefly gliding; usually flies low over the water or shoreline. **HABITAT** Rivers and streams, especially with extensive cobble; also still waters; sea level to high mountains. **STATUS** Common summer resident; fairly common migrant.

Breeding Spotted Sandpiper

Nonbreeding Spotted Sandpiper

Spotted Sandpiper range

Solitary Sandpiper
(*Tringa solitaria*)

As its name implies, the Solitary Sandpiper is most often seen alone. Its distinctive spotted plumage and greenish legs helps separate it from the Greater and Lesser Yellowlegs and Spotted Sandpipers. Found in habitats rarely used by other sandpipers, such as wooded ponds, drainage ditches, and feed-lot drainages, the Solitary Sandpiper nests in trees and will use nests built by songbirds. Nesting is suspected in Oregon and southern British Columbia.

LENGTH 8.5 inches. **WINGSPAN** 22 inches. **SEXES SIMILAR** Dark olive-brown above, patterned with white dots; bright white below; medium-length gray bill with black tip; white eye ring and lores; greenish legs; when at rest, wings extend beyond tail; in flight, dark underwings contrast with white belly; dark rump; white tail has black lateral bars and dark center feathers. **VOICE** Call is a rising, whistled *peep-weep* or *peep-weep-weep*. **BEHAVIORS** Flushes vertically; strong, fast flight; upon landing, briefly raises wings straight upward; acts nervous, often bobbing front part of body. **HABITAT** Breeds in boreal forests and muskeg bogs in Canada. During migration, prefers freshwater, including muddy ponds, ditches, sewage settling ponds and the like. **STATUS** Rare to uncommon migrant; possible very rare summer resident Oregon and southern British Columbia.

Breeding Solitary Sandpiper

Solitary Sandpiper range

Nonbreeding Solitary Sandpiper

Wandering Tattler (*Tringa incana*)

The region's only shorebird that is unmarked gray above, with no black or white accents in the tail or wing, the Wandering Tattler expertly navigates surf rocks and jetties in search of prey, bobbing its rump and tail nonstop, and often ducking in and out of crevices and pockets, even on steep inclines. The rugged Pacific Northwest coast provides the Wandering Tattler ample habitat in the form of rocky beaches and jetties.

Breeding Wandering Tattler

LENGTH 11 inches. **WINGSPAN** 26 inches. Yellow legs, fairly long bill, distinct black eye line borders narrow white brow stripe; white eye ring; In flight, solid gray above, gray tail, gray wings. **NONBREEDING** (October–March) Unmarked steely gray above, grayish breast, white below. **BREEDING** (April–September) Partially to heavily black-and-white-barred below. **VOICE** Call is a series of 3 to 6 identical, plaintive, high-pitched whistles. **BEHAVIORS** Usually solitary rather than in flocks; bobs rump almost nonstop while foraging; walks and climbs on rocks, including steep rock faces near water line. Rapid flight on fairly stiff wingbeats, often while calling. **HABITAT** Intertidal rocks, jetties. **STATUS** Fairly common late summer and spring migrant (August–September and mid-April–mid-May); rare winter visitor. **KEY SITES** Oregon: Barview jetty (Tillamook Bay); Coquille River jetties; Seal Rock State Wayside; Yachats State Park. Washington: Point Brown jetty (Ocean Shores); Westport jetty. British Columbia: Iona Island (Vancouver); Ogden Point Breakwater, Clover Point, Cattle Point (Victoria).

Wandering Tattler range

Nonbreeding Wandering Tattler

Greater Yellowlegs
(Tringa melanoleuca)

Noticeably larger and more robust than the very similar Lesser Yellowlegs, the Greater Yellowlegs is a widespread migrant throughout the Northwest, and also a winter visitor along the coast and in the valleys west of the Cascades. While bill size is generally reliable in separating the 2 species of yellowlegs, the length for each species can vary enough to make this field mark troublesome at times. Generally the Greater Yellowleg's bill is substantially longer and heavier than that of the Lesser Yellowlegs, and is also usually very slightly upturned. With some practice in listening to both species of yellowlegs, voice becomes a fairly reliable identifying feature, and if the 2 yellowlegs are seen together, the size difference is marked.

LENGTH 14 inches. **WINGSPAN** 28 inches. Tall, with long, bright yellow legs; thin bill, grayish olive at its base and black at the tip, is longer than total length of head, and is typically very slightly upturned. **NONBREEDING** (July–February) Dark gray to gray-brown above, patterned with white spots and specks; whitish neck and face lightly speckled with gray-brown; white belly. **BREEDING** (March–June) Black chevrons on breast and extensive black bars along flanks; boldly patterned above in gray, white, and black. **VOICE** Generally a series of 3 or more high, rapid notes: *pi-pi-pew*, or *tu-tu-tu*, or *per-it per-it per-it*. **BEHAVIORS** Feeds by walking or wading rapidly and quickly grabbing prey; will wade deep, sometimes deep enough to swim. Swift, direct flight. **HABITAT** Marshes, lakeshores, flooded and wet fields, coastal mudflats, estuaries. **STATUS** Common fall migrant, fairly common spring migrant; uncommon winter resident. **KEY SITES** Idaho: Deer Flat NWR; Market Lake Wildlife Area. Oregon: Fern Ridge Reservoir; Willamette Valley NWR Complex; Malheur NWR; Summer Lake Wildlife Area. Washington: Columbia NWR and Potholes Reservoir; Eide Road on Leque Island (Stanwood); Grays Harbor area; Skagit Wildlife Area; Toppenish NWR. British Columbia: Boundary Bay; Oak Bay (Victoria); Reifel Migratory Bird Sanctuary (Vancouver).

Breeding Greater Yellowlegs

Nonbreeding Greater Yellowlegs

Greater Yellowlegs range

Lesser Yellowlegs (*Tringa flavipes*)

Smaller and daintier than the very similar Greater Yellowlegs, the Lesser Yellowlegs is a widespread migrant through the Northwest. While bill size is generally reliable in separating the 2 species of yellowlegs, the length for each species can vary enough to make this field mark troublesome at times. Generally the Lesser Yellowleg's slender blackish bill is typically about the same length as the total length of the head measured from the base of the bill. With some practice in listening to both species of yellowlegs, voice becomes a fairly reliable identifying feature, and if the 2 yellowlegs are seen together, the size difference is marked.

LENGTH 10–11 inches. **WINGSPAN** 24 inches. Tall, with long, bright yellow legs; thin, straight, mostly blackish bill is about the same length as the length of the head. **ADULT NONBREEDING** (September–February) Dark gray to gray-brown above, patterned with white spots and specks; whitish neck and face lightly speckled gray-brown; white belly. **ADULT BREEDING** (March–August) Black chevrons on breast and limited black barring along upper flanks; boldly patterned above in gray, white, and black. **JUVENILE**

Similar to nonbreeding adult. **VOICE** Typically a series of 2 (sometimes one) high, sharp notes: *ti-ti ti-ti ti-ti*, or *tu-tu tu-tu tu-tu*; sometimes a more excited, faster, rolling *pip-pip-pip-pip-pip.* **BEHAVIORS** Feeds by walking or wading rapidly and quickly grabbing prey; will wade deep, sometimes deep enough to swim. Swift, direct flight. **HABITAT** Mudflats, saltmarshes, estuaries, lakeshores, marshes, sewage lagoons, flooded and wet fields. **STATUS** Fairly common fall migrant (July–October) and uncommon spring migrant east of the Cascades, uncommon to fairly common west of the Cascades in Oregon and Washington; uncommon fall migrant and rare spring migrant in Idaho; common fall migrant and uncommon spring migrant in coastal British Columbia. **KEY SITES** Idaho: Deer Flat NWR; Market Lake Wildlife Area. Oregon: Fern Ridge Reservoir; Malheur NWR; Summer Lake Wildlife Area. Washington: Columbia NWR and Potholes Reservoir; Grays Harbor area; Skagit Wildlife Area. British Columbia: Boundary Bay; Iona Island.

Adult nonbreeding Lesser Yellowlegs

Adult breeding Lesser Yellowlegs

Lesser Yellowlegs range

Willet (*Tringa semipalmata*)

When foraging and probing on marsh edges or wet fields, the Willet appears as a robust plain-gray shorebird, but all that changes when the bird takes flight, suddenly flashing its striking black-and-white wings and often noisily announcing its displeasure at being interrupted. These gregarious birds are most common in marshy environs of southeast Oregon and southern Idaho, where they are an iconic member of the wetlands community.

LENGTH 15 inches. **WINGSPAN** 26 inches. Fairly long, stout bill; long grayish olive legs; white eyebrow stripe from top of eye to base of bill; in flight, both wing surfaces are black with a broad, bright white stripe through the primaries and secondaries. **ADULT BREEDING** (April–July) Mottled gray, dark brown, and black above, with lightly barred flanks; speckled neck and upper breast; white belly. **ADULT NONBREEDING** (August–March) Plain gray above, white below. **JUVENILE** Similar to nonbreeding adult. **VOICE** Common call is a loud, ringing *pitiwi will-whit*; other calls, often in flight, include a high, slightly raspy *kee-keeeaa* or more plaintive *peeeur*. **BEHAVIORS** Feeds by walking wetland edges, wet fields, and other such habitats, and by wading in shallow water; in nesting season, pairs are territorial and individuals will feign injury to lure predators away from nest or chicks; Willets will congregate to mob and drive away predators. **HABITAT** Breeds on freshwater marshes and wetlands; also beaches, mudflats, saltmarshes, estuaries in migration and winter. **STATUS** Fairly common summer resident and migrant in southeast Oregon west to Klamath Basin; otherwise uncommon east of Cascades and rare west of Cascades. Uncommon summer resident and migrant in southern Idaho. Uncommon winter resident and migrant in south-west coastal Washington, rare elsewhere in Washington. Rare migrant and winter visitor to southwest coastal British Columbia. **KEY SITES** Idaho: Camas Prairie Centennial Marsh; Market Lake Wildlife Area and Camas NWR. Oregon: Malheur NWR; Silvies River floodplain (Burns). Washington: Tokeland (Pacific County); Willapa Bay.

Adult breeding Willet

Adult nonbreeding Willet

Willet range

Upland Sandpiper
(*Bartramia longicauda*)

The unique and attractive Upland Sandpiper has, in years past, bred at a few locations in the Northwest—mountain valleys in Oregon's Grant County and in the grasslands near Spokane, Washington. But eastern Washington's historically widespread population is now likely extirpated due to habitat loss, and there are no recent breeding records for Oregon, Idaho, or southern British Columbia, so this species is now considered a vagrant in the region.

LENGTH 12 inches. **WINGSPAN** 26 inches. Large head atop slender neck, with pale face and large, prominent dark eye; long yellow legs and slender yellowish bill; handsomely scalloped in black, beige, and white above; bright white below with lightly barred flanks and speckled breast. In flight, white wing linings are check-marked with black; dorsal wing surface shows bright white patch through middle of black primaries. **VOICE** Exotic sounding song begins with a rapid froglike croaking and then rapidly transitions to an accelerating, high-pitched, whistled whinny, and ends with a very high, descending whistle. Calls include a rolling *pip-pip-pip-pip-pip*. **BEHAVIORS** Like Wilson's Snipe (also found in upland wet meadows), often perches on wooden fence posts, phone line poles, and other such perches. **HABITAT** Upland prairie, including wet meadows, pastures, croplands. **STATUS** Vagrant during migration; formerly very rare and local summer resident.

Upland Sandpiper

Whimbrel (*Numenius phaeopus*)

Among shorebirds in the Northwest, only the Long-billed Curlew—a much larger bird—has a long, downward-curved bill like that of the Whimbrel. These handsome members of the curlew tribe are most common during spring migration along the coast, less common in fall migration. They are also found in the valleys west of the Cascades, and a few fall-migrating birds linger into winter. Individuals and small groups are typical, but most years a few large flocks arrive during spring migration in April and May.

LENGTH 17 inches. **WINGSPAN** 33 inches. Long, grayish, downward-curved bill; mottled or speckled grayish brown above; pale gray neck and breast is finely speckled with dark spots; barred flanks, pale belly; dark crown stripes divided by light median stripe, and eye stripe; fairly long gray legs; long, narrow, sharply pointed wings. **VOICE** Flight and alarm call is a sharp, rapid *wip-wip-wip-wip-wip-wip*. **BEHAVIORS** Generally found alone, foraging, but sometimes occurs in small flocks and in loose associations with other shorebirds or gulls. When feeding, it probes with its bill, and often dashes quickly ahead in spurts. Flight is strong and direct, and Whimbrels often glide. **HABITAT** Mudflats, estuary edges, sand beaches, marshes, vegetated headlands, lawns, and fields. **STATUS** Uncommon spring migrant, rare fall migrant, and rare winter resident along coast and valleys west of Cascades; very rare migrant east of Cascades. **KEY SITES** Oregon: Bandon Marsh and Coquille River estuary; Bayocean Spit (Tillamook Bay); Fort Stevens State Park; Yaquina Bay. Washington: Drayton Harbor and Blaine Marina Park; Dungeness NWR; Grays Harbor, Ocean Shores, and Westport; Midway Beach and Grayland Beach (Pacific County). British Columbia: Boundary Bay and Iona Island.

Whimbrel

Whimbrel range

Whimbrel in flight

Long-billed Curlew
(*Numenius americanus*)

Largest of the North American shorebirds, the Long-billed Curlew is aptly named for its distinctively long, downward-curved bill. Though Long-billed Curlews are primarily found in the Northwest during summer and the migration seasons, a few have been known to linger through all or part of the winter along the coast.

Long-billed Curlew

Long-billed Curlew in flight

Long-billed Curlew range

LENGTH 24 inches. **WINGSPAN** 33 inches. Extremely long, decurved bill; long neck and legs; attractive mottled dark brown and pinkish tan above, forming a contrasting barred pattern on wing coverts and tail; lightly streaked neck and pale tan breast and belly; crown stripes and eye stripes generally indistinct. In flight, long, slender, pointed wings with pale cinnamon underwings, and dark outermost wingtips and elbow patch. **VOICE** Call is a loud, ringing *whir-it whir-it*, the last note highest, and sometimes with a slightly raspy first note, *kwir-it kwir-it*; sometimes substantially drawn-out *whiiiiiiir-it*; also a quick, high, piping *wit-wit-wit-wit-wit*, delivered in various speeds. **BEHAVIORS** Curlews will aggressively defend their nests and chicks, to the point of driving away coyotes and hawks, and groups of curlews will even mob raptors flying over nesting areas. In flight, typically alternates flapping on stiff wings with gliding. **HABITAT** Nests on shortgrass habitats, including arid, sparse cheatgrass-dominated steppe and open, low-sage plains; agricultural fallow and crop fields; grassy meadows and pasturage; in migration, also frequents shallow wetlands, wet fields and prairies, wet desert playas, estuary mudflats, and saltmarshes. **STATUS** Fairly common summer resident and migrant in southeast Oregon and southern Idaho; locally common summer resident of Columbia Basin in Morrow and Umatilla counties of Oregon; uncommon summer resident and migrant in Washington and British Columbia; rare west of Cascades in migration and winter. **KEY SITES** Idaho: Blacks Creek Bird Reserve (southeast of Boise); Camas Centennial Marsh; Camas NWR. Oregon: Malheur NWR and Burns area; Summer Lake Wildlife Area; Umatilla NWR and Boardman area. Washington: Black Rock Valley (north of Sunnyside); Columbia NWR. British Columbia: Road 22 north of Osoyoos Lake.

Marbled Godwit (*Limosa fedoa*)

A large, long-legged shorebird that breeds on the northern prairies, the Marbled Godwit migrates through the Pacific Northwest in modest numbers spring and fall, primarily along the coast, but also—in lesser numbers—inland as far east as eastern Idaho. In the interior Northwest, they are seen regularly, with density seemingly dependent on water levels at key wetlands used as stopovers during migration. The long, slightly upturned, bicolored bill is diagnostic.

LENGTH 18 inches. **WINGSPAN** 32 inches. Very long, slightly upturned pinkish bill with black outer half; long gray legs, fairly long neck; mottled dark brown and pale cinnamon-tan above, with bold dark bars on lower coverts; rufous-colored tail; in flight, rufous wings show black outer primary tips; lightly barred pale rufous flanks; light buff breast. Juvenile and winter adult lacks strong barring on flanks. **VOICE** Typical calls, especially flush and flight calls, include a rapid clucking *chukaa* and squeaky repeated and more excited *aa-ak* or *arr-AK arr-AK*, often in combination. **BEHAVIORS** Frequently forages by wading deep, and often submerges entire head. **HABITAT** During migration, typically found on coastal estuaries, beaches, mudflats, and saltmarshes; and shallow wetlands, fields, lake margins of the interior Northwest. **STATUS** Fairly common fall and spring migrant (July–October and April–May) and rare winter resident along Washington coast, uncommon Oregon coast; uncommon migrant inland. **KEY SITES** Idaho: American Falls Reservoir; Deer Flat NWR; C.J. Strike Wildlife Area (Jacks Creek arm). Oregon: Bay Ocean Spit and Tillamook Bay; Fort Stevens State Park; Malheur NWR and Burns area. Washington: Dungeness Bay and Dungeness NWR; Grays Harbor and Westport; Ocean Shores beaches; Willapa Bay (esp. Tokeland). British Columbia: Boundary Bay; Tofino and Pacific Rim National Park.

Nonbreeding Marbled Godwit in flight

Nonbreeding Marbled Godwit

Breeding Marbled Godwit

Marbled Godwit range

Hudsonian Godwit
(*Limosa haemastica*)

A rare visitor to the Northwest, the Hudsonian Godwit's primary migration route takes it through the center of the continent. In the Northwest, most sightings occur in coastal British Columbia and Washington, far fewer in Oregon, and Idaho only has a few accepted records.

LENGTH 15.5 inches. **WINGSPAN** 28 inches. **SEXES SIMILAR** Fall adult and juvenile pale gray to pale gray buff overall, lightest below, lightly mottled above; long, slightly upturned bill is pinkish or orangey at base, black at tip; in flight, black wing linings are distinctive (Marbled Godwit has pale buff wing linings); dorsal wing surface is gray with a narrow white stripe at base of flight feathers; white tail with broad black terminal band; prominent white-buff eyebrow stripe. **VOICE** Common call is a high, bright *too-wit*. **BEHAVIORS** Generally seen feeding on coastal mudflats in the company of other shorebirds. **HABITAT** Saltwater and freshwater mudflats; lakeshores, flooded fields; beaches, estuaries. **STATUS** Vagrant. Rare fall migrant; accidental in spring. Most records are for birds in nonbreeding plumage, and most are coastal for August–September.

Nonbreeding Hudsonian Godwit

Ruddy Turnstone
(*Arenaria interpres*)

The Ruddy Turnstone, generally a fall and spring migrant and only rarely an over-wintering species, often forms mixed flocks with the closely related Black Turnstone, which winters along the Northwest coast. The 2 species frequent intertidal rocks, often foraging on steep faces within the spray zone. Thorough examination of turnstone flocks may be required to pick out the occasional Ruddy among numerous Black Turnstones.

Breeding Ruddy Turnstone

LENGTH 9.5 inches. **WINGSPAN** 21 inches. **NONBREEDING** Soft mottled brown-gray above, white below, with dark breast partially divided by white at the center; white-tan mottled face; fairly stout black bill; dull orange legs. Birds from late spring through late summer usually show breeding plumage of intricate black-and-white face, neck, and breast pattern and rich rufous-brown and black upperparts. In flight, flashy black, white, and gray above; white tail with black subterminal band. **VOICE** Variable; alarm call typically a high, bubbling *pip-pip-pip-pip*, somewhat woodpeckerlike; flight call a high, slightly raspy *paa-pipipipi*. **BEHAVIORS** Often forages frenetically but blends in well with rocky habitat. Flight is swift and highly maneuverable. **HABITAT** Rocky intertidal zones, jetties, tidal flats, and sand beaches. **STATUS** Fairly common spring and fall migrant (July–May) along coast; rare but regular migrant inland. **KEY SITES** Oregon: Barview jetty (Tillamook Bay); Seaside Cove. Washington: Ediz Hook (Port Angeles); Ocean Shores; Tokeland (Willapa Bay); Westport. British Columbia: Boundary Bay; Whiffin Spit Park (Sooke).

Nonbreeding Ruddy Turnstone

Ruddy Turnstone in flight

Ruddy Turnstone range

Black Turnstone
(*Arenaria melanocephala*)

Generally found in flocks wintering along the coast, Black Turnstones feed within the intertidal zone with a decided preference for rocky habitat, including jetties. They blend in well with the dark coastal rocks until they take flight and show off their flashy black-and-white dorsal pattern. The Northwest offers so much prime winter and migration habitat—rocky shorelines and jetties—that Black Turnstones are ubiquitous and almost always easy to find from September through April.

LENGTH 9.5 inches. **WINGSPAN** 21 inches. Dark slate-gray or brownish gray back, breast, and head; white belly; short, blackish bill; dark gray legs. In flight, flashy black-and-white pattern, with white rump, white wing stripes, and white tail with black subterminal band. Late spring (breeding plumage) birds show white cheek spots at base of the bill. **VOICE** Alarm and flight calls include a high, fast, rolling chatter, and an even higher fast, rolling whistle. **BEHAVIORS** Forages energetically on intertidal rocks, rocky shores. Flight is swift and highly maneuverable. **HABITAT** Rocky intertidal zone, jetties, rocks outside intertidal zone, occasionally tidal flats. **STATUS** Common migrant and winter resident. **KEY SITES** Oregon: Barview jetty (Tillamook Bay); Chetco Point and Chetco jetties; Coquille River jetties; Seal Rock beach; Siuslaw River jetties; Yaquina Bay jetties. Washington: Ediz Hook; Fort Flagler State Park (Marrowstone Island); Neah Bay and Shipwreck Point; Salt Creek County Park (Clallam County); Semiahmoo Spit (Drayton Harbor); Tokeland (Willapa Bay); Westport. British Columbia: Cattle Point, Oak Bay, and Clover Point (Victoria); Klootchman Park (West Vancouver); Neck Point Park and Pipers Point (Nanaimo); Sunshine Coast; Tsawwassen Ferry Terminal jetty (Vancouver).

Breeding Black Turnstone

Nonbreeding Black Turnstone

Black Turnstone in flight

Black Turnstone range

Red Knot (*Calidris canutus*)

A fairly large *Calidris* sandpiper, the Red Knot nests in the high Arctic, and primarily migrates along both coasts. Though common nowhere in the Northwest, Red Knots occur regularly along the coast in Washington and British Columbia, and are sometimes found along the Oregon coast. Spring migrants, as well as early fall adult migrants, display at least partial rufous-colored breeding plumage. These birds often intermix with other sandpipers at feeding areas, and in juvenile and nonbreeding plumage can be distinguished from similar species by their larger, chunkier appearance, and relatively short bill.

LENGTH 10.5 inches. **WINGSPAN** 23 inches. **SEXES SIMILAR** Chunky and short billed, with relatively short legs; in flight, long, slender wings with white stripes; grayish or white underwing coverts; gray rump and tail. **ADULT BREEDING** Rufous colored overall, with contrasting gray back; white undertail coverts with black barring; greenish legs; during molt, varying amounts of patchy rufous coloration, with gray showing on head. **ADULT NONBREEDING** Light gray above with some white edging to larger feathers on back; breast and flanks with small, light chevrons against light grayish background. **JUVENILE** Similar to nonbreeding adult but

with pale buffy wash on breast and back. **VOICE** Usually silent. **BEHAVIORS** Often seen in groups of other shorebirds. Uses bill to probe or pick at food items. Flocks of knots are seen primarily in spring migration. **HABITAT** Strongly coastal during migration; prefers sand beaches and rock ledges. **STATUS** Uncommon spring migrant (April–May) on northern Washington coast and British Columbia's Boundary Bay and surrounding areas; increasingly rare south through Oregon; uncommon to rare during fall migration (July–September). Extremely rare inland throughout the Northwest.

Adult nonbreeding Red Knot

Juvenile Red Knot

Red Knot range

Adult breeding Red Knots

Sanderling *(Calidris alba)*

Found primarily on gently sloping sand beaches, this common light-colored sandpiper alternately sprints and runs or walks, poking and prodding and chasing prey, right along the edge of the surf line. Its wave-chasing behavior is an excellent identification clue even before further field marks can be studied. Widespread, Sanderlings typically occur in flocks ranging from a few individuals to hundreds of birds. In the Northwest, breeding-plumage and transitional-plumage Sanderlings occur during migration in spring and fall.

LENGTH 7.5 inches. **WINGSPAN** 16 inches. Black legs and bill; in flight, flashy white wing stripes separate dark gray rear wing edges from black leading edges; larger than common peeps, such as Western and Least Sandpipers. **ADULT BREEDING** (April–August) Scalloped gray above with rufous tones and feathers edged with white; finely speckled face, neck, and breast with variable rust tones (increasingly rust-toned in prime breeding plumage); clean white belly. **ADULT NONBREEDING** (September–April) Pale gray above, slightly darker on wing coverts; snowy white below; black bill and legs; black wingtips; black shoulders range from bold to subtle. **JUVENILE** (July–November) Striking pattern of black-on-white chevrons above, gleaming white below; white throat and forehead; white face with faint gray eye line; light gray flecks forming faint incomplete

Adult nonbreeding Sanderling

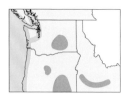
Sanderling range

Adult breeding Sanderling

necklace. **VOICE** A high, sharp *whit whit*, in series or singly. **BEHAVIORS** Feeds along the surf line, rapidly retreating as a wave flows up the beach, then following the draining water back out. Swift and maneuverable in flight; they typically flush as a flock and twist and turn in unison. **HABITAT** Open sand beaches; occasionally shell beds, mudflats, jetties, tide-pool zone. **STATUS** Common winter resident and migrant (August–mid-May). **KEY SITES** Oregon: Bayocean Spit (Tillamook Bay); Coos Bay North Spit; Fort Stevens State Park; Ona Beach (Lincoln County); Oregon Dunes Overlook. Washington: Dungeness NWR area; Ediz Hook (Port Angeles); Fort Flagler State Park (Marrowstone Island); Grayland Beach and Midway Beach (Pacific County); Long Beach Peninsula; Ocean Shores; Semiahmoo Spit (Drayton Harbor); Westport. British Columbia: Boundary Bay; Clover Point (Victoria); Iona Island (Vancouver); Jericho Park and Stanley Park (Vancouver); Whiffin Spit Park (Sooke).

Adult nonbreeding Sanderling showing wing pattern

Juvenile Sanderling

Dunlin (*Calidris alpina*)

An abundant migrant and winter resident in the Pacific Northwest, Dunlins often occur in large flocks along the coast and in the valleys of western Oregon and western Washington. Numbers peak in spring when the Northwest's over-wintering Dunlins are joined by flocks migrating northward from wintering grounds farther south. Though less common inland, flocks can show up even on remote mudflats and lakeshores east of the Cascades.

LENGTH 8.5 inches. **WINGSPAN** 17 inches. Heavy, fairly long, decurved black bill; black legs; in flight shows distinctive white wingbar and black stripe down the center of light tail. **ADULT NONBREEDING** (August–March) Drab brownish above; brownish breast; white below. **ADULT BREEDING** (April–August) Mottled rufous-gray above, with black belly patch bordering black-speckled white breast. **JUVENILE** (July–September) Variable black spotting on belly (sometimes none) and dark speckling on breast; brownish wash on head and neck; brownish gray mantle (sometimes with rust tones) with lightly scaled appearance. **VOICE** Call is a raspy, reedy *kreep*. **BEHAVIORS** Can form large flocks in the Northwest. In flight, flocked birds twist and turn in unison, with the flock flashing bright white when bellies and underwings are turned toward observers, and almost disappearing when backs are facing observers; such flight is especially pronounced when the birds are pursued by falcons, which commonly target shorebirds. **HABITAT** Coastal mudflats, beaches, estuaries; muddy lakeshores; flooded fields and pastures. **STATUS** Locally common to abundant on coastal wintering grounds; fairly common in western Oregon and Washington valleys; common spring (March–May) and fall migrant (September–December) west of Cascades; uncommon east of Cascades. **KEY SITES** Oregon: Barview jetty (Tillamook); Fern Ridge Reservoir. Washington: Columbia River Estuary; Grays Harbor. British Columbia: Boundary Bay; Iona Island Beach.

Juvenile Dunlin

Adult nonbreeding Dunlin

Dunlin range

Flock of adult nonbreeding Dunlins in flight

Adult breeding Dunlin

Western Sandpiper
(*Calidris mauri*)

The Northwest's most abundant shore-bird, Western Sandpipers often form massive flocks during migration—some-times totaling tens of thousands of birds. In flight, large flocks are amazing spec-tacles, the tightly bunched birds twist-ing and turning in unison. They nest on the tundra of western Alaska and eastern Siberia and typically winter from Califor-nia to South America, passing through the Northwest in transit; some Western Sandpipers overwinter along the North-west coast. The longer, thin-tipped, droop-ing bill and black legs help distinguish it from other *Calidris* sandpipers such as the smaller peeps.

LENGTH 6.5 inches. **WINGSPAN** 14 inches. Fairly long, drooping black bill, thin at tip; black legs; white stripe on wings and white outer tail feathers; in flight shows white wingbar and white edges on upper tail and rump. **ADULT BREEDING** (March–August) Light rufous crown and auricular; heav-ily streaked and spotted breast, flanks, and back; rufous scapulars. **ADULT NONBREED-ING** (August–March) Pale gray overall with white breast. **JUVENILE** (July–September) Brownish above, with streaked crown and pale brown auricular and breast; upper scapulars edged in rich rufous; molts ear-lier than other peeps. **VOICE** Calls include a high, burry, whistled *chee* or *chew*; and an ascending series of whistled *dew* notes end-ing in a descending buzzy trill. **BEHAVIORS** Swift and erratic flight; flocks constantly twist and turn, causing them to almost disappear from sight when the birds' backs face the viewer. Frenetic when feed-ing, and compared to the Semipalmated

Juvenile Western Sandpiper (August)

Adult breeding Western Sandpiper

Western Sandpiper range

Sandpiper, more likely to feed by probing deeper in standing water. **HABITAT** Intertidal mudflats and sand flats; lake, pond, and alkali lake edges; freshwater mudflats. **STATUS** Common to locally abundant spring migrant (March–May) and fall migrant (late June–October); uncommon winter resident. **KEY SITES** Idaho: American Falls Reservoir; Deer Flat NWR. Oregon: Bandon Marsh NWR; Tillamook Bay. Washington: Grays Harbor NWR; Potholes Reservoir. British Columbia: Boundary Bay.

Adult breeding Western Sandpiper in flight

Adult nonbreeding Western Sandpiper

Least Sandpiper
(*Calidris minutilla*)

The world's smallest shorebird, this little peep is a fairly common migrant throughout the Northwest. Most easily distinguished from other peeps by their leg color, as well as their smaller size and bill shape, Least Sandpipers often intermingle with large flocks of Western Sandpipers and with other shorebirds. Though pure flocks of hundreds of Least Sandpipers occur along the coast, most flocks are much smaller.

LENGTH 6 inches. **WINGSPAN** 13 inches. Greenish yellow legs in all plumages (among the other regularly occurring *Calidris* species in the Northwest, only the Pectoral Sandpiper has yellowish rather than blackish legs); smaller than all other peeps and appears hunched; in all plumages, darker above than other peeps; thin, dark, short, slightly drooping bill; in flight, shows white wingbars and white edges on upper tail and rump. **ADULT BREEDING** (April–September) Black-and-brown mantle with pale streaks; rufous-and-buff wing coverts; rufous-and-buff-edged scapulars with pale stripes; brown streaks on white breast; pale eye line extends to the bill, creating a pale forehead. **ADULT NONBREEDING** (October–March) Mostly uniform brown above; eye line less pronounced than in breeding plumage. **JUVENILE** (July–October) Bright, crisp plumage; dark feathers edged in rufous and buff. Juvenile maintains its crisp plumage even after adults have molted to their drab winter plumage. **VOICE** Call is a high, trilled *brrreet*. **BEHAVIORS** Moves slowly while feeding and rarely runs (Western Sandpiper frequently darts and runs while foraging). **HABITAT** Prefers freshwater, so often found inland, on margins of lakes, ponds, and marshes; also coastal mudflats, estuary edges, and tidal creeks. **STATUS** Common spring and fall migrant; uncommon coastal winter resident; rare winter resident inland to the Rockies. **KEY SITES** Oregon: Bay Ocean Spit; Fern Ridge Reservoir. Washington: Grays Harbor NWR; James T. Slavin Conservation Area (Spokane County); Nisqually NWR; Potholes State Park; Samish Flats; Whiffin Spit Park (Sooke). British Columbia: Boundary Bay; Pacific Rim National Park.

Least Sandpiper range

Adult breeding Least Sandpiper (mid-July)

Adult nonbreeding Least Sandpiper

Juvenile Least Sandpiper (August)

Adult breeding Least Sandpiper (late April)

Baird's Sandpiper *(Calidris bairdii)*

Like all the other *Calidris* sandpipers that occur in the region, the Baird's Sandpiper is a Far North breeder and long-distance migrant. Its normal migration path is east of the Rocky Mountains, but fair numbers of birds wander into the Northwest, especially juveniles, mostly during late summer and fall (adults are comparatively rare in the Northwest). These fairly large, long-winged peeps occur regularly on low-elevation wetlands, but also forage at high elevations—even on alpine snowfields on Northwest peaks.

LENGTH 7.5 inches. **WINGSPAN** 17 inches. Black to blackish olive legs; at rest, wingtips project beyond tip of tail; horizontal profile; straight, thin-tipped bill; faint white wingbars in flight; light supraloral spots; faint light eyebrow line. **ADULT BREEDING** (April–September) Rich silvery grayish mantle appears sharply scaled, with each feather having a white terminal band and black central spot; finely speckled tannish head and neck, fading to white at base of breast. **ADULT NONBREEDING** (September–April) Similar to juvenile, but lacks scaly pattern, and more solid brown above. **JUVENILE** Brown back gives scaly appearance due to whitish-edged feathers; dark rump; buff-brown head extending down to breast; white below. **VOICE** Call is a low, raspy *kreep*. **BEHAVIORS** Does not occur in large flocks in the Northwest. Swift, direct flight. **HABITAT** Sand beaches, coastal and freshwater mudflats, lake margins, mountain snowfields. Prefers the drier parts of a wet habitat and when present with other peeps, often forages in the drier parts of the feeding area. **STATUS** Uncommon fall migrant, rare spring migrant. More common east of the Cascades. **KEY SITES** Idaho: American Falls Reservoir; Deer Flat NWR. Oregon: Bay Ocean Spit (Tillamook County); Summer Lake Wildlife Area. Washington: Ocean Shores; Walla Walla River Delta. British Columbia: Boundary Bay; Sidney Spit (Gulf Islands National Park).

Adult nonbreeding Baird's Sandpiper

Juvenile Baird's Sandpiper

Juvenile Baird's Sandpipers

Baird's Sandpiper range

Pectoral Sandpiper
(Calidris melanotos)

Often described as a large lookalike of the diminutive Least Sandpiper, the Pectoral Sandpiper is indeed substantially bigger, and also tends to stand more upright—a valuable clue in identification. It is primarily an autumn migrant in the Northwest, passing through on its way south from its breeding grounds on the Arctic tundra to its faraway wintering regions in South America.

LENGTH 8.75 inches. **WINGSPAN** 18 inches. Heavily streaked breast sharply delineated from white belly (unlike Sharp-tailed Sandpiper); fairly long neck for a peep; upright posture; bill dark at tip, pale at base, and slightly decurved; yellowish green legs; faint white wingbars in flight. **MALE BREEDING** (April–October) Bold, dark streaking on breast; larger than female. **FEMALE BREEDING** (April–October) Dull brownish above; brown-streaked breast (lighter than on male); some rufous edging on upper body feathers; sparsely streaked flanks. **ADULT NONBREEDING** (October–March) Overall plain brown above; lighter, more nondescript streaks on breast than in breeding plumage. **JUVENILE** (July–November) Crisply patterned like adult breeding, but with less extensive speckling on breast; rows of white-edged feathers along the back form distinctive white "braces." **VOICE** Call is a sharp, reedy *chert* or *trik*. **BEHAVIORS** Walks slowly when foraging. Swift, erratic flight when flushed; otherwise casual, seemingly unhurried flight, with steady, even wingbeats. **HABITAT** Wet grassy areas near saltmarshes, lake or pond edges, flooded agricultural lands. **STATUS** Uncommon fall migrant, rare spring migrant. **KEY SITES** Idaho: American Falls Reservoir; Deer Flat NWR. Oregon: Fern Ridge Reservoir; Tillamook Bay. Washington: Skagit Wildlife Area; Ocean Shores; Grays Harbor. British Columbia: Boundary Bay.

Adult breeding Pectoral Sandpiper (May)

Juvenile Pectoral Sandpiper (September)

Pectoral Sandpiper range

Semipalmated Sandpiper
(*Calidris pusilla*)

Although most of the Northwest lies just outside its normal range, the Semipalmated Sandpiper is a regular migrant in the region, arriving in small numbers during late summer. Adults arrive first, followed by juveniles making their first southward migration from the Far North breeding grounds. They are rare in spring. Most are seen along the coast, but these birds are seen in the interior Northwest annually. Adults in nonbreeding plumage are very similar to other black-legged *Calidris* sandpipers, but most adults migrating through the Northwest are still in breeding plumage.

LENGTH 6.25 inches. **WINGSPAN** 14 inches. Black, blunt-tipped bill (shorter than that of Western Sandpiper and does not droop at the tip like the Western); black legs (Least Sandpiper has yellowish legs); in flight shows white wingbar and white edges on upper tail and rump. **ADULT BREEDING** (March–September) Mottled black and grayish brown above, with little if any rufous. **ADULT NONBREEDING** (October–March) Pale gray above, with brownish wash on some individuals. **JUVENILE** (July–November) Buffy brown above with strong, uniform scaled pattern (unlike Western Sandpiper); bold eyebrow contrasts with dark crown

and a dark ear patch (chestnut shades in the crown). **VOICE** Calls include a short, barking *chup*, and a wavering, *ghe-ee-ee-ee-ee* of varying length and tone. **BEHAVIORS** Usually found with flocks of other *Calidris* sandpipers. Unlike Western Sandpiper, feeds less frenetically and does not probe as deeply. Flight is swift and erratic. **HABITAT** Mudflats, shorelines, sparsely vegetated saltwater flats. **STATUS** Uncommon to rare regular late summer migrant; rare spring migrant. **KEY SITES** British Columbia: Boundary Bay area.

Adult nonbreeding Semipalmated Sandpiper

Juvenile Semipalmated Sandpiper

Semipalmated Sandpiper range

Adult breeding Semipalmated Sandpiper

Rock Sandpiper
(*Calidris ptilocnemis*)

Squat and stocky, the Rock Sandpiper, which breeds in the Far North of Alaska, winters along the Northwest coast. Often they are found near or among other shorebirds, especially Black Turnstones and Surfbirds, less often Black Oystercatchers. Full breeding-plumage birds are seldom seen in the Northwest.

LENGTH 9 inches. **WINGSPAN** 16 inches. **ADULT NONBREEDING** Gray above, with gray head, throat, neck; lighter breast streaked with rows of dark chevron spots, extending down flanks; white supercilium; short, dull yellowish legs; slightly decurved dark bill with yellowish base. In flight, white tail with dark center; white trailing edge on inner wings; dark throat and breast. **JUVENILE** Rich mix of dark gray, rust, and white above, strongly scalloped appearance; white below with fine dark speckling; grayish face. **VOICE** Flush and flight call is a high, scratchy *screeep*. **BEHAVIORS** Frequently feeds within the splash zone, scampering about on rocks and cobble; often roosts and rests near splash zone. Very fast and highly maneuverable in flight. **HABITAT** Rocky intertidal zone, including jetties; less commonly, sand or cobble beaches. **STATUS** Uncommon winter resident and migrant (October–April). **KEY SITES** Oregon: Barview jetty (Tillamook Bay); Seal Rock State Wayside; Yaquina Bay jetties. Washington: Ediz Hook (Port Angeles); Point Brown Jetty and Westport Jetty. British Columbia: Cattle Point and Clover Point (Victoria).

Adult nonbreeding Rock Sandpiper

Juvenile Rock Sandpiper

Rock Sandpiper range

Surfbird *(Calidris virgata)*

During winter, when it lives in the Northwest, the stocky little Surfbird rarely strays from the intertidal zone. Its nests primarily on rocky domes in the mountains of Alaska, but spends the rest of the year on the coast. Most coastal birds are in winter plumage, but in spring, transitional-plumage and breeding-plumage specimens occur, and in fall, juvenile birds also occur, similar in appearance to winter-plumage adults.

LENGTH 10 inches. **WINGSPAN** 26 inches. Dark and stocky, with short legs; yellowish legs; short, stocky bill with pale yellowish base; in flight shows white wing stripes, and white tail with black terminal band. **NON-BREEDING** (August–March) Gray above, gray breast, white belly. **BREEDING** (April–July) Mottled gray above with rufous wash on shoulders and back; gray-white breast patterned with black spots, white flanks patterned with black chevrons. **VOICE** Subdued, squeaky, high-pitched chatter when feeding or flying, *chir chir chir chir-chir*; sometimes slightly rolling *chirip chirip*. **BEHAVIORS** Feeds in the intertidal zone by foraging on rocks; often occurs in small groups and mixes with turnstones and other sandpipers. Swift in flight, and often makes quick fluttering flights between rocks. **HABITAT** Rocky intertidal zone, jetties, breakwaters. **STATUS** Fairly common winter resident and migrant, most common October–March. **KEY SITES** Oregon: Barview jetty (Tillamook Bay); Seal Rock State Wayside; Siuslaw South Jetty; Yaquina Bay jetties. Washington: Neah Bay; Port Townsend area; Westport area. British Columbia: Cattle Point and Clover Point (Victoria); Klootchman Park and Lighthouse Park (West Vancouver).

Adult nonbreeding Surfbird

Adult breeding Surfbird

Surfbird in flight

Surfbird range

Ruff (*Calidris pugnax*)

The Ruff, an Old World species, is a rare visitor to North America. Almost annually a few show up in the Northwest, primarily in fall; most are juveniles. Though extremely rare in the region, the adult male in its breeding plumage is unmistakable with a dense ruff of fluffy plumes—brown, tan, black, white, barred black and white, or even mixed shades—adorning its neck and head. On their Eurasian breeding grounds, males display on leks during spring, puffing up their shaggy manes to impress the smaller females, which are called Reeves.

LENGTH 11 inches. **WINGSPAN** 21 inches. **JUVENILE** (July–November) Short, slightly drooping bill is dark at the tip and pale gray at the base; small head, long neck, and plump body; scaly brown and buff above; pale buff face, lores, and breast; white belly; long, pale yellowish olive legs. **ADULT NONBREEDING** (August–March) Sexes similar except that most males are larger, with white on head and neck; orange, yellow, or pinkish bill with or without black tip; grayish brown above; some birds may have a white face, neck, and breast; orange-yellow legs (some females have greenish legs).

MALE BREEDING (March–June) Unmistakable, with diagnostic ruff; mottled above; white belly. **FEMALE BREEDING** (March–August) Similar to nonbreeding, but breast can be blotched with black, and tertials are barred. **VOICE** Generally silent. **BEHAVIORS** Flight is steady with measured wing strokes, with long legs trailing well beyond tail; in flight, white feathers outline dark center of rump and tail, but white can meet on the tail forming a U; white underwings; narrow white wingbars. **HABITAT** Estuaries, marshes, wet grasslands, pastures. **STATUS** Vagrant, but nearly annual fall visitor to coast; very rare inland; very rare in spring.

Male nonbreeding Ruff

Juvenile Ruff

Sharp-tailed Sandpiper
(*Calidris acuminata*)

Sharp-tailed Sandpiper are rare fall visitors to the Northwest during migration from their breeding grounds in Siberia. They usually pass through the region offshore, on the Pacific side of their migratory routes. A few show up along the coast. Almost all sightings are of juveniles.

LENGTH 8.5 inches. **WINGSPAN** 18 inches. **JUVENILE** Similar to Pectoral Sandpiper, but generally brighter plumage; slightly drooping bill is pale at base, with a dark, thin tip; yellowish green legs; white eye stripe; rufous cap and rufous stripe running from base of bill through eye to ear; white eye ring; lightly streaked orange-buff breast, not as sharply delineated from white belly as in Pectoral Sandpiper; faintly streaked flanks; rufous-edged dark brown feathers on back; white belly; white wingbar and uniform streaks on white upper tail coverts visible in flight; dark, pointed tail. **VOICE** Call is similar to the *chert* of the Pectoral Sandpiper but higher pitched. **BEHAVIORS** Often seen in the company of similar Pectoral Sandpipers. **HABITAT** Saltmarshes and tidal mudflats. **STATUS** Vagrant. Rare fall migrant along the coast; very rare elsewhere.

Juvenile Sharp-tailed Sandpiper

Stilt Sandpiper
(*Calidris himantopus*)

The handsome Stilt Sandpiper's migratory path takes it through the Great Plains, but a few individuals stray into the Northwest. Most sightings are of juveniles during the fall, but the northward spring migration also produces a few sightings. They are generally seen alone, but small groups of Stilt Sandpipers occasionally show up in the Northwest, usually in eastern Washington, eastern Oregon, or southern Idaho. They are smaller, with shorter bills than the somewhat similar dowitchers and yellowlegs.

LENGTH 8.5 inches. **WINGSPAN** 18 inches. **SEXES SIMILAR** Slender and similar in profile to a yellowlegs, with long, yellowish green legs; dark, long, heavy, slightly drooping bill; at rest, wingtips extend beyond tail. **JUVENILE** (July–September) Brown above, with scaled appearance; lightly streaked breast and flanks. In flight, all plumages have white wing linings outlined in pale gray; dorsal wing surface gray with white stripe extending through flight feathers; white rump and dark tail; legs extend beyond tail. **ADULT BREEDING** (April–August) Mottled dark brown, black, and white above; heavily black-and-white barred below; white neck peppered with black; bright white eye line; rufous ear patch, lores, and nape. **ADULT NONBREEDING** (September–August) Pale gray above; whitish below, with faintly streaked breast and flanks; gray crown and white eyebrow. **VOICE** Call is a descending *keer*. **BEHAVIORS** Feeds like a dowitcher, with a sewing machine motion; in deeper water, is a capable swimmer; often wades belly deep.

Flight is swift and direct. **HABITAT** Shallow lake and pond margins, shallow wetlands; coastal lagoons and estuary shallows. **STATUS** Vagrant, but annual fall migrant; very rare spring migrant.

Juvenile Stilt Sandpipers

Adult nonbreeding Stilt Sandpiper

Buff-breasted Sandpiper
(Calidris subruficollis)

The handsome, ploverlike Buff-breasted Sandpiper migrates through the Great Plains, but a few (primarily juveniles) make it to the Northwest, with most sightings along the coast during fall. On its breeding grounds in the Arctic tundra, males perform intricate courtship displays on leks.

LENGTH 8.25 inches. **WINGSPAN** 18 inches. **SEXES SIMILAR** Short, straight bill and steep forehead; prominent black eyes; scaled brown above, with streaked crown and plain, buffy face, breast, and belly; white undertail coverts; dark rump and tail; yellow legs; long wings with bright white linings. Adults—rarely seen in the Northwest—are lighter, more richly colored overall, and less strongly scaled above than juveniles. **VOICE** Usually silent but will utter a clicking note. **BEHAVIORS** Flight is swift with twists and turns. Pigeonlike walk or run. **HABITAT** Most sightings are coastal, typically on lawns, golf courses, vegetated beaches, and shortgrass locations near ponds and lakes. **STATUS** Vagrant, but nearly annual during fall migration (August–October).

Juvenile Buff-breasted Sandpiper

Long-billed Dowitcher
(*Limnodromus scolopaceus*)

The 2 species of dowitcher are nearly identical, and—at least in nonbreeding and partial breeding plumages—best distinguished by their voices, which are quite unalike. To some extent, location and seasonality can aid in determining the likelihood of one species over the other because Short-billed Dowitchers are rare east of the Cascades and rarely overwinter in the Northwest, while Long-billed Dowitchers are common east of the Cascades (and west of the Cascades) and some overwinter west of the Cascades. Also, the 2 species will comingle within the same flocks.

LENGTH 11.5 inches. **WINGSPAN** 19 inches. **SEXES SIMILAR** Long, sturdy, dark olive bill with black tip; fairly long olive-yellow legs; typically a distinctive humpbacked appearance when feeding; in flight, back and rump are white, and wings show pale trailing edges. **ADULT BREEDING** (April–August) Tannish above, richly mottled with black and white; warm rufous to nearly brick red below, including belly (unlike Short-billed Dowitcher), with scattered freckling and lightly barred flanks diminishing to light speckling on sides of neck; prominent dark eye line and pale supercilium; dark crown. **ADULT NONBREEDING** (September–March) Pale gray above, white below with light gray neck and usually lightly barred flanks; white belly and chin; white supercilium. **JUVENILE** (July–November) Plain grayish overall with hints of rufous; white belly; dark grayish above, with tertial and scapular feathers edged with rust (unlike richly patterned mantle of juvenile Short-billed Dowitcher). **VOICE** High-pitched, airy *peep* calls given in flight and when feeding; also rapid bubbly trills. **BEHAVIORS** Commonly feeds in what has been termed a sewing-machine motion by repeatedly plunging its bill into

the mud. Strong and direct flight; typically flushes and flies in flocks; in flight, white rump and back are distinctive. **HABITAT** Marshes, mudflats, exposed lakeshores, saltmarshes. **STATUS** Common migrant; uncommon winter resident (primarily west of Cascades). **KEY SITES** Idaho: American Falls Reservoir; Deer Flat NWR; Kuna Sewage Ponds; Market Lake Wildlife Area. Oregon: Fern Ridge Reservoir; Malheur NWR; Summer Lake Wildlife Area; Willamette Valley NWR Complex. Washington: Grays Harbor; McNary NWR and Walla Walla River delta; Nisqually NWR; Columbia NWR and Potholes Reservoir; Samish Flats (Skagit County); Skagit Wildlife Area. British Columbia: Boundary Bay; Burnaby Lake (Vancouver); Reifel Migratory Bird Sanctuary (Vancouver); Robert Lake (Kelowna).

Adult breeding Long-billed Dowitchers

Adult nonbreeding Long-billed Dowitcher

Flock of adult breeding Long-billed Dowitchers in flight

Juvenile Long-billed Dowitcher

Long-billed Dowitcher
range

Short-billed Dowitcher
(*Limnodromus griseus*)

The 2 species of dowitcher are nearly identical, but distinguishable by voice, as well as by minor plumage differences. To some extent, location and seasonality can aid in determining the likelihood of one species over the other because Short-billed Dowitchers are rare east of the Cascades and rarely overwinter in the Northwest, while Long-billed Dowitchers are common east of the Cascades (and west of the Cascades) and some overwinter west of the Cascades. Huge flocks, such as those that occur on the flats of British Columbia's Tofino Inlet and Washington's Grays Harbor, are generally Short-billed Dowitchers, often with a few Long-billed Dowitchers mixed in. Regardless, identification is difficult, though much easier with juveniles than with adults.

LENGTH 11 inches. **WINGSPAN** 18 inches.
SEXES SIMILAR Long, sturdy, dark olive bill; fairly long olive-yellow legs; in flight, back and rump are white, and wings show pale trailing edges; back is flat when feeding (unlike humpbacked appearance of Long-billed Dowitcher). **ADULT BREEDING** (April–August) Tannish above with large black spots; white belly (unlike breeding-plumage Long-billed Dowitcher); warm rufous breast fading to pale rufous flanks; strong barring on flanks and heavy freckling on sides of breast; prominent dark eye line and pale supercilium; dark crown. **ADULT NONBREEDING** (September–April) Pale gray above, white below with light gray neck and usually lightly barred flanks; white belly and chin; white supercilium. **JUVENILE** (July–November) Brightly patterned above in dark gray and cinnamon, with tiger-striped tertials and coverts (unlike far plainer mantle of juvenile Long-billed Dowitcher); bright rust upper breast fading to white belly **VOICE** Sharp, fairly low-pitched 2- or 3-syllable whistled *too-too-too*; also sometimes a wavering *chichichi-pee-purr*. **BEHAVIORS** Commonly feeds in what has been termed a sewing-machine motion by repeatedly plunging its bill into the mud. Strong and direct flight; typically flushes and flies in flocks. **HABITAT** Coastal mudflats, including exposed mud and shallow water; saltmarshes; occasionally freshwater mudflats. **STATUS** Common coastal migrant (April–May and July–October), uncommon interior western Oregon and Washington, rare east of Cascades. **KEY SITES** Oregon: Fort Stevens State Park; Siuslaw River estuary. Washington: Grays Harbor; Willapa Bay. British Columbia: Tofino Inlet.

Adult breeding Short-billed Dowitcher

Adult nonbreeding Short-billed Dowitcher molting to breeding plumage (April)

Juvenile Short-billed Dowitcher

Short-billed Dowitcher range

Adult nonbreeding Short-billed Dowitcher

Wilson's Snipe
(*Gallinago delicata*)

With a flexible bill-tip lined with sensory pits, the Wilson's Snipe feeds by probing its long bill into soft substrate to capture invertebrates. Its spring winnowing flights produce one of the characteristic sounds of highland marshes and wet meadows, where the species nests. Wintering and migrating Wilson's Snipes can form fairly large aggregations in prime feeding habitat and even in flight, especially on coastal estuaries, but they do not flock-flush like similar shorebirds, such as dowitchers.

LENGTH 10.5 inches. **WINGSPAN** 17 inches. Mottled rich brown, black, and tan above, with lines of whitish stripes running down the back; speckled breast, black-and-white-barred flanks, white belly; pale buff face with brown eye stripe and buff eyebrow stripe; dark brown crown stripes; long, straight, olive bill with black tip; fairly long olive legs. **VOICE** Flush and flight call a nasal *scaipe*, singularly or rapidly repeated. **BEHAVIORS** Very fast and agile flight; explosive gamebirdlike flush followed by zigzagging low-and-away escape flight before rapid climb; often flies a wide, high circle before landing again. Aloft, flight is very fast and direct. Flush is usually accompanied by *scaipe* calls. On breeding grounds, winnowing flights occur aloft, typically in evening, with circling flight followed by dive wherein tail feathers produce an eerie whistling. Also on breeding grounds, Wilson's Snipe often perches on wooden fence posts, fence anchors, and other flat-topped objects; otherwise secretive and retiring. **HABITAT** Breeding: wet upland meadows and marshes. Migration: ditches, pastures, tidal creeks and estuaries, marshes, wet fields. **STATUS** Locally common summer resident of interior, uncommon west of Cascades; common migrant and winter resident.

Wilson's Snipe

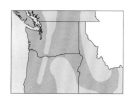

Wilson's Snipe range

Wilson's Snipe

Wilson's Phalarope
(Phalaropus tricolor)

The only phalarope that nests in the Northwest, the Wilson's Phalarope is especially at home in the marshes of the Great Basin, Columbia Plateau, and Snake River Plain. As with the smaller Red and Red-necked Phalaropes, the female is brighter colored than the male, and averages somewhat larger; as in all phalaropes, the males alone care for the young. Phalaropes are good swimmers, and a swimming (rather than wading) shorebird is a solid clue to its identity as a phalarope (although others, especially yellowlegs, will occasionally swim as well). Wilson's Phalaropes—a must-see species for birders visiting the inland wetlands during late spring and early summer—winter in South America and generally depart the Northwest by early September, by which time they have molted to their pale fall and winter plumage. Traditionally many thousands stage at Oregon's Lake Abert in late summer.

LENGTH 9 inches. **WINGSPAN** 17 inches. **FEMALE BREEDING** Pale gray mantle bordering rufous wing coverts, with broad dark stripes along sides of neck from eyes to upper back, black at eyes and upper neck and becoming chestnut-colored along lower neck; white below, with pale cinnamon wash on throat; silvery gray crown and nape; white eyebrow stripe; long, black, needlelike bill; in flight, wings are plain unmarked gray above, whitish below; black legs. **MALE BREEDING** Similar to female, but much paler and less distinctly marked, with longer, more pronounced white eyebrow stripe. **ADULT NONBREEDING** Both sexes pale gray above, white below, with yellow legs; distinct white eyebrow stripe. **JUVENILE** Similar to adult male, but with dark mottling on brownish back; lacks any

Male breeding Wilson's Phalarope

Adult nonbreeding Wilson's Phalarope

Female breeding Wilson's Phalarope

chestnut coloring on neck; nape is pale gray; pale yellowish legs help distinguish it from various sandpipers. **VOICE** Typical call is a surprisingly husky, low-pitched barking *chuk chuk chuk*, more persistent and higher as an alarm call. **BEHAVIORS** Feeds by swimming and wading, and also on land near water's edge, frequenting mudflats; often swims in quick, tight circles to create upwellings that draw prey toward the surface. **HABITAT** Marshes, lakeshores, wet playas, flooded fields, ponds, mudflats. **STATUS** Common summer resident and migrant (May–September) southeast and south-central Oregon; Columbia Plateau, Washington; south and southeast Idaho; fairly common summer resident and migrant central Oregon, northeast Washington, central and eastern British Columbia; rare breeder west of Cascades. Uncommon migrant coastal British Columbia; rare migrant western Oregon/Washington. **KEY SITES** Idaho: Camas NWR and Market Lake Wildlife Area; Deer Flat NWR; Indian Creek Reservoir (Ada County); Kuna sewage ponds; Mann Lake (near Lewiston). Oregon: Fern Ridge Reservoir; Finley NWR; Lake Abert; Malheur NWR; Silvies River floodplain (Burns); Summer Lake Wildlife Area. Washington: Columbia NWR and Potholes area; Sprague sewage lagoon and Sprague Lake (Adams County). British Columbia: Road 22 north of Osoyoos Lake; Robert Lake (Kelowna).

Juvenile Wilson's Phalarope

Wilson's Phalarope range

239

Red-necked Phalarope
(*Phalaropus lobatus*)

The Red-necked Phalarope—small-est of the 3 phalarope species—is a Far North breeder that spends the balance of the year at sea. Its migrations take it through the Northwest in considerable numbers in 2 divergent flyways. Many birds migrate along the coast, mostly over the ocean. They are seen regularly from shore, though the number of such sightings varies annually. Others migrate through the interior Northwest, and while they can be fairly widespread, one of the traditional staging areas is Lake Abert in southeast Oregon, a massive playa that hosts flocks that can number in the hun-dreds and occasionally the thousands at the peak of southward movement in July and August if the playa retains water. In the Northwest, most northward migrat-ing Red-necked Phalaropes occur in May, when they wear full or partial breeding

Juvenile Red-necked Phalarope

Red-necked Phalarope range

Adult breeding Red-necked Phalarope

plumage. As with the other phalaropes, the female is more colorful than the male. In fall and winter plumages, Red-necked and Red Phalaropes are difficult to distinguish unless they are seen together, where the size difference is obvious. Otherwise, bill thickness (very thin in the Red-necked) and back color are the most reliable field marks.

LENGTH 7.5 inches. **WINGSPAN** 15 inches. **ADULT NONBREEDING** Sexes similar, but female brighter and averages slightly larger than male; gray above with contrasting darker and lighter streaks or blotches; white below; white face and throat with dark crown and bold black eye stripe; thin, needlelike bill (much thinner than Red Phalarope). In flight, shows white stripe through upper wing surface and dark leading edge on both wing surfaces (Red Phalarope has white underwings). **ADULT BREEDING** Grayish above, with more extensive mottling in male; dark slate-gray face and head, more completely and darker gray in female; small

white eyebrow; white below; distinctive cinnamon-red neck, richer and more extensive in female; bright white throat. **JUVENILE** (July–October) Similar overall to nonbreeding adult, but mottled black and brownish gray above, usually with distinct pale tan dorsal stripes. **VOICE** Flight call is a sharp *peek*, singly and in series. **BEHAVIORS** Excellent swimmer; when feeding in freshwater, often swims in quick, tight circles to create upwellings that draw prey toward the surface; typically forms flocks in migration, though scattered sightings of individual birds or small groups are common. **HABITAT** Inland: shallow lakeshore margins, wetlands, sewage ponds, flooded fields. Coastal: open ocean, bays and harbors, estuaries. **STATUS** Common fall migrant (July–September); uncommon spring migrant (late April–May). **KEY SITES** (interior) Idaho: American Falls Reservoir; Deer Flat NWR. Oregon: Lake Abert; Silvies River floodplain (Burns). Washington: Sprague Sewage Lagoons and Sprague Lake (Adams County). British Columbia: Okanagan Valley.

Adult nonbreeding Red-necked Phalarope

Red Phalarope
(*Phalaropus fulicarius*)

Like the similar Red-necked Phalarope, the Red Phalarope is more seabird than shorebird; it breeds in the Far North, but then spends up to 9 months at sea. It migrates offshore along the Northwest coast, but is regularly found along the coast, especially during and after storms. They are most easily found on pelagic birding trips, but are less common than the Red-necked Phalarope, even though their fall migration is lengthier. As with the other phalaropes, the female is more colorful than the male. In fall and winter plumages, Red-necked and Red Phalaropes are difficult to distinguish unless they are seen together, where the size difference is obvious. Otherwise, bill thickness (very thin in the Red-necked) and back color are the most reliable field marks.

LENGTH 8.5 inches. **WINGSPAN** 17 inches. Sexes similar, but female brighter in breeding plumage and averages slightly larger than male. **ADULT NONBREEDING** Uniformly pale gray above, white below; irregular gray blotches during fall in transition to winter plumage; white face and throat with dark crown and bold black eye stripe; bill noticeably heavier than that of Red-necked Phalarope. In flight, shows pronounced white stripe through upper wing surface and white wing linings. **JUVENILE** (July–October) Mottled black or dark gray and tan above, with dark feathers edged

in tan; striped upper back. **ADULT BREEDING** Fully to partially bright rufous below, from belly and flanks to neck and throat; rich mottled brown, tan, and black above; black head with large white face patch; yellow or yellowish bill. Male less colorful than female. **VOICE** Call is a high *peep* or *purp*. **BEHAVIORS** Swims buoyantly and lifts on and off the water effortlessly; often swims in quick, tight circles to create upwellings that draw prey toward the surface; typically forms flocks in migration, though scattered sightings of individual birds or small groups occur, and individuals often mix with other phalaropes and shorebirds. **HABITAT** Primarily pelagic; also coastal waters, including bays, inlets, estuaries. **STATUS** Uncommon pelagic migrant (April–May and July–November); uncommon on coastal waters; generally rare inland, primarily west of Cascades; rare December–February. **KEY SITES** Pelagic.

Female adult breeding Red Phalarope

Adult nonbreeding Red Phalarope

Red Phalarope range

Alcids, Gulls, Terns, and Jaegers

Of the same taxonomic order as shorebirds, but representing different families, alcids (Alcidae), jaegers (Stercorariidae), and gulls and terns (Laridae) are primarily coastal and oceanic species, though several species of gulls and terns are freshwater specialists that breed in the interior Northwest. The alcids are stocky, waterfowl-like seabirds ranging in size from the tiny Cassin's Auklet to the robust and showy Tufted Puffin. The jaegers are birds of the open ocean that frequently sustain themselves by stealing food from other birds, often harassing gulls, shearwaters, and other species into regurgitating meals. There are 3 Northwest breeding species of terns and several migratory and transient species. Among the variety of gull species in the region are several that can be difficult to distinguish, especially in subadult plumages that change each year for up to 4 years, depending on species. To make matters more challenging, various gulls commonly hybridize.

Tufted Puffin

Common Murre (*Uria aalge*)

The most numerous alcid in the Northwest, the Common Murre nests in densely packed colonies on nearshore sea stacks and headlands. The starkly black-and-white birds (though actually dark brown rather than black in good light) are so tightly clustered that such colonies are visible from considerable distances. During the nonbreeding season, they occur singly and in small groups, wintering both offshore and nearshore.

LENGTH 16 inches. **WINGSPAN** 26 inches. **ADULT BREEDING** (March–August) Black above (or dark charcoal brown at close range), including all-black head and neck; white below, with slight dark striations on flanks (often difficult to see); narrow, straight, black bill. **ADULT NONBREEDING** (September–March) All-black head of breeding plumage replaced by white throat and face, with thin dark line trailing back from eye. **JUVENILE** (July–April) Similar to nonbreeding adult, but with dusky blackish face, lacking the dark line trailing behind the eye; white throat. **VOICE** Nasal grunting chatter on nesting colonies; otherwise silent. **BEHAVIORS** Excellent divers, they use their wings to aid in underwater propulsion, and can reach astonishing depths—at least 300 feet. Perches upright. **HABITAT** Nests on nearshore rocks and headlands;

Common Murre range

Adult breeding Common Murre

feeds both offshore and near shore, including bays, coves, channels, and outer surf zone. **STATUS** Common year-round resident. **KEY SITES** (breeding colonies) Oregon: Three Arch Rocks NWR; Yaquina Head. Washington: Point Grenville (Grays Harbor County); Tatoosh Island (Cape Flattery). British Columbia: Florencia Island (Vancouver Island).

SIMILAR SPECIES Thick-billed Murre (*Uria lomvia*) is a very rare winter vagrant; in winter plumage, throat is white, but head and face are black; bill is shorter and thicker than in Common Murre, with decurved culmen.

Adult nonbreeding Common Murre

Common Murre nesting colony

Pigeon Guillemot
(*Cepphus columba*)

In contrast to the superficially similar and more numerous Common Murre (which nests in elbow-to-elbow colonies and lays its eggs on bare ground) the Pigeon Guillemot is a cavity nester. Generally they nest on islands and headlands, using fissures and pockets in rock, burrows in soil or sand cliffs, enclaves under tree roots, and other such crevices and holes. They form loose conglomerations on prime breeding sites, but also nest solitarily, sometime even in sand-cliff burrows along beaches.

LENGTH 13 inches. **WINGSPAN** 23 inches. **ADULT BREEDING** (March–September) Black overall (dark charcoal gray at close range), with large white wing patches on coverts; bright orange feet. **ADULT NONBREEDING** (October–March) Whitish overall, with clean white throat, breast, and belly; white head, usually with grayish crown, eye patch, and nape; large white wing patches; grayish back. **JUVENILE** (July–April) Similar to adult nonbreeding, but duskier overall, with more grayish head, and reduced white wing patches. **VOICE** Call is a rapid series of high, airy whistles, usually near the nest site. **BEHAVIORS** Feeds by diving, usually in fairly shallow water; typically perches upright, making its bright orange feet obvious. Flight is swift and direct on broad, rounded, rapidly beating wings. **HABITAT** Marine, including bays, sheltered beaches, nearshore seas, inlets, and passages. **STATUS** Fairly common spring and summer resident (March–August); uncommon year-round resident of Oregon and outer Washington coast; common year-round resident of Washington and British Columbia.

Adult breeding Pigeon Guillemot

Pigeon Guillemot range

Adult nonbreeding Pigeon Guillemot

Marbled Murrelet
(*Brachyramphus marmoratus*)

For decades, the nesting strategy of the little Marbled Murrelet was a mystery. Finally, in 1974, the first North American nest was located—in a tree. All other alcids are ground and/or cliff nesters that primarily use coastal islands and headlands. Marbled Murrelets typically nest near the canopy of old-growth conifers, laying their eggs directly onto large, moss-draped branches.

LENGTH 9.5 inches. **WINGSPAN** 16 inches. **ADULT BREEDING** (March–July) Dark brown overall, with rust-brown scapulars; light scaling on throat and breast. **ADULT NON-BREEDING** (August–March) Clean white below, blackish above; dark cap extends down to envelope eye; white neck extends back to form nearly complete white collar; white stripe above base of wings. **JUVENILE** (August–October) Similar to adult nonbreeding, but duskier overall **VOICE** Call is a high, slightly quavering *keeer*. **BEHAVIORS** Departs to feed and returns to nest at or after dawn and dusk. Flight is swift and direct on narrow, pointed, rapidly beating wings. **HABITAT** Shallow, nearshore waters, including bays, inlets, straits, coves; rarely on coastal freshwater lakes and lagoons; nests in coniferous coastal forests as far as 40 miles inland. **STATUS** Uncommon year-round resident; most numerous northern part of range. **KEY SITES** (nonbreeding) Oregon: Boiler Bay; Cape Arago; Yaquina Head. Washington: Point No Point; Port Townsend area; Washington Park and Anacortes Ferry Terminal. British Columbia: Cattle Point (Victoria); Neck Point Park (Nanaimo); Sunshine Coast.

Adult breeding Marbled Murrelet

Adult nonbreeding Marbled Murrelet

Marbled Murrelet range

Ancient Murrelet
(*Synthliboramphus antiquus*)

A burrow and cavity nester, the distinctive Ancient Murrelet is the only seabird whose chicks are reared entirely at sea; adults and their offspring depart the nesting colonies when the hatchlings are just one to 3 days of age. In the Northwest, however, this handsome little auk is a winter visitor; its southernmost well-documented breeding colony is on British Columbia's Queen Charlotte Island.

LENGTH 10 inches. **WINGSPAN** 17 inches. Short, pale bill; black cap, face, and collar; gray above, with black primaries; white below, with white wing linings. **ADULT NONBREEDING** (September–January) Black face contrasts with white throat and breast. **ADULT BREEDING** (February–August) Entire head, including throat, is black, leaving narrow white collar; white plumes over eyes in fresh spring breeding plumage. **JUVENILE** Similar to adult nonbreeding. **VOICE** Calls are whistled *peep* notes. **BEHAVIORS** Wing-propelled divers, Ancient Murrelets sometimes plunge underwater from flight. **HABITAT** Marine, especially well offshore, but regularly occurs close to shore, including Strait of Juan de Fuca, Puget Sound, and British Columbia's Inland Passage. **STATUS** Uncommon winter visitor to Oregon, becoming increasingly common northward; common off British Columbia; uncommon to rare fall and spring transient; rare summer visitor.

SIMILAR SPECIES Scripps's Murrelet (*Synthliboramphus scrippsi*) is a very rare late summer (August–September) pelagic visitor; black above, clean white below, including chin, throat, neck, breast, belly, undertail coverts, and underwings; small, thin, black bill; at close range, partial white eye ring.

Adult breeding Ancient Murrelet

Ancient Murrelet range

Juvenile Ancient Murrelet (October)

Cassin's Auklet
(*Ptychoramphus aleuticus*)

Despite being numerous at a number of breeding colonies on islands off the Washington outer coast and off the outer coast of Vancouver Island, diminutive Cassin's Auklets are troublesome to see from shore because they feed at sea during the day, departing the colonies before dawn and returning to the nests after dark. The best opportunity to see these wide-ranging alcids is via pelagic birding trips.

LENGTH 9 inches. **WINGSPAN** 15 inches. Gray overall, darkest on dorsal surface; small white eyebrow is distinctive and white "eyelid" (technically a nictitating membrane) flashes when the bird blinks; short, stubby bill, with pale spot at base of lower mandible. **VOICE** Rapidly stuttering *chereee* calls at colony. **BEHAVIORS** Commonly forms large feeding flocks at sea. **HABITAT** Pelagic, except breeds at colonies, primarily on islands, but also at a few headlands. **STATUS** Locally common summer resident at breeding colonies on the outer coasts of Washington and British Columbia (but largely nocturnal); common year-round resident and transient at sea.

Cassin's Auklet

Cassin's Auklet

Cassin's Auklet range

Rhinoceros Auklet
(*Cerorhinca monocerata*)

Nearly as large as its relatives, the puffins, the Rhinoceros Auklet is named for the white hornlike projection that grows from its culmen, somewhat reminiscent of a rhinoceros horn. Though typically crepuscular or nocturnal around the breeding colonies, the adults, their bills stuffed with fish, often stage in the water near their colony around dusk, then carry food back to their nesting burrows, which can extend into the ground for several yards.

LENGTH 14 inches. **WINGSPAN** 22 inches. **ADULT BREEDING** (April–September) Dark gray mantle, nape, and crown; slate-gray face, throat, breast, and flanks; dirty white belly; wispy white facial plumes trailing rearward from eye and from base of bill; large yellowish bill with light upright "horn" at the base of the culmen; dark underwings in all plumages. **ADULT NONBREEDING** (October–March) Similar to adult breeding, but lacking white facial plumes and culmen horn. **JUVENILE** Similar to adult nonbreeding, but with grayish bill. **VOICE** Low growling calls on breeding colonies; generally silent otherwise. **BEHAVIORS** Nests in deep burrows on vegetated headlands and islets; adults depart and return to burrows under cover of darkness. **HABITAT** Marine, including bays, inlets, straits, coves, channels, and open ocean. **STATUS** Fairly common year-round resident.

SIMILAR SPECIES Parakeet Auklet (*Aethia psittacula*) is a very rare winter and spring transient, with most Northwest records from offshore; about 50 records total for the region. Smaller than Rhinoceros Auklet; dark gray above, white below, with short, roundish, slightly upturned orange bill (dark in juvenile); slender white plume extending back from white eye (reduced in juvenile).

Adult nonbreeding Rhinoceros Auklet

Adult breeding Rhinoceros Auklet

Rhinoceros Auklet range

Tufted Puffin (*Fratercula cirrhata*)

The showiest of Northwest seabirds, the Tufted Puffin, in its extravagant breeding plumage, is unmistakable. These duck-sized alcids nest on sea stacks, islands, and headlands with sufficient soil for digging burrows and sufficient slope or launch points to facilitate easy takeoff. Throughout the year, Tufted Puffins are one of the most wide-ranging members of the auk family, frequenting waters far offshore.

LENGTH 15 inches. **WINGSPAN** 25 inches. **ADULT BREEDING** (March–September) All black except bright white triangular face adorned with long, dense, pale yellow plumes curling back over each side of the head and neck; massive, orange, laterally compressed parrotlike bill; bright orange feet; white eyes. **ADULT NONBREEDING** (October–March) Similar overall to adult breeding, but with gray face; bill appears smaller, with dark gray base; head plumes greatly reduced. **JUVENILE** similar to adult nonbreeding, but paler overall. **VOICE** Generally silent. **BEHAVIORS** Tufted Puffins wander widely in search of fish to feed their chicks, and typically carry the prey crossways in their bills when they return to the nesting colony. Active during the day, they are often seen standing outside their burrows. **HABITAT** Nests on nearshore sea stacks and islands, as well as headlands; mostly pelagic during nonbreeding season. **STATUS** Locally fairly common spring and summer resident (April–August); balance of year

spent at sea, but rare sightings occur near shore. **KEY SITES** Oregon: Haystack Rock (Cape Kiwanda State Park); Three Arch Rocks. Washington: Cape Flattery; Jagged Island; Protection Island. British Columbia: Cleland Island (northwest of Tofino).

SIMILAR SPECIES Horned Puffin (*Fratercula corniculata*) is a rare spring and summer transient and very rare winter transient, primarily well offshore, but occasionally near shore. Black above, white below; massive yellow and orange bill; clownlike face with large white face patches surrounded by black, and with a black checkmark over eye; in winter and juvenile plumages, face and base of bill are gray.

Adult breeding Tufted Puffin

Adult nonbreeding Tufted Puffin

Tufted Puffin range

251

Bonaparte's Gull
(*Chroicocephalus philadelhia*)

The Northwest's smallest gull is a common spring and fall migrant in the region. Interestingly, the Bonaparte's Gull almost always nests in trees, but its Canadian and Alaskan breeding range dips only as far south as south-central British Columbia.

LENGTH 13.5 inches. **WINGSPAN** 33 inches. Adult plumage acquired in 2 years. Pinkish legs; long, thin bill. **ADULT BREEDING** (April–August) White primaries form a wedge shape with black tips; sooty black head with white partial eye ring forming crescent shape; white neck, breast, belly, and tail; uniform pearl-gray mantle. **ADULT NONBREEDING** (August–April) White head with dark ear patch. **FIRST-YEAR SUMMER** (April–August) Similar to adult breeding with varying amounts of white on black head. **FIRST-YEAR WINTER** (August–April) Similar to adult nonbreeding, with black bar across lesser wing coverts; more black in primaries and trailing edges of wing; black bar across tip of tail. **VOICE** Call is a grating, ternlike *weer*. **BEHAVIORS** Buoyant and graceful in flight. Sometimes forms large flocks during migration. Takes food from the surface of the water and also takes

Adult breeding Bonaparte's Gull

Bonaparte's Gull range

Adult breeding Bonaparte's Gull in flight

insects on the wing; sometimes spins on the water similar to phalaropes to bring food items to the surface. **HABITAT** Coastal areas, open ocean, marshes and lakes. **STA-TUS** Common spring (March–May) and fall (August–December) migrant; rare summer and winter resident. **KEY SITES** Idaho: American Falls Reservoir; C.J. Strike Reservoir. Oregon: Smith and Bybee Lakes Wetlands Natural Area (Portland); Upper Klamath Lake. Washington: Grays Harbor; Tacoma Narrows. British Columbia: Ambleside Park (West Vancouver); Creston.

Adult nonbreeding Bonaparte's Gull

Adult nonbreeding Bonaparte's Gull in flight

First-winter Bonaparte's Gull in flight

Franklin's Gull
(*Leucophaeus pipixcan*)

A small gull of the inland Northwest, the Franklin's Gull—the only gull that molts completely twice per year—breeds at a handful of sites in eastern Oregon and eastern Idaho, with its primary nesting range in the northern Great Plains of the United States and Canada.

Adult breeding Franklin's Gull

First-winter Franklin's Gull

Franklin's Gull range

LENGTH 14.5 inches. **WINGSPAN** 37 inches. Adult plumage acquired in 3 years. **ADULT BREEDING** (April–August) Black hood with bold white partial eye ring forming crescent shape; red bill and reddish legs; warm pinkish tinge to the normally white breast and belly; dark gray mantle; wingtips have some black with a white band that separates the tips from the rest of the gray wing, with the white band continuing down the trailing edge of the wing; white tail. **ADULT NON-BREEDING** (August–March) Black hood with white streaks; thin, black bill. **FIRST SUMMER** (April–August) Similar to adult nonbreeding, but with more black in the wingtips and little or no white band separating black wingtips from gray wing. **SECOND WINTER** (September–April) Similar to adult nonbreeding. **FIRST WINTER** (September–April) Similar to adult nonbreeding, but with dusky tinged gray mantle; black wingtips and white trailing-edge on wings; black bar on tail that does not extend to the tips. **VOICE** Call is a loud, harsh *huh*. **BEHAVIORS** Highly insectivorous; agile, swallowlike flight as it pursues flying insects or plucks prey from the water's surface. **HABITAT** Marshes, lakes, flooded fields. **STATUS** Locally common summer resident. Rare along coast during migration; elsewhere uncommon during migration and post-breeding dispersal. **KEY SITES** Idaho: Grays Lake NWR; Market Lake Wildlife Area. Oregon: Malheur NWR; Upper Klamath Lake.

Mew Gull (*Larus canus*)

One of the Northwest's most common coastal gulls during winter, the Mew Gull is the smallest of North America's white-headed gulls and the only one that nests in trees as well as on the ground. Its breeding range reaches southward from Alaska and western Canada to southwestern British Columbia.

LENGTH 16.5 inches. **WINGSPAN** 43 inches. Adult plumage acquired in 3 years. **ADULT BREEDING** (April–September) White head, breast, belly, and tail; red eye ring; dusky iris; unmarked yellow bill; yellow legs; gray mantle with large, white tertial crescent and smaller white scapular crescent; in flight, wings show large white spots on first 2 primaries bordered by black primary base and black tips; white trailing edge to gray wings. **ADULT NONBREEDING** (September–April) Similar to adult breeding, but with dusky neck and head; dark eyes; dusky spot on the lower mandible of yellow bill. **SECOND WINTER** (August–April) Similar to adult nonbreeding, but pale bill with dark ring; more black on wingtips and a pale black band across tail; pale legs. **FIRST SUMMER** (April–August) Dingy brown overall, with some gray on mantle; pink bill with dark tip; pale, flesh-colored legs; brown underwings; brown, heavily barred tail. **FIRST WINTER** (October–April) Brownish overall, with gray mantle; small, pink bill with black tip; brown wings with darker wingtips. **VOICE** Calls include a harsh, 2-tone *mew*; and a screaming, falsetto *keeah*. **BEHAVIORS** Strong, direct flight, but also graceful and buoyant; sometimes spins on the water like a phalarope to create upwellings to draw small prey to surface. **HABITAT** Rocky shorelines, beaches, pastures, landfills, estuaries, bays and harbors. **STATUS** Abundant winter resident, common summer resident. Rare transient east of the Cascades.

Adult breeding Mew Gull

Second-winter Mew Gull

Mew Gull range

First-winter (front) and adult Mew Gulls

Adult nonbreeding Mew Gulls molting to breeding (April)

Ring-billed Gull
(*Larus delawarensis*)

The fast-food gull of the inland North-west, the Ring-billed Gull has learned the value of hanging out in urban parking lots, waiting for wayward French fries and other treats launched from vehicle windows. Hence, these birds are often easy to find within towns and cities.

LENGTH 17.5 inches. **WINGSPAN** 45 inches. Adult plumage acquired in 3 years. **ADULT BREEDING** (April–September) White head; gray mantle with faint white tertial and scapular crescents; yellow bill with black ring; yellow legs; black primaries with small white spot near tips. **ADULT NONBREEDING** (September–April) Similar to adult breeding, but with brown flecking on head and neck. **SECOND WINTER** (September–April) Similar to adult nonbreeding, but wingtips mostly black, and faint black bar across the tail. **FIRST SUMMER** (April–September) Gray mantle with some brown in upper wing coverts; pink bill with black ring; pink legs; black wingtips; variable dark band on tail. **FIRST WINTER** (September–April) Speck-led brown overall, with gray mantle; pink bill with black tip; pink legs. **VOICE** Call is a shrill, forced, falcetto *ow*. **BEHAVIORS** Often found in parking lots of restaurants and at landfills; also frequents fields with dense rodent populations, and follows plows. **HABITAT** Widespread, including marshes, fields, cities, sewage ponds, landfills, lakes. **STATUS** Locally common summer resident interior; uncommon fall and winter resident and transient interior (though fairly common along Columbia and Snake Rivers); uncommon summer resident and transient west of Cascades; common fall and winter resident west of Cascades, including coast.

Juvenile Ring-billed Gull (early August)

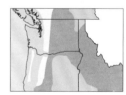

Ring-billed Gull range

Adult breeding Ring-billed Gull

First-winter Ring-billed Gull in flight

Adult nonbreeding Ring-billed Gull in flight

California Gull (*Larus californicus*)

The California Gull is the bird that, according to Mormon folklore, arrived en masse in Utah's Salt Lake Valley through divine intervention to save the crop of 1848 from swarms of what became known as Mormon crickets, a species of katydid. Subsequently the California gull became the official state bird of Utah.

Adult nonbreeding California Gull in flight

LENGTH 21 inches. **WINGSPAN** 54 inches. Adult plumage acquired in 4 years. **ADULT BREEDING** (April–September) Medium-sized gull with white head, breast, belly, and tail; black and red spots on yellow bill; wings with black tips and small white patches near the tips; greenish yellow legs. **ADULT NON-BREEDING** (October–April) Similar to adult breeding, but with brown streaks on neck and head. **THIRD WINTER** (August–April) Similar to adult nonbreeding, but partial or complete black bar on the tail; black wingtips. **SECOND SUMMER** (April–September) Gray mantle, with brown mottling on head, neck, and wings; bluish gray legs and bill; black on wingtips; black tail and white rump. **SECOND WINTER** (August–April) Gray

Juvenile California Gull (July)

Adult breeding California Gull

California Gull range

mantle; mottled brown head, neck, flanks, and belly; bluish gray legs and bill. **FIRST SUMMER** (April–August) Mottled brown overall, with pale head and breast; pinkish bill with black tip; black wingtips; pink legs. **FIRST WINTER** (September–April) Pale brown face but darker than in second-winter plumage; pink bill with black tip; mottled brown overall; black wingtips. **JUVENILE** (August–September) Brown overall; dark bill; black tail; pink legs. **VOICE** Calls include a squealing *keeh*; a scratchy, hoarse *oh*; and a trumpeting *keeaha* followed by repeated *uh-uh-uh-uh-uh*. **BEHAVIORS** An opportunistic feeder that frequents parking lots, and is often found with Ring-billed Gulls; also feeds in freshly plowed fields and follows raptors to feed on any rodent leftovers. **HABITAT** Breeds in freshwater marshes, rivers, and alkaline lakes; frequents flooded or plowed fields. Ranges outside of breeding season, and sometimes occurs along the coast and offshore. **STATUS** Abundant summer resident, common winter resident.

Second-winter California Gull

First-winter California Gull (October)

Herring Gull (*Larus argentatus*)

A common, large gull with circumpolar distribution, the Herring Gull was once nearly extirpated in North America by plumage hunters and egg collectors. They are winter residents in the Northwest, where adult-plumage Herring Gulls outnumber subadults, which vary substantially in plumage. Note that Herring Gulls are easily confused with the paler northern subspecies of the Western Gull.

LENGTH 25 inches. **WINGSPAN** 58 inches. Adult plumage acquired in 4 years. **ADULT BREEDING** (February–September) Large gull with relatively flat crown; pale yellow eyes; white head, neck, breast, and belly; heavy yellow bill with red mark on lower mandible; light gray mantle and dorsal wing surfaces; black wingtips with small white spots on the ends of the outer primaries white tail and rump; pink legs. **ADULT NONBREEDING** (September–April) Similar to adult breeding, but heavily streaked head and neck. **THIRD WINTER** (August–April) Gray mantle with brown tinge to the dorsal wing surfaces, and less white in the wingtips than in adult; yellow bill with dark spot; yellow

Adult nonbreeding Herring Gull

Juvenile Herring Gull

Adult breeding Herring Gull

First-winter Herring Gull

Second-winter Herring Gull

Third-winter Herring Gull, molting to adult, in flight

Herring Gull range

eye; some black on the tail. **SECOND SUMMER** (February–September) Brown body and head; no white wingtips; yellow bill with dark spot; yellow eye. **SECOND WINTER** (August–April) Brown overall, with some gray in the back and wings; black tail and mostly white rump; brown barring on undertail coverts; pink bill with a black tip; pink legs. **FIRST SUMMER** (April–August) Pale brown and worn-looking overall; pink bill with black tip; pale legs; brown barring on undertail coverts; dark eyes. **FIRST WINTER** (September–April) Variable brown with mottled back and wings; clean brown breast and belly, sometimes with light streaks; dark eyes; dark bill with a pink base. **JUVENILE** (August–November) Brown overall, with barred undertail coverts; black bill; dark eyes; pink legs; black wingtips; brownish tail. **VOICE** Call is a loud, trumpeting *wuuck, wuuck, wuuck* **BEHAVIORS** Sometimes makes shallow dives from height for prey, but buoyant and cannot dive too deep. **HABITAT** Rocky shorelines, sand beaches, bays, estuaries, rivers, lakes, fields, landfills, city parking lots. **STATUS** Common fall, winter, spring visitor (August–May), though less common southward into Oregon, where Western and Glaucus-winged Gulls are more common.

Thayer's Gull *(Larus thayeri)*

Gull identification challenges perhaps reach a pinnacle in the Northwest with the various plumages of Thayer's and their near-lookalike cousins, Herring Gulls. Thayer's Gulls are subject to taxonomical debate because they are closely related to, or even conspecific with, a non-Northwest species, the Iceland Gull, and hybridize with similar species, adding to the difficulty in identification.

LENGTH 23 inches. **WINGSPAN** 55 inches. Adult plumage acquired in 4 years. **ADULT BREEDING** (April–September) Rounded crown as opposed to flattened crown of Herring Gull; white head, breast, and belly; rather small, yellow-green bill with red mark on lower mandible; dark eye with deep reddish orbital ring; gray mantle; white primaries with black edges; wings mostly white below. **ADULT NONBREEDING** (September–April) Similar to adult breeding, but with streaks on head and neck.
THIRD WINTER (August–April) Whitish head with brownish smudges; greenish yellow bill with dark ring near tip; blackish wingtips with minimal white; some black on tail; dark pink legs. **SECOND YEAR** (August–April) Gray mantle with brown mottling on wings; whitish head, breast, and belly, with brown smudges; light barring on undertail coverts; dark tipped bill with pale base; pale pink

Adult nonbreeding Thayer's Gull

Second-winter Thayer's Gull

Thayer's Gull range

Adult breeding Thayer's Gull

legs; outer primaries may show darker tips; dark tail. **FIRST SUMMER** (April–August) pale brown overall; wingtips and tail may be darker than rest of body; dark bill and pale pink legs. **JUVENILE** Dark bill; mottled brown back and wings with slightly darker wingtips; pale pink legs. **VOICE** Calls include a hoarse *huh*, and a drawn-out, trumpeted,

hiyuu huh huh huh huh. **BEHAVIORS** Opportunistic feeder often found at landfills and sewage ponds; Columbia River smelt runs draw concentrations of Thayer's Gulls. **HABITAT** Bays and estuaries, agricultural fields, landfills, and sewage ponds. **STATUS** Common migrant and winter visitor; uncommon to rare into the inland Northwest.

Juvenile Thayer's Gull

Glaucous Gull (*Larus hyperboreus*)

Very light colored—appearing almost all white at times—the rare Glaucous Gull stands out when seen with other gull species, all of which are darker. Younger Glaucous Gulls are most easily confused with Glaucous-winged Gulls, but their paler coloration and their bill color helps distinguish between the 2 species.

LENGTH 27 inches. **WINGSPAN** 60 inches. Adult plumage acquired in 4 years. **ADULT BREEDING** (March–September) Large pale gull with all-white primaries; light gray mantle and dorsal wing surfaces; white underwings; yellow bill with red spot on lower mandible; yellow eyes with yellow orbital ring; pink legs. **ADULT NONBREEDING** (September–April) Similar to adult breeding, but with limited streaking on the head and neck. **THIRD WINTER** (August–April) Similar to adult, but with dark ring on bill. **SECOND WINTER** (August–April) White overall, with pale brownish smudging; pink bill with dark tip. **FIRST SUMMER** (April–August), White overall, sometimes with brownish mottling on wing coverts; dark eye; pink legs. **JUVENILE** Varies from light brown to off-white; pink bill with dark tip; pink legs. **VOICE** Call is a hoarse, barking *qwaa*. **BEHAVIORS** Opportunistic feeder, and will steal food from other gulls; sometimes preys on small birds. **HABITAT** Bays, estuaries, beaches, lakes, rivers, landfills, parking lots. **STATUS** Uncommon winter visitor (October–March).

Glaucous Gull range

Adult breeding Glaucous Gull in flight

Adult nonbreeding Glaucous Gull

First-winter Glaucous Gull

Glaucous-winged Gull
(*Larus glaucescens*)

One of the more common gulls in the Northwest, this large, white-headed gull breeds in coastal locations as far south as Oregon, and winters as far south as the southern tip of Baja California. Precise population estimates for Glaucous-winged Gulls have proved elusive because they readily hybridize with other gull species, especially Western Gulls in the Northwest.

LENGTH 26 inches. **WINGSPAN** 57 inches. Adult plumage acquired in 4 years. **ADULT BREEDING** (February–September) Large, pale gull with dark eye and pink orbital ring; white head, neck, breast, belly, and underwings; light gray mantle and dorsal wing surfaces, with slightly darker gray wingtips with white spots; yellow bill with red spot on lower mantle; pink legs. **ADULT NONBREEDING** (September–March) Similar to adult breeding, but with light brown barring on head and neck. **THIRD WINTER**

First-winter Glaucous-winged Gulls

Second-winter Glaucous-winged Gull

Third-winter Glaucous-winged Gull in flight

Adult breeding Glaucous-winged Gull

(August–March) Similar to adult nonbreeding, but pale yellow bill with dark tip; pink legs. **SECOND WINTER** (August–March) Off-white overall, with grayish back; dark bill with pink base; pink legs. **FIRST SUMMER** (April–August) Pale overall with light brown mottling on scapulars; all-dark bill; pink legs. **JUVENILE** Uniform pale brown, with dark bill and pink legs. **VOICE** Calls include a trumpeting *keeah keeah keeah*, and a soft, quacklike *ka ka ka ka*. **BEHAVIORS** Opportunistic feeder that readily steals food from other gulls; preys on eggs and chicks of other birds; scavenges at landfills and sewage ponds; nests on flat, sandy or gravely surfaces, including rooftops. **HABITAT** Coastal and freshwater wetlands, including bays, harbors, estuaries, beaches; also towns, parking lots, parks. **STATUS** Common year-round and winter resident.

Adult nonbreeding Glaucous-winged Gull in flight

Adult breeding Glaucous-winged Gull in flight

Glaucous-winged Gull range

Western Gull (*Larus occidentalis*)

This large, handsome, white-headed gull is one of the most common gulls found year-round along the Northwest coast. Western Gulls exhibit variation in plumage and eye color owing to extensive hybridization with Glaucous-winged gulls, and the northern subspecies is paler than the birds found throughout Oregon, which can lead to misidentification as Herring Gulls. As the name implies, the Western Gull is purely a western species, ranging from British Columbia to Baja California.

LENGTH 25 inches. **WINGSPAN** 58 inches. Adult plumage acquired in 4 years. **ADULT BREEDING** (February–September) Dark gray mantle and dorsal wing surfaces; black wingtips with one white spot; white head, breast, and belly; large, thick-tipped, yellow bill with red spot on lower mandible; pink legs; eyes variable from dark to yellow, with orange orbital ring. **ADULT NONBREEDING** (September–March) Similar to adult breeding, but neck lightly streaked with brown. **THIRD WINTER** (August–March) Dark gray back with variable amounts of brown; yellow bill has dark ring near tip; black wingtips with little or no white; limited brown mottling and streaking on neck and around eyes. **SECOND WINTER** (August–March) Brown patches on dark gray mantle; brown wash on head and neck; black tail and white rump; pale yellow bill with black tip; dark eye; pink legs. **FIRST SUMMER** (April–September) Brown overall, with blackish wingtips and tail; barred brown rump; dark bill with pale base. **JUVENILE** Similar to first summer, with heavily mottled mantle. **VOICE** Call is a 2-tone *waha*, the second note falsetto, usually in a series, *waha waha waha*. **BEHAVIORS** Nests in colonies along the coast; like other gulls, an opportunistic predator; preys on eggs and chicks of other birds; drops mollusks from height onto rocks to break open the shells. **HABITAT** Coastal; inhabits bays, estuaries, harbors, beaches, headlands, agricultural areas, landfills. **STATUS** Abundant year-round resident.

Adult breeding Western Gull

Adult breeding Western Gull in flight

First-winter Western Gull

First-winter Western Gull in flight

Second-winter Western Gull

Western Gull range

Third-winter Western Gull in flight

Heermann's Gull
(*Larus heermanni*)

The attractive and distinctively colored Heerman's Gull breeds in Mexico, but then disperses northward, reaching the Northwest in summer and lingering well into autumn. Its overall dark color (and its habits) are reminiscent of the jaegers, with which Heerman's Gulls can be confused.

LENGTH 19 inches. **WINGSPAN** 51 inches. Adult plumage acquired in 4 years. **ADULT BREEDING** (December–August) Gray body with white head; red bill with black tip; black legs; black wingtips; black tail with white terminal band. **ADULT NONBREEDING** (August–February) Grayish head, lighter toward front of face, with entire head often speckled; black-tipped red bill. **THIRD WINTER** (August–February) Like adult nonbreeding. **SECOND WINTER** (September–March) Charcoal-gray overall; red bill with black tip; black legs; weak white terminal band on tail. **FIRST SUMMER** (May–August) Sooty brown overall; pale orangish bill with black tip.

VOICE Calls include a nasal *ow*; loud trumpeting call similar to other gulls, but more nasal. **BEHAVIORS** Aggressive and often seen chasing other gulls or mobbing pelicans and diving birds to steal food. Often glides just over the surface of the water in the troughs between waves. **HABITAT** Coastal; rocky shorelines, mudflats, kelp beds, bays, estuaries. Rarely found in freshwater. **STATUS** Fairly common summer and fall visitor (June–October).

Adult nonbreeding (rear) and juvenile Heermann's Gulls

Second-winter Heermann's Gull

Adult breeding Heermann's Gull

Heermann's Gull range

Sabine's Gull *(Xema sabini)*

A unique, largely pelagic gull whose sig-
nature wing pattern allows for identification even at long range, the Sabine's Gull
migrates to and from its arctic breeding
grounds off the Northwest coast, although
the occasional individual shows up far
inland. Feeding on zooplankton and small
fish, these small gulls are often targeted
by kleptoparasitic (food stealing) jaegers.

LENGTH 13 inches. **WINGSPAN** 34 inches.
Adult plumage acquired in 2 years. Distinctive white triangle on mid-wing borders
black primaries and black leading wing
edge. **ADULT BREEDING** (April–September)
Black hood; gray mantle; white breast
and belly; black bill with yellow tip;
slightly forked tail. **ADULT NONBREEDING**
(October–March) Similar to adult breeding, but with partial hood. **FIRST WINTER**
(January–April) Rarely seen in Northwest; mottled gray-brown mantle; brown
patches on head; black bill. **JUVENILE**
(August–December) Brown back, neck,
and sides of breast; black terminal band on
tail; dark bill. **VOICE** Call is a raspy, ternlike chattering. **BEHAVIORS** Buoyant, ternlike flight; plucks zooplankton and small
fish from near the surface. **HABITAT** Pelagic;
occasionally seen from shore, including
estuaries and bays. **STATUS** Common spring
migrant (March–May) and fall migrant
(July–November); rare coastal migrant; rare
but nearly annual inland.

Adult breeding Sabine's Gull in flight

Juvenile Sabine's Gull in flight

Sabine's Gull range

Black-legged Kittiwake
(*Rissa tridactyla*)

This small pelagic gull is a bonus to the birders' checklist when seen from shore, but such appearances are uncommon, often occurring during and after raging Pacific storms and high winds that push birds toward the coast. The name "kittiwake" derives from the bird's call, rarely heard off the Northwest coast, a shrill *kittee-wa-aaaake*.

LENGTH 17 inches. **WINGSPAN** 36 inches. Adult plumage acquired in 2 years. **ADULT BREEDING** (April–September) Small white-headed gull; gray mantle and dorsal wing surfaces; black wingtips; yellow bill; black legs. **ADULT NONBREEDING** (August–March) Similar to adult breeding, but with gray smudges behind eyes. **JUVENILE AND FIRST YEAR** Black markings on dorsal wing surfaces form an M in flight; black to gray collar; black terminal band on tail; black legs (or rarely varying in color from red to yellow). **VOICE** Generally silent. **BEHAVIORS** Takes food from the water's surface and dives for prey just under the surface; flies with stiff wingbeats. **HABITAT** Pelagic; rarely seen from shore. **STATUS** Fairly common migrant and winter visitor (September–May).

Adult Black-legged Kittiwake in flight

Juvenile Black-legged Kittiwake

Juvenile Black-legged Kittiwake in flight

Adult Black-legged Kittiwake

Black-legged Kittiwake range

Caspian Tern (*Hydroprogne caspia*)

The Caspian Tern—the world's largest tern—is a major predator of salmon and steelhead smolt on the Columbia River system, and has thus garnered the ire of anglers and the consternation of fisheries managers. But this bird's penchant for swarming to waters dense with fish also has an upside: in some places, such as Oregon's Malheur National Wildlife Refuge, manmade islands created as nesting habitat for the terns are aimed at bolstering their numbers so they can feed on common carp, an invasive species whose proliferation can devastate entire ecosystems.

LENGTH 21 inches. **WINGSPAN** 50 inches.
ADULT BREEDING (February–October) Gull-like in size; heavy red bill; black cap; dark primaries when viewed from below.
ADULT NONBREEDING (October–February) Similar overall to adult breeding, but black crown reduced to leave grayish forehead and forecrown. **JUVENILE** (June–October) Similar to adult nonbreeding, but pale silvery gray mantle is patterned with darker chevrons, to form variable scalloping. **VOICE** Adult call is a harsh, raspy *arrr* or *arryayum*; juvenile call is a whistled *weeup*. **BEHAVIORS** Flies with steady wingbeats, often with bill pointed downward as the tern looks for food in the water. Nests on sand or gravely islands, isolated beaches, dikes, and levies, as well as manmade islands built specifically for them. **HABITAT** Lakes, wetlands, large ponds, estuaries, large rivers. **STATUS** Common summer resident (April–September).

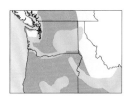

Caspian Tern range

Adult breeding Caspian Tern

Adult breeding Caspian Tern in flight

Adult nonbreeding Caspian Tern in flight

Juvenile Caspian Tern

Elegant Tern (*Thalasseus elegans*)

The slender, graceful Elegant Tern is a late summer and fall visitor to coastal locations, and a fairly recent addition to the avifauna of the Northwest, first appearing in Oregon in the early 1980s. Numbers can be sporadic; some years the Northwest hosts flocks of more than 400 birds, while other years, Elegant Terns are very scarce.

LENGTH 17 inches. **WINGSPAN** 34 inches. **ADULT NONBREEDING** (August–February) Long, slender, slightly decurved orange bill; long, black crest, often lowered; white fore-crown and forehead; narrow, pointed wings with pale undersides. **ADULT BREEDING** (March–July) Crown is completely black. **VOICE** Adult call is a raspy *kedeeah*; juvenile call is a high-pitched whistle. **BEHAVIORS** Often congregates on pilings and docks; feeds by flying over water, hovering, then plunge-diving to capture fish; highly social and often feeds in flocks. **HABITAT** Bays, harbors, estuaries. **STATUS** Common to rare fall visitor (August–October) Oregon to southern Washington; rare farther north. **KEY SITES** Oregon: Brookings area.

Nonbreeding Elegant Tern in flight

Elegant Tern range

Breeding Elegant Tern in flight

Common Tern (*Sterna hirundo*)

The Common Tern, small and balletic, occurs along the Northwest coast during migration and post-breeding dispersal from its breeding range across central and eastern Canada and the upper Midwest. Common Terns also turn up inland, primarily in autumn and rarely in spring. They resemble Arctic Terns and can be differentiated by their bill color during breeding season and the dark wedge in the primaries of the Common Tern.

LENGTH 12 inches. **WINGSPAN** 30 inches. **ADULT BREEDING** (April–November) Red bill with black tip; black cap; at rest, wingtips extend beyond tail; dark wedge on primaries (lacking in all other small terns); dark outer edging on deeply forked tail. **ADULT NONBREEDING** (October–March) White forehead; dark red bill; may have dark leading wing edges. **JUVENILE** Brownish to grayish back may show faint barring. **VOICE** Calls are similar to Arctic and Forster's Tern, typically a rattling *kerrrr* and a high *kit* or *keet*. **BEHAVIORS** Typically seen in flocks, hunting for prey located on or just below the surface of the water; hovers to spot prey and then plunge-dives, sometimes completely immersing. Buoyant flight with strong, steady wingbeats. **HABITAT** Open ocean, bays, estuaries; inland rivers, marshes, and lakes. **STATUS** Common to abundant migrant offshore during spring (May–June) and fall (August–October), fairly common along the coast; rare to uncommon inland during fall, and rare in spring.

Breeding Common Tern

Common Tern range

Nonbreeding Common Tern

Arctic Tern (*Sterna paradisaea*)

Famous for its incredible yearly 25,000-mile migratory journey between the Arctic and Antarctica, the Arctic Tern is the world's long-distance champion, racking up a lifetime total of some 850,000 miles, not including foraging flights. A circumpolar species, its migration in the Pacific Northwest generally occurs over the open ocean.

LENGTH 12 inches. **WINGSPAN** 31 inches. Very short legs when seen perched. **ADULT BREEDING** (March–October) Black cap; small, dark red bill; pale gray back and wings, with translucent primaries; white, deeply forked tail extends beyond wingtips at rest.

ADULT NONBREEDING (November–February) Same as adult breeding, but forehead is white. **VOICE** Calls similar to Common and Forster's Tern, typically a rattley *kerrrr* and a high *kit* or *keet*. **BEHAVIORS** Most often seen in flocks, hunting for prey located on or just below the surface of the water; hovers to spot prey and then plunge-dives, sometimes completely immersing. Buoyant flight with strong, steady wingbeats. **HABITAT** Pelagic; also surf zone, bays, and estuaries. **STATUS** Uncommon offshore spring (April–May) and fall (July–September); occasionally seen from land along the coast; very rare inland.

Nonbreeding Arctic Tern in flight

Breeding Arctic Tern in flight

Arctic Tern range

Forster's Tern (*Sterna forsteri*)

A freshwater tern of the inland Northwest, the Forster's Tern plunge-dives for fish like other terns, but also snatches flying insects and even plucks insects from the water's surface. Though similar in color and pattern to the Caspian Tern, which is also found on inland waters, the Forster's Tern is substantially smaller and daintier.

LENGTH 13 inches. **WINGSPAN** 31 inches. **ADULT BREEDING** (March–August) Black cap; large red bill with black tip; pale gray back and wings, including primaries; deeply forked tail extends beyond wingtips when bird is at rest. **ADULT NONBREEDING** (August–February) Dark mask replaces black cap; dark bill; dark primaries. **JUVENILE** (July–August) Dark mask; rich brown markings on back. **VOICE** Calls similar to Common and Arctic Terns, typically a rattley *kerrrr* and a high *kit* or *keet*. **BEHAVIORS** Often found on small ponds, creeks, and marshes. **HABITAT** Wetlands and marshes, lakes and ponds, creeks and rivers, vegetated flooded fields. **STATUS** Fairly common summer resident (April–September) east of the Cascades, rare west of the Cascades. **KEY SITES** Idaho: C.J. Strike Wildlife Area; Market Lake Wildlife Area. Oregon: Malheur NWR; Upper Klamath NWR. Washington: Potholes Reservoir; Yakima River delta. British Columbia: Creston Valley Wildlife Center.

Adult breeding Forster's Tern in flight

Juvenile Forster's Tern

Adult breeding (left) and nonbreeding Forster's Terns

Forster's Tern range

Black Tern (*Chlidonias niger*)

At first glance, the small, dark Black Tern might be confused with a large swallow as it acrobatically catches insects on the wing. Buoyant and agile in flight, these denizens of freshwater environs twist and turn, dart and dive in pursuit of insects, sometime even deftly plucking them from the surface of the water. The Black Tern rarely plunge-dives like other terns.

LENGTH 9.5 inches. **WINGSPAN** 24 inches. **ADULT BREEDING** (March–August) Black head, breast, belly, and flanks; slate-gray back, wings, and tail. **ADULT NONBREEDING** (August–February) Dark ear patch and hindcrown, with the rest of the head, and underparts, dingy white; slate-gray back, wings, and tail. **JUVENILE** (July–November) Similar to nonbreeding adult, but with scaly pattern on back. **VOICE** Calls include a raspy, scolding *keee*, and a high, complaining *kip*. **BEHAVIORS** Flight is buoyant and swallow-like. Nests on floating vegetation in loose colonies. Preys largely on insects during breeding season, with small fish making up the balance of the diet. **HABITAT** Marshes, small lakes, ponds, flooded fields, canals, and rivers. **STATUS** Common summer resident (May–August). **KEY SITES** Idaho: Camas Centennial Marsh; Coeur D'Alene Chain Lakes; Lake Lowell. Oregon: Fern Ridge Reservoir; Klamath Marsh NWR; Malheur NWR. Washington: Potholes Reservoir; Ridgefield NWR; Turnbull NWR. British Columbia: Creston Valley Wildlife Center; Osoyoos Lake.

Juvenile Black Tern

Adult breeding Black Tern in flight

Adult breeding Black Tern

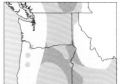

Black Tern range

South Polar Skua
(*Stercorarius maccormicki*)

The gull-like South Polar Skua is a pelagic species rarely seen from shore, but as with all offshore birds, they are occasionally driven close to shore by raging storms in the Pacific. Generally rare off the Northwest coast, skuas breed in Antarctica. Aggressive predators, they often steal meals by forcing shearwaters to disgorge food by harassing and pursuing them, sometimes going so far as to grab the shearwater by the head and shaking until the smaller bird disgorges its food.

LENGTH 21 inches. **WINGSPAN** 52 inches. **ADULT** Color ranges from uniform light brown to dark brown (most Skuas seen in the Northwest are dark); hunchbacked and pot-bellied appearance; white wing patches; wedge-shaped tail. **VOICE** Rarely vocalizes. **BEHAVIORS** Direct flight on steady, even wingbeats, only rarely gliding; harasses other birds in order to steal fish, both by directly stealing prey and by forcing a bird to vomit its meal. **HABITAT** Pelagic. **STATUS** Uncommon to rare summer and fall visitor (June–October).

South Polar Skua in flight

South Polar Skua range

Molting adult South Polar Skua in flight

Long-tailed Jaeger
(*Stercorarius longicaudus*)

A pelagic species rarely seen from shore, the Long-tailed Jaeger is a graceful exotic treat for birders who join offshore excursions to look for open-ocean species. As with the other jaeger species, Long-tailed Jaegers have been recorded inland, but normally they occur even farther out to sea than the Parasitic and Pomarine Jaegers.

LENGTH 15 inches. **WINGSPAN** 43 inches. **ADULT BREEDING** (March–October) Long central tail feather distinctive; slender body and narrow wings; black cap and rounded head; short stout bill; yellowish buff nape and neck; white breast and gray belly; pale gray back and wing coverts contrast with black wing feathers. Only 2 white shafts on outer primaries. Adult nonbreeding plumage seldom seen in the Northwest. **JUVENILE** Color phases in juveniles range from light gray to dark brown but most in Northwest are light gray; short, blunt-tipped central tail feather extends beyond tail; whitish head; gray wash on breast and whitish belly;

barred back, wings, undertail coverts, and rump; white patch at base of underwing primaries. **VOICE** Generally silent. **BEHAVIORS** Buoyant, graceful flight; less aggressive than other jaegers and skuas. **HABITAT** Pelagic. **STATUS** Common to uncommon well offshore during spring (April–May) and fall migration (July–October); extremely rare inland.

Juvenile Long-tailed Jaeger in flight

Long-tailed Jaeger range

Adult breeding Long-tailed Jaeger in flight

Parasitic Jaeger
(*Stercorarius parasiticus*)

Of the 3 species of jaegers, the Parasitic Jaeger is the one most likely to be seen from shore. The odd-sounding name—Parasitic—derives from this jaeger's habit of harassing other seabirds to make them disgorge their prey. They even follow migrating terns (especially Arctic Terns), routinely feeding by this means, known as kleptoparasitism.

LENGTH 16.5 inches. **WINGSPAN** 46 inches. Color morphs range from dark to light; central tail feathers are narrow and extend just beyond the tail; bill is long and narrow; outer primaries have white shafts. **ADULT BREEDING** (March–November) Narrow central tail feathers extend beyond the tail; long, slender bill; several white shafts in outer primaries. Dark morph ranges from uniformly dark brown to intermediate brown with a lighter neck and a dark cap. Light morph features a dark cap; yellowish neck and white throat; brownish breast band; white belly; brown back, tail, and undertail coverts. Adult nonbreeding seldom seen in the Northwest. **JUVENILE** Outer third of bill is dark; dark morph all dark brown; light morph barred above and below. **VOICE** Generally silent. **BEHAVIORS** Harasses terns and other seabirds until they disgorge their meals, then deftly dives and catches the food in midair. Falconlike flight with rapid wingbeats and frequent gliding; highly acrobatic when pursuing terns and other seabirds. **HABITAT** Pelagic, but its habit of following migrating terns brings it closer to shore than other jaegers. **STATUS** Rare spring migrant (March–June), uncommon to common fall migrant (July–October). Rare inland during fall.

Parasitic Jaeger range

Adult breeding Parasitic Jaeger in flight

Pomarine Jaeger
(*Stercorarius pomarinus*)

Bulky and gull-like, the Pomarine Jaeger is the largest and most common jaeger found off the Northwest coast. Pelagic and rarely seen from shore, the Pomarine Jaeger is kleptoparasitic like the other members of the genus, meaning it routinely feeds by stealing fish and other prey from various other seabirds, harassing them until they disgorge their meals. This jaeger, however, will also catch its own fish and even preys on small seabirds, such as phalaropes.

LENGTH 20 inches. **WINGSPAN** 51 inches. **ADULT BREEDING** (March–November) Heavy, 2-toned bill; long, rounded, and twisted central tail feathers; white shafts in outer primaries. Morphs range from dark to light with most being light. Dark morph all dark. Light morph features dark cap; yellowish neck; dark breast band; broader wings than other jaegers; white belly. Adult nonbreeding seldom seen in the Northwest. **JUVENILE** Heavily barred in all plumages. **VOICE** Generally silent. **BEHAVIORS** Harasses other birds for food and also hunts fish and small seabirds; falconlike flight. **HABITAT** Pelagic. **STATUS** Fairly common migrant and summer visitor (March–September). Rare inland.

Pomarine Jaeger range

Adult breeding Pomarine Jaeger in flight

Doves and Pigeons

The Americas are home to several dozen species of pigeons and doves, but they reach their greatest diversity in Central and South America; the Pacific Northwest is home to only one native pigeon, the Band-tailed Pigeon, and one native dove, the Mourning Dove. In addition to the natives, however, are 2 species introduced to North America from Europe; both have been wildly successful. The Rock Pigeon is the familiar pigeon of city streets, and in the Northwest, the Eurasian Collard-Dove is commonplace throughout much of the region and rapidly expanding its range.

Mourning Dove

Rock Pigeon (*Columba livia*)

The familiar pigeon of city streets and parks, the introduced Rock Pigeon is native to Europe, but has been wildly successful in North America. It frequently becomes tame enough to forage under outdoor dining tables on city streets and flocks to people in parks who toss food to the birds. Racing pigeons come from Rock Pigeon stock and sometimes escape or get lost, explaining why birders sometimes see pigeons with leg bands.

Rock Pigeon

LENGTH 12–14 inches. **WINGSPAN** 26–28 inches. Most common plumage is soft gray overall, darkest on the mantle and head, with 2 broad black bars or blotches on the lower wings; usually whitish rump patch; in good light, neck shows green iridescence; in all plumages, bill has a white cere (Band-tailed Pigeon has a yellow bill without the white cere). Plumage variations range from nearly all white to nearly black, and include piebald shades, rusty orange shades, and various others. **VOICE** Low-pitched cooing notes. **BEHAVIORS** Easily adapts to human settlement and thrives in urban areas, particularly cities and towns with tall buildings for roosting, perching. **HABITAT** Cities, farmlands and ranch lands; also cliffs and canyons in open country. **STATUS** Common year-round resident.

Rock Pigeon

Rock Pigeon range

Band-tailed Pigeon
(*Patagioenas fasciata*)

The native pigeon of the Northwest, the Band-tailed Pigeon seems to have increased its numbers after populations declined enough to prompt the U.S. Fish and Wildlife Service and state game agencies to shorten hunting seasons and severely curtail traditionally liberal bag limits on this bird during the 1980s. A century ago, migrating flocks of Band-tailed Pigeons could total hundreds

of birds, but such flocks are nowhere evident now. Still, the birds have rebounded substantially and rank among the iconic species of Pacific Northwest forests.

LENGTH 14.5 inches. **WINGSPAN** 26 inches. Bluish gray above, rosy tan below, trending to a purplish tinged head; white crescent on nape at base of head (missing on juvenile), bordered below by a patch of iridescent feathers; long, squared, gray tail with narrow dark subterminal band, then broad, light tan terminal band; yellow bill with black tip; yellow legs; dark primaries. **VOICE** Typical call is a resonant, owl-like *who-whooo*, with the second hoot sometimes slightly wavering. **BEHAVIORS** Often perches high in forest canopy, and also high on standing snags, both in the forest and on coastal estuaries; overhead flight, usually in small flocks, is very rapid; in appropriate habitat, can be attracted to bird feeders. **HABITAT** Generally breeds in mature conifer forests, but occasionally uses second growth, mixed forests, and mature deciduous stands, from sea level to at least 5000 feet, including suburban and rural areas; occurs in a wider variety of wooded habitats in migration. **STATUS** Fairly common summer and year-round resident (March–September), except rare winter resident (October–February) in Cascades and coastal mountain ranges. Rare east of Cascades.

Flock of Band-tailed Pigeons

Band-tailed Pigeon

Band-tailed Pigeon range

Eurasian Collared-Dove
(*Streptopelia decaocto*)

After its initial introduction to the Bahamas in the 1970s, the Eurasian Collared Dove has colonized nearly half the North American continent with staggering rapidity. These large doves continue to spread across the Pacific Northwest, seeming to expand their range annually. They readily eat at bird feeders and adapt well to human settlement, factors thought to have spurred their systematic range expansion.

LENGTH 12 inches. **WINGSPAN** 21 inches. Larger and stockier than Mourning Dove and much paler than most Rock Doves; pale tan overall; tail with white corners is squared when the bird is perched and fanlike in flight; thin black crescent on nape; dark primaries. **VOICE** Common call is a 3-syllable *who-who, who*, higher, more musical, and more hurried than Band-tailed Pigeon. **BEHAVIORS** Readily comes to bird feeders and may aggressively chase away other birds. **HABITAT** Small towns, residential areas in cities of all sizes, farmlands. **STATUS** Fairly common and increasingly numerous year-round resident.

Eurasian Collared-Dove in flight

Eurasian Collared-Dove

Eurasian Collared-Dove range

Mourning Dove
(*Zenaida macroura*)

Named for its familiar and distinctive call, the slender, elegant Mourning Dove is a familiar resident of open areas, ranging from city neighborhoods to remote prairies and sagebrush plains. These incredibly swift fliers—they can zoom overhead at 50 miles per hour—migrate en masse southward from the Pacific Northwest, but many remain all winter in areas with ample food, even in the coldest regions of eastern Washington and eastern Oregon.

LENGTH 11 inches. **WINGSPAN** 16 inches. Grayish tan above, warm light pinkish tan below; black spots on the wing coverts; long, pointed tail showing black-tipped white feathers on sides. **JUVENILE** Similar to adult but lacks black spots and may appear somewhat mottled. **VOICE** Most musical of the Northwest doves and pigeons; song, for which the species is named, is a mournful, drawn-out *whooo-AHH whooo who-who.* **BEHAVIORS** In hot, late summer weather, Mourning Dove flocks routinely seek water sources daily, congregating on cobble river banks, farm ponds, standing water from irrigation systems, and other locales. **HABITAT** Grasslands, high plains, agricultural areas, shrub steppe, residential areas, open and relatively open riparian corridors. **STATUS** Common summer resident (March–September) and migrant; fairly common year-round resident.

SIMILAR SPECIES White-winged Dove (*Zenaida asiatica*) is a rare vagrant to the Pacific Northwest, but seems to be slowly expanding its range northward from the Southwest; tan overall, with dark primaries and broad white wing stripes in flight, showing as white stripes when bird is perched; tail is squared, with white corners.

Mourning Dove

Mourning Dove range

Cuckoos

Represented by a single species in the region (if indeed any Yellow-billed Cuckoos remain in their last-known Northwest holdout in eastern Idaho), the Cuckoo family also includes the iconic Greater Roadrunner of the desert Southwest.

Yellow-billed Cuckoo
(*Coccyzus americanus*)

Historically the Yellow-billed Cuckoo ranged across the West, from southern British Columbia to northern Mexico and from the Rocky Mountains to the Pacific Ocean. But throughout its former range in the West, this beautiful and secretive bird has been victimized by the loss of the large blocks of intact riparian habitat on which it relies. In the Northwest, the only possible remaining breeding population may still persist in southeast Idaho, in the Snake River Valley (possibly only along the South Fork of the Snake River), and the state's entire breeding population, if indeed the birds still nest there, comprises no more than a few pairs.

LENGTH 12 inches. Long, streaming, dark tail, with rows of large white spots on the underside; grayish above, with rust-colored primary flight feathers; white below; fairly long decurved bill with yellow lower mandible; thin yellow eye ring. **VOICE** Distinctive song is a rapid series of clattering *chuk* notes, sometimes followed by a slow, squeaky *cheeaw cheeaw cheeaw*; calls include a reedy, hollow cawing. **BEHAVIORS** Highly secretive and difficult to view because of preference for dense cover, where it gleans foliage and branches for insects. **HABITAT** Large tracts of healthy riparian cottonwood and willow forests with dense understory.

STATUS Probably extirpated, but possible very rare summer resident southeast Idaho in the Snake River Valley, primarily along the South Fork downstream from Palisades Dam. Rare visitor along the Columbia River, and vagrant elsewhere in the region, such as southeast Oregon.

Yellow-billed Cuckoo

Owls

Whereas hawks, eagles, and falcons hunt by daylight, owls rule the nights. In the Northwest, one species—the Spotted Owl—has stolen many a headline, having once become symbolic of the plight of public-land forests in the region, but the most familiar of the tribe is the Great Horned Owl, a deadly predator capable of killing mammals several times its weight. While most owls do indeed hunt by night, several species in the Northwest have no compunction about foraging diurnally, including the Northern Pygmy-Owl and the Short-eared Owl. Most owls are permanent residents, but the Flammulated Owl and the Burrowing Owl migrate south for the winter, and the Snowy Owl and the Northern Hawk Owl periodically over-winter in the region, wandering down from their normal ranges farther north.

Great Horned Owl

Barn Owl (*Tyto alba*)

The Barn Owl has superb night vision, but often locates prey by sound, and is among the most adept animals at doing so. Widespread, though not often seen because they are entirely nocturnal, Barn Owls adapt well to human habitation and often nest in crevices, deep enclaves, and other cavity-type locations on buildings. They readily nest in barns and other structures as well as artificial nest boxes.

LENGTH 14.5 inches. **WINGSPAN** 44 inches. **SEXES SIMILAR** Slender, with long wings; large, heart-shaped white facial disk; male is white below, female tends to be very light cinnamon-tan on breast, and both sexes tend to appear white when seen from below flying over lights or lit up by headlights; softly mottled light gray and light, bronzy tan above; flight feathers have dark bars forming lines down the wings when the bird is seen in flight from above. Long wingtips extend beyond tail when bird is perched. **VOICE** Typical call is an eerie scratchy, hissing screech, *chuuuureeech*. **BEHAVIORS** Strictly nocturnal; often seen at night hunting near roadways that pass through extensive open country, and thus are often struck and killed by vehicles. **HABITAT** Nests and roosts in undisturbed structures, caves, crevices, and cavities in dense tree stands; hunts over open country, such as prairies, farmlands and ranch lands, and extensive marshlands. **STATUS** Fairly common year-round resident.

Barn Owl

Barn Owl in flight

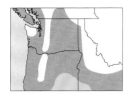

Barn Owl range

Great Horned Owl
(*Bubo virginianus*)

Renowned for its ability to tackle the
large prey it prefers, the Great Horned
Owl has a widely varied diet, making
just about any creature it can kill subject
to grave danger when the owl is on the
hunt. Though mammals—rabbits, hares,
skunks, raccoons, mice, and even the
occasional house cat—make up most of
its normal diet, these owls also kill birds
and will readily kill and eat virtually any
other owl.

LENGTH 22 inches. **WINGSPAN** 48 inches.
Large and robust, with large ear tufts,
bright yellow eyes, and a white bib; mottled
brown, tan, and gray overall, with extensive
dark-on-light barring on the breast, belly,
and flanks; fairly stiff wingbeats in flight.
VOICE Common call, especially frequent
during winter prior to nesting, is a deep,
soft, resonant hooting, typically *who who
who-oo*, or more drawn-out *who who hoo-oo
who who*; also various squeaky and raspy
barking calls and screeches from female
and juveniles. Sometimes hoots during daylight. **BEHAVIORS** Begins nocturnal courtship hooting in midwinter. Though usually
nocturnal, will hunt during daylight, especially morning and evening, in times of
food scarcity. **HABITAT** Widespread, from
open desert (where it roosts in cliffs and
rimrocks) to forest edges; also farmlands,
suburban residential areas and parklike
areas, wooded swamps. **STATUS** Fairly common year-round resident.

Great Horned Owl

Great Horned Owl range

Snowy Owl (*Bubo scandiacus*)

Well-known in the Arctic and subarctic where it normally lives, the aptly named Snowy Owl disperses southward in winter, and in some years is irruptive (unpredictably travels outside typical seasonal range, especially in fall and winter—often related to food supply). Individuals can stray deep into the lower 48 states. At least a few individuals show up in the Northwest annually. Accustomed to the Far North tundra, Snowy Owls occurring in the Pacific Northwest are frequently found perched on the ground.

Snowy Owl

Snowy Owl range

LENGTH 22–24 inches. **WINGSPAN** 50–56 inches. **SEXES SIMILAR** Large and robust; white overall, ranging from nearly pure white (adult male) to strongly barred with black or dark brown (first-year female); yellow eyes; legs and feet heavily feathered. **VOICE** Rarely heard in winter in the Northwest; calls include low grunts when flushed, soft barking calls, and deep hoots. **BEHAVIORS** Usually perches on ground or low structures, such as posts and rooftops; strongly diurnal. **HABITAT** Generally found in open spaces, including coastal dunes, airports, pasturage, prairies. **STATUS** Rare winter visitor (November–March). **KEY SITES** British Columbia: Boundary Bay. Other general locations with fairly regular sightings include the coast from Astoria, Oregon, northward, and the Palouse, in eastern Washington.

Flammulated Owl
(*Psiloscops flammeolus*)

Barely larger than a sparrow, the miniscule Flammulated Owl is seldom seen, not only because it is strictly nocturnal, but also because it tends to feed on insects within forested areas, and then day roosts close to tree trunks within dense branch cover, or in holes. It is far more often heard than seen, but can alter

Flammulated Owl sleeping on roost

Flammulated Owl with ear tufts lowered

the volume of its surprisingly deep call to make locating the bird difficult (it has been described as a ventriloquist). Unlike most owls, the Flammulated Owl is strongly migratory.

LENGTH 6.5 inches. **WINGSPAN** 16 inches. Grayish overall, with highly variable degrees of light rufous, with some specimens appearing more rufous than gray; pale below and finely marked with bars and streaks; darker and finely speckled above, usually with rufous-tipped scapulars forming a pale line on each side of the back; dark eyes (all other small Northwest owls have yellow eyes); facial disk and short ear tufts lend square shape to head, but ear tufts not always visible; wingtips extend beyond tail. **VOICE** Call is a monotonous single, hollow hoot at intervals of several seconds that can continue for hours during the spring courtship period; occasionally, abrupt and somewhat quieter hoot precedes the primary call, making a double hoot; also a variety of loud shrieks and barks. **BEHAVIORS** Gleans insects from trees, takes them in flight, and sometimes captures them on the ground. **HABITAT** Open pine forest, especially ponderosa pine. **STATUS** Uncommon summer resident (May–September). **KEY SITES** Idaho: Craig Mountain Wildlife Area; Rock Creek Road and Magic Mountain area (South Hills south of Twin Falls). Oregon: Idlewild Campground (Harney County); Westside Road area (Klamath County). Washington: Bethel Ridge Road (Highway 12 near White Pass); Liberty area and Blewett Pass. British Columbia: Max Lake (Penticton); Oliver and McKinney Road (Okanagan Valley).

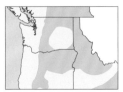

Flammulated Owl range

Western Screech-Owl
(*Megascops kennicottii*)

Nocturnal, but not particularly shy, the handsome little Western Screech-Owl tends to show up in surprising places— it thrives in remote settings far from civilization, but also adapts well to human settlement. It is a secondary cavity nester, often using holes made by woodpeckers, and will use nesting boxes placed in proper habitat. In the Northwest, Western Screech-Owls appear to be declining in numbers owing to direct predation by Barred Owls, which are relatively recent arrivals in the region after expanding their range westward during the 20th century.

LENGTH 8–9 inches. **WINGSPAN** 22–24 inches. Grayish or brownish gray above, with darker gray or brownish streaks; silvery gray below, with bold dark streaks; gray facial disk, bordered by black on both sides, does not wrap around the top of the head; prominent ear tufts (can be laid down);

bright yellow eyes (Flammulated Owl has all-dark eyes); in flight, shows prominent black-and-white barred flight feathers. **VOICE** Common call is an accelerating "bouncing ball" series of high, whistlelike hoots; alarm calls are various high wails and barks. **BEHAVIORS** Western Screech-Owls hunt a wide variety of prey—small mammals, insects, birds, amphibians, and more—but most surprisingly, they even snatch fish from shallow water. **HABITAT** Open mixed woodlands and deciduous woodlands, usually with adjacent open areas; woodland edges; riparian corridors with tree stands, including desert canyons; suburban and residential areas with appropriate habitat; generally found at low to middle elevations. **STATUS** Uncommon year-round resident (generally rare in British Columbia).

Western Screech-Owl

Western Screech-Owl range

Western Screech-Owl with ear tufts lowered

Northern Pygmy-Owl
(*Glaucidium gnoma*)

One of the tiniest birds of prey, not much larger than most sparrows, the Northern Pygmy-Owl is nonetheless a lethal predator, and a daylight hunter that preys heavily on songbirds, even snatching them from bird feeders. In fact, one way to find these tiny predators is to listen and watch for congregations of agitated groups of songbirds, such as chickadees, nuthatches, and kinglets mobbing the owl, perhaps either to drive it away or announce the threat.

LENGTH 6.75 inches. **WINGSPAN** 15 inches. Plump, with proportionately large, rounded head; long tail that extends well beyond wingtips; brown above (darker and more reddish in Pacific slope population; lighter and gray-brown in interior population), with small pale spots; 2 large false-eye spots on back of neck; dark crown heavily patterned with white freckles; dark throat; white breast and belly patterned with dark streaks; bright yellow eyes and narrow white eyebrows. **VOICE** Song is a series of monotonous, high, hollow toots, usually spaced by one to 4 seconds. **BEHAVIORS** Usually perches diagonally rather than upright like other owls; often perches on tiptop of small to medium-sized trees, or on outer branches. Undulating flight, unique among Northwest owls, is a woodpeckerlike alternating series of rapid wingbeats and closed-wing glides. **HABITAT** Coniferous and mixed woodlands with scattered openings. **STATUS** Uncommon year-round resident; rare in lowland valleys west of the Cascades; generally absent from southeast Oregon, southwest Idaho, and the Columbia Plateau.

Northern Pygmy-Owl

Northern Pygmy-Owl range

Northern Pygmy-Owl

Burrowing Owl
(*Athene cunicularia*)

Unusual for birds, the Burrowing Owl, true to its name, nests below ground, using holes dug by mammals, such as badgers, marmots, or ground squirrels. They modify these existing holes, but do not dig their own burrows. They are both diurnal and nocturnal hunters, often active at dawn and dusk, and are highly migratory.

LENGTH 9 inches. **WINGSPAN** 21 inches. Distinctive long legs; brown above, heavily spotted with white; barred brown and white below, with white throat and broad white brow; yellow eyes. **JUVENILE** Plain brown above; brown head and face with prominent light mustache and eyebrow stripes; pale buff below **VOICE** Calls include a high, rapid, 2-note cooing, *coo-cooo*, often ending in an abrupt growl or grunt, *coo-coooo-ur*; also chattering calls. **BEHAVIORS** Frequently perches on the ground, wooden fence posts, fence anchors, or other structures during daylight. In hot weather, tends to remain underground until nightfall. May bob and weave when approached, then flush a short distance and land on the ground or a low perch. **HABITAT** Dry prairie, sagebrush steppe, ranch lands, farmlands. **STATUS** Uncommon to locally common summer resident (March–September); rare winter resident. **KEY SITES** Idaho: Kuna area (South Cole Road, Kuna and Kuna-Mora Road); Snake River Birds of Prey National Conservation Area. Oregon: Catlow Valley; Highway 78 Burns to Crane; Malheur NWR. Washington: Desert Unit, Columbia Wildlife Area; Othello area.

Burrowing Owl

Burrowing Owl in flight

Burrowing Owl range

Spotted Owl (*Strix occidentalis*)

In the 1980s, the Northern Spotted Owl (one of 3 subspecies, and the only one occurring in the Northwest) became an iconic symbol of the habitat devastation wrought by decades of virtually unchecked logging on federal timberland. The 1994 Northwest Forest Plan sought to better manage timber harvests to help ensure survival of the Spotted Owl, a threatened species under the federal Endangered Species Act. This created a system of late-successional forest reserves across the owl's range, but population trend studies suggest that Northern Spotted Owl populations continue to decline range-wide at an average rate of 2.9 percent annually, and past trends suggest that much of the bird's remaining habitat could disappear within 30 years. In addition to declining numbers owing to habitat loss, the Spotted Owl now faces the threat of being outcompeted by (and interbred with) the larger and less habitat-specific Barred Owl, a relative newcomer to the Northwest by way of greatly expanding its range westward beginning in the 20th century.

LENGTH 17 inches. **WINGSPAN** 40 inches. Chocolate-brown overall, heavily peppered with white spots that form oblong bars on the breast (Barred Owl has brown vertical bars on the breast); large rounded head; light brown facial disk framed by dark brown; dark eyes; broad, rounded wings. **VOICE** Common call is a series of 4 low, staccato hoots, *who hoo-hoo who* or *who hoo-hoo whooa*; also sharp barks; rising wail, *aaruup.* **BEHAVIORS** Hunts primarily at night by perching on branches and pouncing on rodents on the ground or on lower branches; chief prey include northern flying squirrels, woodrats, voles, and other rodents. **HABITAT** Mature and old-growth conifer forests with a variety of tree species, a multilayered closed canopy, and fallen logs and snags. **STATUS** Rare year-round resident; regionwide, populations are densest in southwest Oregon and critically sparse in British Columbia.

Spotted Owl

Spotted Owl

Spotted Owl range

Barred Owl (*Strix varia*)

A relative newcomer to the Pacific Northwest and the West in general, the Barred Owl was historically an eastern species until its rapid and dramatic spread westward in the last century. Many experts postulate that humans made the expansion possible by planting shelter belts of trees all across the Great Plains, creating stepping stones for the owls to move westward alongside human development. Whatever the cause of the sudden range expansion, one thing is clear: the Barred Owl puts the already imperiled Northern Spotted Owl in a dire situation. The larger Barred Owl is a habitat and prey generalist, well suited to outcompete, displace, and even kill the habitat- and prey-specific Northern Spotted Owl.

LENGTH 19 inches. **WINGSPAN** 42 inches. Dark brown above, heavily patterned with white; whitish below, with each breast feather having a narrow brown streak down its center, contrasting with pale horizontal barring on throat; gray facial disk outlined by dark brown; dark eyes and yellow bill; large, rounded head with heavily speckled crown, and no ear tufts; dark-and-white barred flight feathers. **VOICE** Call is a low hooting, rising and then tailing off in pitch, often phonetically described as who-cooks-for-you, with the you usually drawn out; often begins with a series of spaced hoots before drawing them together for the final note, who who who who who who-UU-a. Also a loud wail and hollow barking calls. **BEHAVIORS** Generally nocturnal, but sometimes hunts opportunistically by day. **HABITAT** Mixed woodlands and conifer forest. Status Uncommon year-round resident.

Barred Owl

Barred Owl

Barred Owl range

Great Gray Owl (*Strix nebulosa*)

Elusive and secretive, the Great Gray Owl always provides birders a terrific thrill, for this largest of North American owls is almost always a surprise discovery rather than a species specifically sought and found. It is substantially larger in appearance than a Great Horned Owl, but much of its size is due to dense plumage—it actually weighs less than a Great Horned Owl. Key sites listed below represent locations where sightings are most numerous, but finding Great Gray Owls is rarely predictable.

LENGTH 26–28 inches. **WINGSPAN** 56 inches. Large and robust (female averages larger than male); grayish overall; large, bulbous, rounded head, with large facial disk with concentric circles of dark lines within and narrow white arcs between bright yellow eyes (Barred Owl has dark eyes); black chin patch beneath bill, and white "bow tie" pattern on throat. **VOICE** Very deep, resonant, hollow hoots. **BEHAVIORS** Not strictly nocturnal, Great Gray Owls often hunt during daylight hours, generally by perching and waiting for an opportunity to pounce on rodents. **HABITAT** Mature coniferous forest, especially near mountain meadows. **STATUS** Rare year-round resident. **KEY SITES** Idaho: Centennial Range and Island Park area (also Yellowstone and Teton National Parks in adjacent Wyoming). Oregon: east side of Klamath Marsh NWR; Howard Prairie Reservoir; Spring Creek Great Gray Owl Management Area (for information, contact La Grande Ranger District, Wallowa-Whitman National Forest, 541-963-7186). Washington: Okanogan Highlands. British Columbia: Anarchist Mountain.

Great Gray Owl range

Great Gray Owl

Long-eared Owl (*Asio otus*)

Typically seen while roosting, the strongly nocturnal Long-eared Owl appears tall and slender, its long, close-set ear tufts only strengthening that impression. A bird of open country, especially prominent in the interior Northwest, these medium-sized owls are well adapted to locate food by sound, as well as sight, and subsist primarily on small rodents, including voles, gophers, and kangaroo rats.

LENGTH 15 inches. **WINGSPAN** 38 inches. Tall and slender, with very long ear tufts set close together; light cinnamon-colored facial disk, with white V between yellow eyes; strongly streaked and barred below (Great Horned Owl has horizontal bars, not vertical streaks); silky speckled gray and tan above; dark bars across cinnamon-colored flight feathers, with black wrist mark on lower wing surface. **VOICE** Male's call is a series of hollow, slightly drawn-out hoots, spaced by several seconds; female's call is a scratchy, nasal hoot; also raspy barking calls. **BEHAVIORS** Primarily nocturnal; hunts on the wing over open country. During winter, Long-eared Owls sometimes share communal roosts where several to a dozen or more birds congregate. **HABITAT** Hunts over open country, including grasslands, shrub steppe, open conifer forest (especially juniper), ranch lands; usually nests and roosts in dense trees, though also uses crevices in canyon cliffs and rimrocks; in open desert and sage steppe, often roosts and nests in dense riparian thickets (along perennial and seasonal streams). **STATUS** Uncommon year-round resident interior; rare transient and occasional breeder west of Cascades (most sightings late summer through winter).

Long-eared Owl

Long-eared Owl

Long-eared Owl range

Short-eared Owl (*Asio flammeus*)

A distinctive owl of open country, the Short-eared Owl, which is often active during daylight, can be identified at a distance by its slow, deep wingbeats that give the bird a buoyant, mothlike appearance in flight. Because of their diurnal foraging, these handsome owls are sometimes preyed upon by hawks, falcons, and larger owls.

LENGTH 15 inches. **WINGSPAN** 37 inches. **SEXES SIMILAR** Mottled beige, tan, and dark brown above; pale tan breast with fine dark lines; whitish belly; pale tawny facial disk usually outlined with white; bright yellow eyes; short ear tufts seldom visible; in flight, shows white wing linings, dark wingtips, and a dark crescent on the leading edge of the wing above the wrist. **VOICE** Typical call is an abrupt, scratchy *cheeow*; male's breeding-season call is a rapid series of abrupt, deep hoots. **BEHAVIORS** Flight is agile and somewhat mothlike, with soft, deep wing strokes; sometimes hovers. Nests on the ground. **HABITAT** Prairies, grasslands, marshes, agricultural areas, shrub steppe. **STATUS** Uncommon year-round resident and migrant interior; rare winter visitor and migrant west of Cascades.

Short-eared Owl in flight

Short-eared Owl range

Short-eared Owl

Boreal Owl (*Aegolius funereus*)

True to its name, the Boreal Owl is primarily a resident of the boreal forest zone of Canada and Alaska (as well as the northern Eurasia), but its range also takes in similar habitat from the Rocky Mountains to the Cascades, where it occurs primarily at high elevation, up to the subalpine zone. Boreal Owls are cavity nesters, relying largely on holes made by pileated woodpeckers and flickers, as well as natural cavities.

LENGTH 8.5–11 inches. **WINGSPAN** 22–24 inches. **SEXES SIMILAR** Females larger than males; large head; grayish facial disk framed above by dark brown ring that forms inverted check marks above the eyes; bold, white Y between bright yellow eyes; heavily speckled forehead; white with heavy brown streaks below; dark brown above with white spots; juvenile is buff below, with brownish facial disk, and prominent white Y between and above eyes. **VOICE** Song is a rapid series of high, hollow toots. **BEHAVIORS** Flies through forest by flapping quickly, not gliding, unlike smaller Northern Pygmy-Owl, which flaps and glides in undulating flight. **HABITAT** Mature conifer forest, generally at high elevations; also mature aspen stands. **STATUS** Rare year-round resident of Idaho, Washington, and British Columbia; very rare Oregon.

Boreal Owl

Boreal Owl range

Northern Saw-whet Owl
(*Aegolius acadicus*)

The Northern Saw-whet Owl is widespread and by owl standards, relatively common, but it is also tiny, secretive, reclusive, and nocturnal, so is seldom seen. Its preferred daytime roost is in dense forest or dense tree stands, especially conifers, making it very difficult to find during the day; often voice is the best way to confirm its presence.

LENGTH 8 inches. **WINGSPAN** 18 inches. Large, rounded head, lacking ear tufts; bright yellow eyes; light grayish brown facial disk with pale streaks and a white Y between the eyes; warm brown above with white spots; white breast and flanks with rusty brown vertical streaks; in flight, underwings strongly patterned with white spots. **JUVENILE** (May–September) Rich dark brown overall, with solid buffy underparts; white Y between eyes prominent against brown facial disk. **VOICE** Common call is a fairly rapid, monotonous series of hollow, whistled *toot* notes, even or nearly even in pitch. **BEHAVIORS** Secondary cavity nesters, Northern Saw-whet Owls typically use holes made by woodpeckers. **HABITAT** Conifer and mixed forests, including extensive groves in shrub steppe habitats; also deciduous forests in some areas. **STATUS** Fairly common year-round resident and migrant.

Juvenile Northern Saw-whet Owl

Northern Saw-whet Owl range

Adult Northern Saw-whet Owl

Northern Hawk Owl
(*Surnia ulula*)

The Northern Hawk Owl's North American range extends across the boreal forest belt of Canada and Alaska. In winter, these hawklike owls stray southward in limited numbers, and occasionally show up in the northernmost Lower 48 as vagrants, although a few have nested in southern British Columbia and northern Idaho.

LENGTH 16 inches. **WINGSPAN** 29 inches. Very long tail; resembles an accipiter; gray facial disk boldly outlined in black; bright yellow eyes; dark brown above, heavily spotted; uniformly barred with brown and white below, including flight feathers and tail. **VOICE** Calls include a high, raspy *chuuuur-ip*, and a shrill *kee-kee-kee-kee-kee*. **BEHAVIORS** Active during daylight; uses exposed perches, especially the tops of small trees, to watch for prey; flies low and fast, like an accipiter, on stiff wings with rapid wingbeats; sometimes hovers. **HABITAT** Unpredictable as vagrants, but generally open country with sparse trees. **STATUS** Very rare winter visitor (also a few nesting records for southern British Columbia and northern Idaho).

Northern Hawk Owl

Northern Hawk Owl
range

Nightjars and Swifts

These 2 different types of birds are represented in the Northwest by only 5 species in total. Worldwide, however, there are many more. The nightjars, family Caprimulgidae, are represented in the Northwest by the Common Nighthawk and Common Poorwill, crepuscular and nocturnal birds, respectively, that specialize in snatching insects with bills that appear tiny when the mouth is closed, but gaping when open. They are incredibly agile fliers, especially the Common Nighthawk, which forages almost exclusively on the wing; the night-hunting Common Poorwill prefers to sit on the ground and then burst upward to snatch insects out of the air. In many ways, swifts (family Apodidae) are similar to the nightjars—they specialize in snatching insects while flying with great athleticism and dazzling speed.

Common Nighthawk

Common Nighthawk
(*Chordeiles minor*)

In arid regions of the interior North-west, when flying insects are plentiful on warm summer evenings, congregations of feeding Common Nighthawks often put on a remarkable spectacle of acrobatic aeronautics. These slender-winged insectivores dip and dive and dart at high speeds, snatching bugs on the wing with their wide, froglike mouths. The male's dramatic courtship display begins high in the air with a nasal *peeeink* call; he then dives swiftly toward the ground, braking suddenly as the air rushing through his flight feathers produces a buzzy *woooosh*.

LENGTH 9.5 inches. **WINGSPAN** 24 inches. **SEXES SIMILAR** Rich, mottled grayish brown overall; pale grayish breast and belly patterned with thin, dark bars; in flight, long, narrow wings with distinctive white bands across the base of the primaries; white (male) or buff (female) throat stripe. **VOICE** Call is a loud, nasal *peeeink*, given in flight. **BEHAVIORS** Most active around dusk and after dark; flies erratically and acrobatically in pursuit of insects, from near ground level to very high. During the day, roosts on the ground or lengthwise on branches, and can be very approachable. **HABITAT** Open areas, especially near water that produces dense hatches of Chironomids or other insects; fields, prairies, sagebrush steppe; also feeds over suburban and urban areas. **STATUS** Summer resident (May–September); common interior, uncommon west of Cascades.

Common Nighthawk in flight

Common Nighthawk

Common Nighthawk range

Common Poorwill
(*Phalaenoptilus nuttallii*)

The Common Poorwill, like its better-known cousin, the Common Nighthawk, is insectivorous, but unlike the nighthawk, which fly-catches entirely on the wing, this nocturnal predator settles onto the ground or a low perch, watches for insects, then flits up to catch them. Most poorwills in the Northwest migrate south for the winter, but the species is known for its ability to withstand cold temperatures by entering a state of deep torpor, in which its metabolism nearly shuts down completely.

LENGTH 8 inches. **WINGSPAN** 17 inches. Rich, mottled grayish or brownish gray overall; white throat bordering black breast band; pale gray, lineated breast; short, rounded wings; white tips on dark outer tail feathers. **VOICE** More often heard than seen; call is a repeated, ringing, hollow, whistled *poor-will*, or *poor-will-up*. **BEHAVIORS** Most commonly seen sitting on roads on summer nights, especially in remote areas; flight is mothlike; hikers occasionally flush Common Poorwills from their daytime roosts on the ground (usually rocky slopes), upon which the bird flutters away a short distance and lands on the ground again. **HABITAT** Arid, open country, especially with rocky slopes and rimrocks, including sagebrush steppe, juniper woodlands, open ponderosa forest with scrub understory, chaparral, desert scrub and desert playas with scrub fringe. **STATUS** Uncommon summer resident (May–September).

Common Poorwill

Common Poorwill range

Common Poorwill on a gravel road at night

Vaux's Swift *(Chaetura vauxi)*

Common throughout the Northwest during breeding season, the Vaux's Swift often uses chimneys as roosting sites during fall migration. Thousands of swifts may occupy a single chimney, creating a spectacle enjoyed by birders and nonbirders alike during evenings, when the birds come back from feeding. They begin swirling around the entrance to the chimney, forming a tornadolike spinning mass until the first few birds decide to enter. This signals the entire flock to begin darting into the chimney until all the swifts are inside.

LENGTH 4.25 inches. Small, cigar-shaped body and sicklelike wings; gray-brown overall, with dingy pale brown throat and breast; short, squared tail. **VOICE** Calls include high twittering and a buzzy *cheer.* **BEHAVIORS** Flies rapidly and acrobatically on stiff wings, with shallow wingbeats, and without the folded-wing glides of a swallow. In addition to chimneys, Vaux's Swift roosts in hollow trees in mature forests and stands. Migrating flocks in fall can number in the thousands; spring migrants tend to arrive in smaller groups. **HABITAT** Widespread, but prefers old-growth conifers for nesting. **STATUS** Common summer resident and migrant (April–October). **KEY SITES** (Chimney locations, September) Idaho: downtown Moscow. Oregon: Agate Hall, University of Oregon (Eugene); Chapman Elementary School (Portland). Washington: Frank Wagner Elementary (Monroe); Selleck Old Schoolhouse.

Vaux's Swift in flight

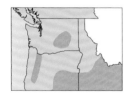

Vaux's Swift range

Vaux's Swifts swarm into the chimney at Chapman Elementary School in Portland.

Black Swift (*Cypseloides niger*)

The largest North American swift, the robust Black Swift is a cliff specialist, nesting behind waterfalls, on riverine canyon walls, and on sea cliffs. Such locations are difficult to study because they are so perilous to reach, so details about nesting and brood behavior remain under-documented. Moreover, Black Swifts tend to make their feeding flights high over mountains, so cloudy days, which force them to fly-catch at lower elevations, are often the best times to see them. British Columbia hosts the largest breeding population of this purely western species.

LENGTH 7.25 inches. Large, black swift, with cigar-shaped body and sicklelike wings; slightly forked tail. **VOICE** Call is a seldom-heard, high-pitched twitter. **BEHAVIORS** Acrobatic flight on stiff wings, with shallow wingbeats, and without the folded-wing glides of a swallow; slower wingbeats than other swifts; tail often fanned in flight. **HABITAT** Mountains and coastal areas; nests on cliffs; feeding flights occur over mountains. **STATUS** Uncommon to rare summer resident and migrant (May–September). **KEY SITES** Idaho: Shadow Falls (Shoshone County). Oregon: Salt Creek Falls (Lane County). Washington: Mount Hardy (Skagit County); Mount Rainier National Park. British Columbia: E.C. Manning Provincial Park; Okanagan Falls Provincial Park.

Black Swift in flight

Black Swift in flight

Black Swift range

White-throated Swift
(Aeronautes saxatalis)

A common swift of the inland Northwest, the White-throated Swift nests and roosts in rock crevices, riparian canyon walls, and rimrocks. These birds can be very vocal, and are often first detected by their call. At a distance, the White-throated Swift's contrasting black-and-white pattern may not be visible, but the long, thin tail distinguishes the bird from other swifts.

LENGTH 6.5 inches. Large swift with cigar-shaped body, sicklelike wings, and long, thin, forked tail; black overall, with white throat and belly; white patches on sides of rump; white tips on black secondary flight feathers. **VOICE** Call is a raspy, descending *kee-kee-kee-kee-kee*. **BEHAVIORS** Fast, acrobatic flight is stiff-winged, with shallow wingbeats, and without the folded-wing glides of a swallow. Nests and roosts in rock crevices and occasionally manmade structures, such as overpasses, bridges, and buildings. **HABITAT** Mountains, rimrocks, and canyons. **STATUS** Common summer resident (April–September). **KEY SITES** Idaho: Malad Gorge State Park; Snake River Birds of Prey National Conservation Area. Oregon: Peter Skene Ogden State Scenic Viewpoint (Jefferson County); Succor

Creek State Natural Area (Malheur County). Washington: Palouse Falls State Park (Franklin County); Sun Lakes State Park (Grant County). British Columbia: Syringa Provincial Park (Central Kootenay); Vaseux Lake (Okanagan).

White-throated Swift in flight

White-throated Swift range

White-throated Swift in flight

Hummingbirds

The tiniest of birds, hummingbirds are remarkable and unique in their frenetic, acrobatic abilities on the wing. These incredible little bundles of energy beat their wings about 50 times per second and can hover in place and change direction with incredible speed. They are nectar specialists, feeding on a wide range of flowers, and also take tiny insects, typically gleaning them from plants. Of course they readily visit hummingbird feeders, to the delight of birders and nonbirders alike. One of the half-dozen Northwest species, the Anna's Hummingbird, often overwinters in the valleys of western Oregon and western Washington; the others are entirely migratory.

Anna's Hummingbird

Anna's Hummingbird
(*Calypte anna*)

Only in recent decades has the Anna's Hummingbird become the most common hummingbird in some parts of lowland western Oregon and western Washington, vying with the Rufous Hummingbird for that distinction. Moreover, it is the only hummingbird that remains year-round in the Northwest. This colorful recent arrival has greatly expanded its breeding range, which was once limited to the mountains of southern and central California. So new to the Northwest, the Anna's Hummingbird has not had time to evolve a migration pattern southward for winter; so it simply stays put, adapting as need be to Northwest winters.

Adult male Anna's Hummingbird

Adult male Anna's Hummingbird (side view)

Adult female Anna's Hummingbird

Juvenile Anna's Hummingbird

LENGTH 4 inches. Fairly large humming-bird; wingtips do not reach the end of the tail. **MALE** Green above, dingy white below, with slightly greenish flanks; gorget and crown are iridescent rosy red in good light, and can beam like a colorful flashlight when the bird turns its head to the right angle and catches the light; in poor light, gorget and crown can appear blackish or deep crimson; in prime plumage, irides-cent gorget and crown feathers envelope the base of the bill; tail is all dark; juve-nile male has partial gorget. **FEMALE** Green above, pale gray below with pale greenish flanks; usually a small partial gorget cen-tered on the throat; green tail has black band and white tips on outer feathers. **VOICE** Song is scratchy, high chatter that can continue for long periods of time; sings all times of year; calls include simple high *tsip* notes, and an abrupt *brrrt* chase call. **BEHAVIORS** Individual males often stake out one or more feeders and defend them aggressively, including occasional fairly aggressive encounters with likewise pugna-cious Rufous Hummingbirds. **HABITAT** Res-idential areas, parks, farmlands, foothills, valleys, coastal lowlands. **STATUS** Common year-round resident.

Anna's Hummingbird range

Costa's Hummingbird
(*Calypte costae*)

The striking male Costa's Hummingbird is unmistakable, but the female is generally very difficult to distinguish from female Anna's and Black-chinned Hummingbirds, which may explain why the bulk of Northwest sightings have been males.

LENGTH 3.5 inches. **MALE** Iridescent purple gorget extends down to long tapers on both sides of the neck; iridescent purple crown; juvenile male is paler overall with minimally formed gorget. **FEMALE** Green above, clean white below, but often with flecks of purple on throat; short tail; wingtips extend to end of tail. **VOICE** Abrupt, high *tic* notes, sometimes in a rapid series. **BEHAVIORS** Often pumps tail up and down while hovering at feeders. **HABITAT** Most Northwest sightings occur at hummingbird feeders. **STATUS** Rare vagrant; most records for spring and summer.

Male Costa's Hummingbird

Rufous Hummingbird
(*Selasphorus rufus*)

Capable of utilizing a wide range of habitat types, the Rufous Hummingbird is the most widespread hummer in the Pacific Northwest, equally at home on the coast and on inland mountain slopes. They can occur in the same general or even specific locations as Anna's, Calliope, Black-chinned, and Allen's Hummingbirds, but they don't particularly care to share feeders and will aggressively chase away, or attempt to chase away, interlopers of their own kind or any other species. In coastal southwest Oregon, where both Rufous and Allen's Hummingbirds occur, females and juveniles are not safely separable in the field.

LENGTH 3.5 inches. **MALE** Bright rusty orange overall, often with flecks of green on upper back; whitish belly; shimmering orange gorget in good light, but can also appear gold, greenish gold, or reddish, depending on light; green crown; pointed rusty tail feathers with dark tips, and the tail feathers next to the 2 center tail feathers have a diagnostic notch on the inside tips (unlike adult male Allen's); juvenile male has rows of iridescent feathers on white throat and usually a central iridescent spot. **FEMALE** Green above, dingy white below, with pale rust flanks; white throat with rows of small green flecks; outer tail feathers are rusty at the base, with black subterminal bands and white tips. **VOICE** Calls, often in response to territorial disputes, include an excited, high, squeaky *zee zee chippity chippity*, scratchy chattering *chip* notes, and high *zewee* notes. **BEHAVIORS** Male's courtship display is a steep dive from high above the female and ends in a very rapid J-hook at the bottom of the dive, with rapid *chu-chu-chu* calls; also zips back and forth in a rapid three-dimensional figure eight waggle flight in front of perched female. **HABITAT** Coniferous and mixed woodlands from sea level to at least 6000 feet, especially edges, clearings, meadows, brushy regrowth from clear-cuts and fires; also riparian habitats, residential areas, city parks; widespread in migration. **STATUS** Common summer resident (March–September).

Male Rufous Hummingbird

Female Rufous Hummingbird

Female Rufous Hummingbird in flight

Rufous Hummingbird
range

Allen's Hummingbird
(*Selasphorus sasin*)

Almost identical to the much more wide-spread Rufous Hummingbird, the Allen's Hummingbird has a very small breeding range extending from coastal Southern California north to coastal southwest Oregon. Only the adult males are safely separable in the field from Rufous Hummingbirds, and even then, careful study is required for proper identification.

LENGTH 3.5 inches. **MALE** Largely identical to adult male Rufous Hummingbird except back is mostly green rather than all or mostly orange; Allen's outer tail feathers are narrower than those of adult male Rufous, and lack the notch found on the inside tip of the second tail feathers on each side of the tail (the feathers adjacent to the center tail feathers). **FEMALE AND JUVENILE** Essentially identical to Rufous Hummingbird and not separable in the field. **VOICE** Similar to Rufous Hummingbird. **BEHAVIORS** Males make noisy diving courtship displays. **HABITAT** Residential areas, parks, brushy coastal lowlands and nearby foothills primarily in western Curry County, Oregon. **STATUS** Fairly common summer resident (March–August) of coastal southwest Oregon.

Male Allen's Hummingbird

Allen's Hummingbird range

Female Allen's Hummingbird

Calliope Hummingbird
(*Selasphorus calliope*)

The smallest bird in North America that breeds north of Mexico, the dainty Calliope Hummingbird is a mountain specialist, breeding as high as timberline from the crest of the Cascades to the Rocky Mountains. When subalpine meadows burst into color with blooming wildflowers, look for these unobtrusive little hummers hover-feeding at flowers that may not grow more than a few inches high.

LENGTH 3 inches. **MALE** Light, bright green above; white below, with light green scaling on flanks; sparse, streaked magenta gorget; short, stubby, all-dark tail can appear very slightly forked unless flared; wings extend to end of tail; juvenile male has only partial gorget. **FEMALE** Green above, pale buff-tan below; white throat lightly flecked with green or bronze-green; short, stubby tail has black subterminal band and white tips on outer feathers; juvenile female similar to adult female. **VOICE** Typical call is a very high, thin *tsip*, often doubled or in series; in courtship display dive, male utters a high spiraling *spizeeee* along with insect-like wing trills. **BEHAVIORS** Because many mountain flowers grow low to the ground, Calliope Hummingbirds often hover-feed near the ground; they also feed from sap wells drilled in trees by Sapsuckers. **HABITAT** Shrubby mountain meadows and riparian margins, especially open ponderosa forests and mixed conifer forests, including 8- to 15-year-old regrowth from fire or logging; up to and above timberline; generally breeds above 3000 feet elevation, but often found at lower elevations near breeding range upon arrival from wintering grounds; nests and visits feeders in mountain towns with appropriate habitat. **STATUS** Uncommon summer resident (April–August). **KEY SITES** Idaho: Idaho City; Kootenai NWR;

University of Idaho Arboretum and Botanical Garden (Moscow). Oregon: Calliope Crossing (Sisters); Shevlin Park (Bend). Washington: Leavenworth Fish Hatchery and Leavenworth area; Little Pend Oreille NWR; Mount Spokane area. British Columbia: Creston Valley Wildlife Area; Okanagan Valley.

Male Calliope Hummingbird in flight

Female Calliope Hummingbird in flight

Calliope Hummingbird range

Black-chinned Hummingbird
(*Archilochus alexandri*)

The most widespread hummingbird east of the Cascades in the Northwest, this green-backed species is found in a variety of habitats and is easily attracted to backyard feeders. The male is aptly named for its dark throat, which in ideal light shows a rich blue-violet band, forming the gorget.

LENGTH 3.25 inches. Long, very slightly decurved bill; at rest, the wingtips extend past tail. **MALE** Metallic green back; black chin and throat with blue-violet band at lower edge (not always visible); dull white underparts, slightly darker flanks brushed with metallic green; dark tail feathers, central pair metallic green, some others with a glossy purple tinge. **FEMALE** Subdued metallic green upper body and drab gray underparts with light metallic green flanks; outer 3 tail feathers tipped with white, central pair of tail feathers metallic green. **JUVENILE** Similar to female. **VOICE** High-pitched warbling song rarely heard; chipping calls when chasing intruders; male's dive display produces a trill at the bottom of the dive. **BEHAVIORS** Pumps and fans tail feathers often while feeding. Display flights are dives from as high as 100 feet and also a flat figure eight shuttle display. **HABITAT** Varies widely; open juniper woodlands, riparian zones in ponderosa forest, willow- and cottonwood-lined streams in sagebrush steppe and desert; highly adaptable as long as there are food sources. **STATUS** Common summer resident (May–September).

Male Black-chinned Hummingbird

Female Black-chinned Hummingbird in flight

Male Black-chinned Hummingbird

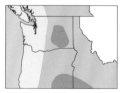

Black-chinned
Hummingbird range

Broad-tailed Hummingbird
(*Selasphorus platycercus*)

The common hummingbird of the central Rockies, the comparatively robust Broad-tailed Hummingbird reaches nearly to the northern extent of its breeding range in the mountains of eastern and central Idaho, and presumably in far eastern Oregon. Sightings in Oregon occur almost annually, but owing to both the scarcity of this species in the state and its high-elevation habitat, the question of whether it nests in the state remains unanswered.

LENGTH 4 inches. **MALE** Green above, including crown; rose-colored gorget; clean white below, except for greenish scaling on flanks. **FEMALE AND JUVENILE** Very difficult to distinguish from female Rufous Hummingbird, but larger and more robust, with less rufous in the base of the outer tail feathers. **VOICE** Typical calls are high, rusty squeaks; produces loud wing noise in flight, which can help differentiate it from Black-chinned Hummingbird. **BEHAVIORS** In his courtship display, the male flies rapidly high into the air, where he then hovers, his wings producing a loud trill; he then dives speedily, headlong, down toward the female sitting below on a low perch, checking his dive at the last moment and whisking by her to again climbing high in the air to start anew. **HABITAT** Open mountain meadows, generally near stands of conifers or aspens; brushy subalpine riparian areas; also open,

high-elevation juniper woodlands associated with deciduous riparian corridors. **STATUS** Uncommon summer resident of southeast and south-central Idaho; very rare summer resident or transient in eastern Oregon and southwest Idaho. **KEY SITES** Idaho: Castle Rock State Park and City of Rocks National Reserve; Magic Mountain and Pike Mountain (south of Twin Falls); Scout Mountain, Cherry Springs, and upper Mink Creek drainage (south of Pocatello).

Male Broad-tailed Hummingbird

Female Broad-tailed Hummingbird in flight

Broad-tailed
Hummingbird range

Kingfishers

The Belted Kingfisher lives up to its name as the king of fishers; these handsome, noisy birds plunge-dive for fish, often from perches overhanging the water, and by hovering to watch for prey beneath the surface. Worldwide there are 90 species of kingfisher, all of them built similarly, but many of them remarkably colorful.

Belted Kingfisher
(*Megaceryle alcyon*)

Loud, conspicuous, and frequently heard before seen (announcing its displeasure at being disturbed with a piercing dry rattle often given in flight), the Belted Kingfisher is fun to watch, especially when actively fishing. These robust kingfishers launch themselves headlong into the water, snatching fish in their bills. Often they perch above favorite fishing grounds, watching the water below for fish to wander close to the surface. They repeatedly use the same perches—limbs, standing snags, and even overhead wires.

LENGTH 12.5 inches. **SEXES SIMILAR** Bluish gray above, white below, with bluish gray breast band; adult female has rust-orange belly band and flanks (lacking in male); ragged crest along entire crown; large chisel-like bill. **JUVENILE** Similar to adult female but with incomplete bands. **VOICE** Call is a distinctive loud, rolling rattle. **BEHAVIORS**

Female Belted Kingfisher

Hunts (primarily fish) by perching above water or hovering, then plunge-diving head first. **HABITAT** Both freshwater and saltwater environs with clear water, so they can see prey, and sufficient over-water perch sites; rivers, ponds, lakes, reservoirs, canals, sloughs, and estuaries. Requires sand and soil banks in which it can excavate nesting burrows. **STATUS** Fairly common year-round resident; withdraws from breeding areas where water freezes.

Male Belted Kingfisher in flight

Belted Kingfisher range

Woodpeckers

With 12 different species breeding within the region, woodpeckers are a major attraction for birders in the Pacific Northwest. These excavating specialists are strikingly patterned, colorful, and energetic—always fun to observe. Species include the familiar and beautiful Northern Flicker, the diminutive Downy Woodpecker, and the giant of the species, the spectacular Pileated Woodpecker. Woodpeckers exhibit a variety of foraging and feeding habits: some species feed heavily on ants; some are post–forest fire specialists, arriving in freshly burned areas to prod and pry for insects under the bark of burned trees; some (the sapsuckers) drill sap wells in trees and feed on the sap that drains out. Some woodpeckers even readily come to suet feeders set up to attract them.

Pileated Woodpecker

Lewis's Woodpecker
(Melanerpes lewis)

In flight, this subtly beautiful western woodpecker, with some un-woodpeckerlike habits, may resemble a small crow. Lewis's Woodpecker often forms sizable flocks. It fly-catches from telephone poles and tall snags during spring and summer. While largely migratory throughout its range, this woodpecker is a year-round resident in parts of the Northwest.

LENGTH 10.25 inches. Iridescent greenish black head and dark red face; gray breast and collar; pinkish red lower breast and belly; iridescent greenish black back, wings, tail. **VOICE** Calls include a raspy, raucous, rattling *peeerrrr*. Drumming is a single tap followed by a series of fast taps. **BEHAVIORS** Commonly fly-catches from tall perches; caches pieces of acorns in bark crevices and other nooks in trees and wooden posts; eats fruits and berries during fall. Flight is level and crowlike, without the undulation characteristic of other woodpeckers. **HABITAT** Open oak woodlands; ponderosa pine forest edges and transition zone between ponderosa forest and shrub steppe habitats; riparian habitats. **STATUS** Uncommon summer resident, locally uncommon winter resident. **KEY SITES** Idaho: Hot Springs Island and Salmon area; Lucky Peak State Park. Oregon: Agate Lake and Emigrant Lake (Jackson County); Tygh Valley and Wamic. Washington: Oak Creek Wildlife Area (Yakima County); Wenas Campground and Umtanum Road; Balch Lake (near Lyle). British Columbia: Anarchist Mountain (Osoyoos); McKinney Road (Oliver).

Lewis's Woodpecker

Lewis's Woodpecker range

Acorn Woodpecker
(*Melanerpes formicivorus*)

A colony bird (and the inspiration for the Woody Woodpecker cartoon character), the clown-faced Acorn Woodpecker is famous for caching winter food stores by chiseling holes into wooden power poles and trees, and then hammering acorns into the cavities. These storage areas are called granaries, and are used year after year by the same woodpecker colony, resulting in trees and poles riddled with holes.

LENGTH 9 inches. **SEXES SIMILAR** Red crown, with white forehead that extends down to throat; female has a black margin between forehead and crown; white eye surrounded by black that extends to the nape and continues down the back; black around bill; black bib separates white underparts; white rump, white wing patch. **VOICE** Calls include a raucous, nasal *wakeup-wakeup-wakeup*, and a raspy, nasal *waa-waa-waa*. **BEHAVIORS** Forms communities that share in acorn caching, rearing young, and colony protection, as well as guarding colonial granaries, which may store hundreds of acorns, from other birds and mammals. Commonly catches flying insects during spring and summer. **HABITAT** Closely tied to oaks; often found in urban areas. **STATUS** Locally common southwest OR, uncommon northward. **KEY SITES** Oregon: TouVelle State Park and Denman Wildlife Area; Ashland Pond and Lithia Park (Ashland); Royal Avenue near Fern Ridge Reservoir.

Female Acorn Woodpecker

Acorn Woodpecker range

Male Acorn Woodpecker

Williamson's Sapsucker
(*Sphyrapicus thyroideus*)

Largest of the 4 sapsucker species, Williamson's Sapsuckers are so sexually dimorphic that for years after the striking bird was first described for science, the female was actually thought to be a different species of woodpecker entirely. Finally, in 1873, the case was solved when a nesting pair was observed in Colorado.

LENGTH 9 inches. **MALE** Black back, head, and breast; white rump and white wing patch (appears as white shoulder when bird is perched); white stripe behind the eye and white mustache stripe; red chin; yellow belly. **FEMALE** Ladderback appearance with black-and-white bars above, gray-black barring below; unlike all other sapsucker species, female lacks white wing patches; brown head, black breast band, yellowish belly, white rump. **VOICE** Calls include a harsh, screeching *keeaww-keeaww-keeaww*; a harsh, raspy *haaaa*. Drumming is a series of fast, almost rebounding-type taps, followed by slower repeated taps. **BEHAVIORS** Drills rows of holes (called wells) into trees to collect sap and insects that feed on sap; sometimes fly-catches. **HABITAT** Mixed conifer forest. **STATUS** Uncommon summer resident.

Female Williamson's Sapsucker

Williamson's Sapsucker range

Male Williamson's Sapsucker

Red-breasted Sapsucker
(*Sphyrapicus ruber*)

The only sapsucker of the far-western parts of the Northwest, this beautiful bird is fond of orchards, where it has been implicated in damaging fruit crops by girdling trees. It and the Red-naped Sapsucker were long considered a subspecies of the Yellow-bellied Sapsucker, but eventually both gained full species status. Where their ranges overlap, the Red-breasted and Red-naped Sapsuckers readily hybridize.

LENGTH 8.5 inches. Solid red head and breast, though red is variable, sometimes extending only down the throat and onto the upper breast, and in other birds extending well down the breast; white mustache stripe; pale yellow or whitish spots on black back; white patch on black wings appears as white shoulder when bird is perched; pale yellow belly; dusky white flanks with black streaks; white rump. **VOICE** Common calls include a descending medium-pitch *hey*, and a squealing *weep-weep-weep-weep*. Drumming is a series of fast taps followed by slower taps. **BEHAVIORS** Drills rows of holes called wells into trees to collect sap and insects that feed on sap; sometimes fly-catches. **HABITAT** Coniferous and mixed forests, including forest burns; deciduous woodlots and orchards. **STATUS** Common year-round resident.

Red-breasted Sapsucker range

Red-breasted Sapsucker

Red-breasted Sapsucker

Red-naped Sapsucker
(*Sphyrapicus nuchalis*)

A highly migratory sapsucker found from the eastern slope of the Cascades through the Rocky Mountains, the Red-naped Sapsucker was long considered a subspecies of the Yellow-bellied Sapsucker (see Similar species), but gained full species status in 1983. Where their ranges overlap, Red-naped and Red-breasted Sapsuckers readily hybridize.

LENGTH 8.5 inches. **SEXES SIMILAR** Red forehead, nape, and throat; female has less red on the throat and sometimes a white chin; white line extends from eye back to nape area; white mustache stripe; black bib on breast; black back spotted with white; yellowish tinged underparts; large white wing patch appears as white shoulder when bird is perched; white rump. **VOICE** Common calls include a nasal squealed *wee*, and a higher squealing *weep-weep-weep-weep*. Drumming is a series of fast taps followed by slower taps. **BEHAVIORS** Drills rows of holes (called wells) into trees to collect sap and insects that feed on sap; sometimes fly-catches. **HABITAT** Riparian habitats, especially aspen and conifer-deciduous mixed forests; coniferous forests, including forest burns. **STATUS** Fairly common summer resident (March–October); rare winter resident. Rare west of Cascades.

SIMILAR SPECIES Yellow-bellied Sapsucker (*Sphyrapicus varius*) is a rare vagrant in the Northwest (mostly late fall through early spring), and very similar to the Red-naped Sapsucker. The male's red throat is fully bordered by black, and he

Male Red-naped Sapsucker

Female Red-naped Sapsucker

Red-naped Sapsucker range

lacks the red patch on the nape as in the Red-naped (also usually present on the female Red-naped); the female lacks the red throat patch of the female Red-naped. In both sexes, the black-and-white back pattern is less precise and more speckled than that of the Red-naped, which is a series of black-and-white bars in 2 rows. Juveniles of the 2 species are very similar, but the Yellow-bellied is lighter overall, especially on the back, and has a speckled light brown crown rather than dark brown crown. Significantly, Yellow-bellied Sapsuckers retain juvenile plumage as late as March, while the Red-naped Sapsucker achieves adult plumage by late fall and early winter, so juveniles seen in the Northwest in later winter are almost certainly Yellow-bellied Sapsuckers.

Similar species: Yellow-bellied Sapsucker

Downy Woodpecker
(*Picoides pubescens*)

The diminutive Downy Woodpecker's small, stubby bill and black barring on the white outer tail feathers, coupled with its smaller size, help distinguish it from the otherwise nearly identical Hairy Woodpecker. North America's smallest woodpecker, this bird commonly visits suet feeders.

LENGTH 6.75 inches. **SEXES SIMILAR** Black-and-white facial pattern, male with a red spot on the back of the crown; short, stubby bill compared to longer bill of the Hairy Woodpecker; underparts white in the interior west and pale grayish west of the Cascades (but the difference can be subtle); wings spotted with white, less in the coastal form; white to brownish gray back bordering black rump; white outer tail feathers with black barring. **VOICE** Common call is a squeaky, sharp *pik* and a squeaky, whinnying rattle. Drumming is a series of fast taps. **BEHAVIORS** Probes for insects on trunks and branches of dead and living trees. **HABITAT** Deciduous forest, mixed forest, riparian areas; common in urban and suburban areas. **STATUS** Common year-round resident.

Female Downy Woodpecker

Downy Woodpecker range

Male Downy Woodpecker

331

Hairy Woodpecker
(*Picoides villosus*)

A common bird in nearly all forested areas, the Hairy Woodpecker is a regular visitor to backyard bird feeders in the Northwest. Named for the hairlike feathers in the middle of its back, it comes in 2 regional variations: the Rocky Mountain form, found in the interior Northwest, is bright white on the back and underparts, while the Pacific form, found west of the Cascades, is slightly duller, more smoky gray than white, and with less white on the wings.

LENGTH 9.25 inches. **SEXES SIMILAR** Black and white overall, with distinctive white back (shared only by Downy Woodpecker); black rump and wing coverts, white breast and belly are dingy white to smoky gray in Pacific form, bright white in interior form; black crown; bold white eyebrow stripe, black eye stripe wrapping around nape, and white mustache stripe that extends back to form white nape spots; male has bright red patch on rear of crown; long bill com-

pared to smaller Downy Woodpecker; wings spotted with white, less in the Pacific form; black tail with white outer feathers. **VOICE** Calls include a loud, sharp, squeaky *peek*; a squeaky, whinnying rattle. Drumming is a series of fast taps. **BEHAVIORS** Probes for insects on trunks and branches of dead and living trees. **HABITAT** Most forest and woodland habitats, but rare in juniper woodlots; common in forest burns. **STATUS** Common year-round resident.

Male Hairy Woodpecker

Female Hairy Woodpecker

Hairy Woodpecker range

White-headed Woodpecker
(*Picoides albolarvatus*)

One of the most striking of the wood-peckers and an iconic must-see bird of the Northwest, the aptly named White-headed Woodpecker thrives in ponderosa pine forests. It sometimes picks a stump or fallen tree in which to excavate a nest cavity just a few feet off the ground. During winter and spring (generally in search of a mate or during years of poor pine cone crops) these wood-peckers may venture away from the pon-derosa forest and visit other forest-type habitats, including urban areas.

LENGTH 9 inches. **SEXES SIMILAR** White head contrasts with black body; black tail; males have red patch on the nape; white wing patches visible in flight. **VOICE** Calls include a sharp *pee-dink* or *pee-dee-ink*; rat-tle is an extended *pee-dink*. Drumming is 15 to 30 beats, usually on a hollow tree or limb. **BEHAVIORS** Heavily dependent on ponderosa pine seeds; also eats other seeds, insects, and sap. Often forages low on trees in search of insects and will chisel bark away to find them; quietly visits water seeps

and springs. **HABITAT** Ponderosa pine for-ests or mixed coniferous forests dominated by ponderosa. Prefers open areas in these forests. **STATUS** Fairly common year-round resident. **KEY SITES** Oregon: Camp Sher-man area; Idlewild Campground (Harney County). Idaho: Idaho City. Washington: Leavenworth Fish Hatchery; Little Pend Oreille NWR. British Columbia: Oliver area, especially along McKinney Road.

Female White-headed Woodpecker

Male White-headed Woodpecker

White-headed
Woodpecker range

333

Black-backed Woodpecker
(*Picoides arcticus*)

Often hard to find, the aptly named Black-backed Woodpecker prefers coniferous forests impacted by fire, where burnt-out or scorched trees still stand, as well as trees infested with wood-boring beetles. The birds feed by probing loose bark, peeling and knocking off outer layers. Substantial patches of bare tree trunks in dead forests are indicators of the possible presence of Black-backed (and other) woodpeckers.

LENGTH 9.5 inches. **SEXES SIMILAR** All black above, white below; black face with white mustache stripe; males have solid yellow crown patch, females have black crown; flanks and flight feathers are black with fine white barring; 3 toes instead of 4,

like all other woodpeckers except American Three-toed Woodpecker. **VOICE** Calls include a sharp, rubbery, descending *che-che-che-cherrrr*, and a chirpy *pet-pet-pet*. Drumming is one series of taps. **BEHAVIORS** Probes burnt trees for wood-boring beetle larvae and other insects; less commonly, live trees with wood-boring beetle infestation. **HABITAT** Conifer forest. **STATUS** Uncommon year-round resident.

Female Black-backed Woodpecker

Black-backed Woodpecker range

Male Black-backed Woodpecker

American Three-toed Woodpecker (*Picoides dorsalis*)

The quiet and seldom-seen American Three-toed Woodpecker, like the closely related Black-backed Woodpecker, is a forest-fire and beetle-killed-forest specialist. It feeds on insects by searching under the bark of standing burnt and beetle-killed conifers, often moving into burns almost before the smoke has settled. Spring drumming and begging calls of the young can reveal the presence of this species.

LENGTH 8.75 inches. **SEXES SIMILAR** Black and white overall, with white-edged black tail, black wings with fine white speckles, and black rump; back ranges from mostly white in the Rockies to barred black and white farther west; male has a yellow cap bordered by white-and-black flecking; black face with narrow white stripe extending back from eye, and white mustache; white throat and breast, barred flanks and wings. **VOICE** Calls include a squeaklike *pik* and a squeaking, rattled *pikpikpikpikpik*. Drumming a series of accelerating taps. **BEHAVIORS** Beetle infestations seem to be more important for feeding than forest fire alone, although burns provide nesting opportunities. Works its way up tree trunks, prying and flaking off bark in search of beetles and beetle larvae. **HABITAT** Diseased or burned conifer forests. **STATUS** Rare year-round resident. **KEY SITES** Fresh forest burns throughout its range.

Female American Three-toed Woodpecker

American Three-toed Woodpecker range

Male American Three-toed Woodpecker

Northern Flicker
(*Colaptes auratus*)

The most common woodpecker in the Northwest, the Northern Flicker was, until 1995, considered 2 species, the Red-shafted Flicker of the West and Yellow-shafted Flicker of the East. The Red-shafted subspecies abounds in the Northwest, but a few Yellow-shafted Flickers occur in the region in fall and winter. Where their ranges overlap east of the Rocky Mountains, the 2 subspecies readily hybridize and hybrids occasionally show up in the Northwest.

LENGTH 12.5 inches. **SEXES SIMILAR** Except female lacks moustache stripe. **RED-SHAFTED SUBSPECIES** Brown above, with black bars; black breast band and black-spotted belly; bright coppery orange wing and tail linings; prominent white rump; brown forehead and nape; gray face, male with red moustache. **YELLOW-SHAFTED SUBSPECIES** Yellow wing and tail linings; gray forehead and crown with a red crescent on nape; brown face, male with black moustache. **VOICE** Typical calls include a loud, nasal, fast *wika-wika-wika-wika* or *wik-wik-wik-wik*; a loud, bold, descending, squeaky *ewwww*; a high, resonant *chee-ear*. Drumming is a series of rapid, even taps **BEHAVIORS** Often forages on the ground for ants and other insects; eats fruits and berries during cold months and frequents feeders. **HABITAT** Coniferous, deciduous, and mixed woods and woodlots, from urban and suburban areas to remote mountains forests; avoids densely forested areas. **STATUS** Common year-round resident.

Male Northern Flicker

Northern Flicker range

Female Northern Flicker

336

Pileated Woodpecker
(Dryocopus pileatus)

Most at home in mature forests, the striking, crow-sized Pileated Woodpecker is the largest woodpecker in the Northwest. It needs large trees for nesting and roosting holes, and to find its favorite food, large carpenter ants. Its heavy drumming and loud, exotic sounding calls are audible over considerable distances.

LENGTH 16.5 inches. Mostly black, with large red crest; prominent white neck stripes angling across face to base of bill; white eyebrow stripes; white chin; long black tail; in flight, shows white wing linings and small white wing stripe. **MALE** Red crest, crown, and forehead; red mustache stripe. **FEMALE** Red crest and crown, blackish forehead. **VOICE** Variety of calls; commonly a Flickerlike, steady pitched *hut-hut-hut-hut-hut*; a fast, higher-pitched *wik-wik-wik-wik-wik*; a rising then dropping series of *wuk notes*; flight call is a rapid series of *wuk-wuk* calls. **BEHAVIORS** Often forages on dead or diseased trees, including fallen timber; nesting and roosting holes, excavated in dead wood, are vertically oblong. **HABITAT** Mature and old-growth coniferous and mixed forests with large dead trees and snags; urban and suburban areas with mature woodlands. **STATUS** Uncommon year-round resident.

Female Pileated Woodpecker

Pileated Woodpecker range

Male Pileated Woodpecker

Flycatchers

Ranging from the colorful kingbirds to the handsome phoebes to the lookalike empids, flycatchers (family Tyrannidae) live up to their names by feeding almost exclusively on insects, typically catching them on the wing, and occasionally gleaning them from foliage. Most flycatchers in the Northwest are summer (breeding) residents, migrating south for the winter, and 7 species comprising the genus *Empidonax* are so similar in appearance that in many cases, appearance alone is not always reliable in separating them, forcing birders to rely on songs and calls, habitat and range, and behavior to identify the different types.

Ash-throated Flycatcher

Olive-sided Flycatcher
(*Contopus cooperi*)

A common bird of Pacific Northwest forests often detected by its distinctive song, the Olive-sided Flycatcher perches in the tops of tall trees or on bare snags and hawks for flying insects. Its lengthy spring migration to the region stretches from early May into the first week of June. Population declines throughout its range may be linked to habitat loss in its South American wintering grounds.

Olive-sided Flycatcher

LENGTH 7.5 inches. Large, dark, slightly crested head; lightly streaked olive-gray flanks contrast strongly with white breast and belly, giving a vestlike appearance; brownish olive above; dark, stout bill; white throat; pale wingbars on dark, pointed wings are not always noticeable. **VOICE** Distinctive song is an enthusiastic, whistled *quick-three-BEERS*; alarm and aggression call a sharp *pit-pit-pit*. **BEHAVIORS** Perches high on top of dead branches or tallest trees; all insects it takes are winged, does not glean like other flycatchers; very aggressive to intruders of nesting area. Strong, direct flight; acrobatic when pursuing insects. **HABITAT** Conifer forests from sea level to timberline. During migration can be found in virtually any habitat. **STATUS** Common summer resident (May–mid-September).

Olive-sided Flycatcher

Olive-sided Flycatcher range

Western Wood-Pewee
(*Contopus sordidulus*)

A common breeding flycatcher in the Northwest, the Western Wood-Pewee picks a favorite perch from which to make sudden swift flights to snatch flying insects, then returns to the same perch each time. Larger than all the *Empidonax* flycatchers, it is considerably easier to identify by sight, as well as by its distinctive call, often heard before the bird is seen. Declining numbers have been attributed to habitat loss in the bird's South American wintering range.

LENGTH 6 inches. Pale gray throat and underparts become darker on the flanks and belly, giving a vested appearance; brownish gray above, with 2 pale wingbars, the uppermost being indistinct; peaked crest; no eye ring; long, mostly dark bill; long wings indicated by long primary projection; long tail; short legs. **VOICE** Commonly a slurred, nasal, down-pitching *beerer*; also a whistled, raspy, rising *beewee*; a variety of sharp, whistled *whit* calls that can indicate alarm or aggression. Calls throughout the day, and at night if disturbed. **BEHAVIORS** Chooses a high perch among trees to hawk insects from midair, using the perch repeatedly. Does not flick tail as do *Empidonax* flycatchers, but may quiver wings after returning to perch. **HABITAT** Open, mature woodlands and conifer forests, especially edges, recently logged areas, forest burns, wooded parks; also clusters of trees near waterways or human habitation. **STATUS** Common summer resident (mid-May–mid-September).

Western Wood-Pewee

Western Wood-Pewee range

Dusky Flycatcher
(*Empidonax oberholseri*)

Common in the interior Northwest, the Dusky Flycatcher is often the most visible bird in aspen-filled drainages and open conifer forests with a brushy understory. It is easily confused with Hammond's and Gray Flycatchers, but the totality of habits, physical characteristics, and song can aid in identification.

LENGTH 5.75 inches. Rounded head; intermediate-length bill mostly dark and narrow; narrow white eye ring and pale lores; pale throat and pale belly, sometimes with an olive or yellowish wash; upperparts gray to grayish brown, sometimes slightly olive-brown; narrow whitish wingbars; short primary projection. **VOICE** A cheery, whistled *whidit chree whidit whee* with the last note higher in pitch; also a *whee-hick* with a rising emphasis on the last part of the call; and a single sharp, short *whit*. **BEHAVIORS** Flicks tail in an upward motion, unlike the Gray Flycatcher, which pumps its tail down. Watches for prey from lower branches of trees and shrubbery, then darts out to catch flying insects, often hovering, and will also fly to the ground to pounce on insects. **HABITAT** Open dry conifer forest with shrub understory, aspen-filled drainages, clear-cuts, forest burns. During migration, occurs in a variety of habitats. **STATUS** Common summer resident (late April–early September).

Dusky Flycatcher

Dusky Flycatcher range

Hammond's Flycatcher
(*Empidonax hammondii*)

A hardy flycatcher of mature conifer forests and mixed woodlands, the Hammond's Flycatcher typically hunts insects from the forest canopy, especially early in the season. As the summer progresses, it will forage lower and also pounce on insects on the ground. The Hammond's Flycatcher molts earlier in the fall than other empids, and tends to be more colorful, with a marked contrast between gray head, greenish back, and yellowish underside. Still, its similarity to other *Empidonax* flycatchers makes detailed observation of plumage, voice, habits, and habitat critical to identification.

LENGTH 5.5 inches. Slightly crested grayish head; tiny, narrow, dark-colored bill; white eye ring that appears bolder behind the eye; throat, upper breast, and flanks are grayish to dark gray, creating a vestlike appearance in some birds; belly and undertail coverts range from white to slightly yellowish; upperparts darker gray with olive tones; narrow whitish wingbars; longer primary projection (the extent to which the tips of the primary flight feathers extend when the bird is at rest with wings folded) than very similar Dusky Flycatcher. **VOICE** Three-part song, any part of which may be sung singly, in 2 parts, or in any order: a crisp, sweet, 2-syllable *seenit*; a buzzy, lower-pitched *zeeit*; and a sharp monotone *zeet*. Call notes include a sharp, whistled *whit* often followed by a whistled *whit-wee* with the *wee* dropping in pitch; alarm calls are a short, sharp *whit* and trilled *zeets*. **BEHAVIORS** Flicks tail upward. Often darts for insects from bare branches high on trunks of mature trees below the canopy. Weak, shallow wingbeats in flight; acrobatic; often hovers near foliage tips to glean insects. **HABITAT** Mature to old-growth conifer forests and forest edges; mature mixed montane woodlands and aspen groves. **STATUS** Common summer resident (mid-April–September).

Hammond's Flycatcher

Hammond's Flycatcher

Hammond's Flycatcher range

Willow Flycatcher
(*Empidonax traillii*)

Common and widespread in the Northwest from late spring through summer,
the Willow Flycatcher—considered conspecific with the eastern Alder Flycatcher
under the name Traill's Flycatcher until
1973, when the two were granted species
status—is somewhat easier to identify
than the other *Empidonax* species. It has
a very faint, almost indiscernible eye ring
instead of the pronounced white eye ring
of the other members of the genus.

LENGTH 5.75 inches. Lack of eye ring and
long, wide bill differentiate Willow from
other *Empidonax* flycatchers; flat forehead
sloping to noticeably peaked crown; white
throat bordered by pale, olive-brown breast;
white belly, sometimes with faint yellowish
wash; back mostly brownish gray, sometimes with a hint of olive; dingy white to
buff wingbars; primary flight feathers do
not extend past the undertail coverts. **VOICE**
Song is a warbled *fitzabew*, or a sharp,
strong, burry *fitzabew*; call is a sharp, whistled *whit* and a chirpy *waa*. **BEHAVIORS**
Often sings from a high perch on smaller
willows and shrubbery. Forages for insects
low in trees and shrubs. Unlike the similar
Western Wood Pewee, does not often return
to the same perch when feeding. Erratic
flight, often following the darting flight of
insects it is preying on. **HABITAT** Brushy
riparian areas with small trees, especially willows; open deciduous woodlands,
especially bottomlands; shrubby clear-cut
edges near springs, seeps, and streams.
STATUS Fairly common summer resident
(mid-May–mid-September).

Willow Flycatcher

Willow Flycatcher

Willow Flycatcher range

Least Flycatcher
(*Empidonax minimus*)

The aptly named Least Flycatcher is the smallest member of its tribe in the Northwest, though relative size among the Empidonax flycatchers, or empids, is essentially useless for identification. Common throughout the eastern half of the continent, it is rare in the Northwest. Unique among empids, it sometimes nests in loose colonies. Its distinctive call is the best identifying feature.

LENGTH 5.25 inches. Short bill gives appearance of large head that appears rounded with a slight peak at crown; complete white eye ring; brownish olive to gray above; white throat; dusky gray breast and flanks; belly can have a yellowish wash; 2 dusky white to yellowish wingbars; short, rounded wings with short primary projection; short, slightly forked tail. **VOICE** Both sexes sing; song is a sharp, explosive, 2-part *chee-bek*, with emphasis on both parts. Calls and alarm calls include a sharp, staccato *whit-whit-whit*,

a single sharp *whit*, and a short *burr*. **BEHAVIORS** Feeds in the uppermost part of shrubs, and in forest canopy; may land on ground to bathe, but rarely to feed. Weak, direct flight, with rapid wingbeats; insect-catching flights from perches can be acrobatic. Most easily detected by song. **HABITAT** Deciduous forest with a shrub understory, especially riparian strips. During migration occurs in a variety of habitats, but prefers deciduous trees. **STATUS** Rare breeder and migrant (May–mid-August) **KEY SITES** Idaho: Kootenai NWR. Washington: Okanogan County. British Columbia: Okanagan Valley; Creston Valley Wildlife Centre.

Least Flycatcher

Least Flycatcher range

Least Flycatcher

Cordilleran Flycatcher
(*Empidonax occidentalis*)

This flycatcher of the Rocky Mountain west was formerly considered conspecific with the Pacific-slope Flycatcher; together, they were the Western Flycatcher. The 2 birds are identical in appearance, making breeding range the best identifier. However, to make matters worse, their ranges appear to overlap to some extent, and in these areas, identification requires great care and may be largely impossible. Only a minor difference in their songs makes identification in the field possible, though even this difference is marginal and uncertain, and best summarized by the cautionary statement of birdsong expert Nathan Pieplow: "...decent recordings would be essential to document any occurrence of either species outside its normal breeding range."

Cordilleran Flycatcher

Cordilleran Flycatcher range

LENGTH 5.5 inches. Peaked, slightly ragged crest; wide, medium-length bill; oblong whitish eye ring often broken on top of the eye and extends behind eye to form a point; pale brownish olive below; olive-brown above; whitish wingbars; long tail and short primary projection. **VOICE** Song is a cheery, sweet, broken *see-pwee-pitit*; variations of these given as calls; also a high-pitched, questioning *seee*. **BEHAVIORS** Forages from low, exposed branches; gleans insects from leaves and the ground; prefers to forage in shaded areas. Weak flight with shallow wingbeats; acrobatic when pursuing insects. **HABITAT** Cooler, dry coniferous forest with brushy waterways that open the forest canopy. **STATUS** Common summer resident (mid-April–mid-September) of Rocky Mountains, rare in other suspected breeding areas.

Pacific-slope Flycatcher
(*Empidonax difficilis*)

Formerly considered conspecific with the slightly larger but otherwise identical Cordilleran Flycatcher under the name Western Flycatcher, the Pacific-slope Flycatcher is a denizen of moist Pacific slope forests. Range is the only sure way to separate the 2 species, a task that is very nearly impossible in the areas where their ranges appear to overlap. A very minor difference in songs makes identification in the field a slight possibility, although even this difference is marginal and uncertain.

LENGTH 5.5 inches. Slightly ragged, peaked crest; wide, medium-length bill; oblong whitish eye ring (often broken on top of the eye) extends behind eye to form a point; pale brownish olive below; olive-brown above; whitish wingbars; long tail and short primary projection. **VOICE** Song a cheery, sweet, broken *see-pwehee-pitit*; variations of these notes given as calls; also a high-pitched, questioning *seee*. **BEHAVIORS** Forages from low, exposed branches; gleans insects from leaves and limbs, and from the ground and low understory; prefers to forage in shade. Weak flight with shallow wingbeats; acrobatic when pursuing insects. **HABITAT** Moist, dense, coniferous and mixed forests below 4000 feet. **STATUS** Abundant summer resident (late April–mid-September) in coastal forests and mountains; common in the rest of its range.

Pacific-slope Flycatcher

Pacific-slope Flycatcher

Pacific-slope Flycatcher range

Gray Flycatcher
(*Empidonax wrightii*)

Gray Flycatcher

An inhabitant of arid regions of the Pacific Northwest, this plain gray fly-catcher pumps its tail downward—a key to its identification. Especially common in mixed sagebrush and juniper steppe and sagebrush steppe, as well as other desert scrublands, the Gray Flycatcher typically hunts insects from the low perches prevalent in such habitats and even lands on the ground to catch prey.

LENGTH 6 inches. Pale gray overall, with lighter gray throat, breast, belly, and under-tail coverts; rounded head; long, narrow bill with dark upper mandible and pink to yellowish lower mandible with a dark tip; narrow whitish eye ring; dark wings with 2 narrow whitish wingbars; long tail with noticeable outer white edges; primary flight feathers do not extend beyond the under-tail coverts. **VOICE** Two-part song in various combinations: a raspy, forceful, 2-syllable *chiwit*, and a higher pitched, imploring *chee*. Call note is a sharp, whistled *whit*, similar to other species of *Empidonax*. **BEHAVIORS** Downward tail pump is a key to identifica-tion (although other empids rarely occur in open desert scrub, where Grays are often found). Highly territorial, and aggres-sively defends nest and territory from other flycatchers. Males sing from prominent perches in the tops of sagebrush and other shrubs. Acrobatic in the air, with shallow wingbeats and darting, short flights. **HAB-ITAT** Arid regions of sagebrush, saltbush, juniper, mountain mahogany, and grassy pine forest. **STATUS** Common summer resi-dent (mid-April–late September).

Gray Flycatcher

Gray Flycatcher range

Black Phoebe (*Sayornis nigricans*)

Historically a Southwest species, the handsome and energetic Black Phoebe has been steadily expanding its range northward into Oregon and Washington, and not only do they now breed in western Oregon and possibly western Washington, but many individuals remain throughout the winter, unlike other flycatchers. Availability of nesting sites that human habitation provides seems to be a key to their range expansion, and these pioneering flycatchers may even raise 2 broods per year.

LENGTH 7 inches. Distinctive black-and-white pattern; slate-black overall, with a white belly, undertail coverts, and outer tail feathers; long tail. **VOICE** Song is a whistled, descending *pee-hee*; call is a single *peet*, repeated often. **BEHAVIORS** Hawks insects from low perches, often over water, and lands on the ground to graze on insects and spiders; may eat small fish from the water's surface; dips and fans tail. Strong, direct flight; acrobatic when pursuing insects; hovers while gleaning insects or distracting intruders in nesting area; display flight of male is a vertical zigzag and will also show female potential nesting sites by hovering near them. **HABITAT**

Natural and manmade riparian areas, with open areas and availability of mud from which to build nest; benefits from nesting sites created by human habitation, such as buildings and bridges, much like the Barn Swallow. **STATUS** Uncommon to locally common year-round resident. **KEY SITES** Oregon: Coquille Valley (Coos County); Touvelle State Park and Ken Denman Wildlife Area; Kirtland Road sewage ponds; North Mountain Park and Ashland Pond; Emigrant Lake; Bandon Marsh; Talking Waters Gardens (Albany).

Black Phoebe

Black Phoebe range

Black Phoebe

Say's Phoebe (*Sayornis saya*)

This hardy flycatcher of arid regions arrives in the Northwest in late winter and begins nesting as early as late March. Relying on insects as a food source in freezing weather (and perhaps with snow-covered ground) would seem tenuous, but the Say's Phoebe has perfected the strategy. When the weather is at its winter worst, Say's Phoebes raid old spider webs, especially on buildings, presumably for freeze-dried insects they may contain. A few individuals even forgo migration and stay year-round. During the hot, dry summers, the Say's Phoebe gets the moisture it needs from the insects it eats.

LENGTH 7.5 inches. Grayish brown overall, with a warm cinnamon belly and undertail coverts; pale gray wingbars (sometimes tannish on juveniles); slightly peaked head and dark, stout bill; long, dark tail. **VOICE** Song is a melancholic *pe-wheeew*, with the last part of the song a long, sliding change of pitch; also a rattling, rising, *breet*. **BEHAVIORS** Frequently hunts and fly-catches on or near the ground, and often perches on fences and other low objects in open areas. Strong, direct flight; acrobatic when pursuing insects; hovers, often just above the ground when searching for insects, and also when investigating nooks for a potential meal. **HABITAT** Sagebrush steppe; dry, rocky canyons and rimrock; especially fond of low, remote buildings and corrals with appropriate overhangs to build nest. **STATUS** Common spring and summer resident east of the Cascades, rare but regular west of the Cascades.

Say's Phoebe

Say's Phoebe

Say's Phoebe range

Ash-throated Flycatcher
(*Myiarchus cinerascens*)

Unlike most other flycatchers, the stately Ash-throated Flycatcher prefers to hunt the foliage and glean insects directly from trees, shrubs, and even from the ground. It often perches quietly on overhead lines and in trees before actively searching for insects by hovering near foliage and above the ground. It is a secondary cavity nester, often choosing abandoned woodpecker holes, and readily uses nest boxes.

LENGTH 8.5 inches. Crested gray head; thick, stout black bill; light gray throat and breast; pale yellow belly and undertail coverts; gray above, with pale wingbars and, when wings are folded, a rufous-colored patch on the primaries; rufous-colored tail may have a thin, dark terminal band. **VOICE** Song is a musical, slurred *kabrrr kabrik*; calls include a slurred *breek*, a sharp *pik*, and individual parts of the song. **BEHAVIORS** Makes short foraging flights; perches quietly in an upright posture; acrobatic when defending territory and in courtship flights; often hovers when gleaning insects from foliage; hops when on the ground. **HABITAT** Open oak woodlands and open juniper woodlands; can utilize a variety of other habitats that provide suitable cavities for nesting; generally found in arid areas; prefers slopes. **STATUS** Uncommon summer resident (late April–mid-August). **KEY SITES** Idaho: Big Cottonwood Wildlife Area (Oakley); Stone Hills (Black Pine Road, Oneida County). Oregon: Lower and Upper Table Rocks (Jackson County); Hatfield Lakes (Bend). Washington: Fort Simcoe State Park; Catherine Creek and Balch Lake (Lyle area).

Ash-throated Flycatcher range

Ash-throated Flycatcher

Western Kingbird
(*Tyrannus verticalis*)

This large, handsome flycatcher is a familiar summer resident throughout much of the interior Northwest, often seen perching on exposed branches, utility wires, and fences. Territorial and protective of their nesting area, Western Kingbirds will dive-bomb humans who get too close.

LENGTH 8.5 inches. **SEXES SIMILAR** Pale gray above, with darker wings; bright yellow below; white throat; dark tail with white outer feathers; slightly peaked gray head; stout black bill; dark mask from base of bill to past the eye; male has a reddish tuft of feathers on the crown that is usually concealed. **VOICE** Dawn song is a sharp *kip-kip-kip-keeweeee-kee-kee* with the *kee* series rising then dropping in pitch; call is a series of rapidly delivered staccato *kip* notes; also a single *kip*. **BEHAVIORS** Takes most of its insects on the wing with occasional forays to the ground. Very aggressive to intruders in nesting area, whether the threat is from above or below. Strong, acrobatic flight when pursuing insects or chasing predators such as hawks, jays, and magpies from its nesting area. **HABITAT** Open, grassy areas with suitable perches, including hay fields, farmlands, roadsides, and large parks and other open areas with manmade structures. **STATUS** Common summer resident (mid-April–early September); rare to uncommon summer resident west of the Cascades.

Western Kingbird range

Western Kingbird

Eastern Kingbird
(*Tyrannus tyrannus*)

This striking flycatcher is a summer resident of drier environs in the eastern half of the Northwest. The Eastern Kingbird migrates over the Rocky Mountains before turning south to its wintering grounds in South America, reversing the course in the spring. Because of its nonstandard migration route, it is a late arrival in the region, and migratory birds usually form small flocks—unique among flycatchers. Eastern Kingbirds readily adapt to manmade structures for perches and nesting sites.

LENGTH 8.5 inches. Black above and white below; distinctive white terminal band on the dark tail; red or orange feathers on crown are concealed unless the bird is alarmed or irritated. Sits upright and male may raise crest slightly. **VOICE** Dawn song is a rapid, high-pitched, buzzy *ke-ke-ke-ke kekwee-wee*; call is a buzzy slurred *bazeer* or *zeer*. **BEHAVIORS** Forages from a high perch, often near habitat edges; forages near ground in cool weather before insects begin flying. Very aggressive to other birds intruding into nesting area or near young. Strong, direct flight; acrobatic when pursuing insects or chasing intruders; greeting flights and display flights on fluttering wingbeats, causing a slow approach to a perch. **HABITAT** In breeding range, generally tied to water. Nests in riparian deciduous groves, especially in canyons and gorges. **STATUS** Uncommon summer resident interior; rare west of the Cascades.

Adult Eastern Kingbird

Eastern Kingbird range

First-fall juvenile Eastern Kingbird

Tropical Kingbird
(*Tyrannus melancholicus*)

Primarily found in the deep Southwest and south through Mexico, Central America, and South America, Tropical Kingbirds—primarily juveniles—make lengthy post-breeding-season dispersals and regularly show up in very small numbers along the Pacific Northwest coast from October into December. They are now regular visitors in Oregon and Washington.

LENGTH 9.25 inches. Pale ashy gray throat, pale olive-green breast, bright yellow belly and undertail coverts; olive-gray mantle, brownish wings; large gray head; stout black bill; brown tail lacks the white outer feathers of Western Kingbird. **VOICE** Calls include a fast, buzzy *tee-ee-ee-ee-ee*; often silent. **BEHAVIORS** Sits on fence posts, fence wires, and overhead utility lines, making flights to pursue insects. **HABITAT** Open or fairly open coastal areas, including pastures, fields, estuaries, towns; very rare inland. **STATUS** Rare fall visitor along coast (October–December); vagrant elsewhere. **KEY SITES** Oregon: Tillamook Bay area; Mark O. Hatfield Marine Science Center (Newport). Washington: Neah Bay; Ocean Shores.

Tropical Kingbird range

Tropical Kingbird

353

Shrikes and Vireos

Shrikes (family Laniidae) are unique songbirds that hunt prey as large as rodents, lizards, and smaller songbirds. Two species occur in the Northwest, with the Northern Shrike a winter visitor and the Loggerhead Shrike a summer resident. These striking birds are known for their habit of impaling prey on thorns and barbed-wire fences.

Vireos (family Vireonidae) are musically inclined, fairly drab-colored songbirds that specialize in gleaning insects from the foliage of trees and shrubs. Five species of vireo breed in the Northwest; one species, the Hutton's Vireo, is a year-round resident whereas the others migrate south for the winter.

Loggerhead Shrike

Loggerhead Shrike
(*Lanius ludovicianus*)

A predatory songbird of sagebrush steppe and other desert scrub and grassland habitats in the interior Northwest, the Loggerhead Shrike, though mostly insectivorous, hunts prey as large as lizards and small mammals. Because its feet are not strong enough to hold prey like a raptor, the bird impales its prey on sharp objects—commonly thorns or barbed wire—to assist in eating.

LENGTH 9 inches. Gray above, with broad black face mask; white throat and light gray breast and belly; black wings with white patches, especially flashy in flight; black tail with white outer tail feathers; short, stout bill; flashy black and white in flight. **VOICE** Song includes a variety of notes, typically a high-pitched *purt-a-eee*, or a lower-pitched, guttural *word-it*, repeated often; calls include a loud, harsh, scolding, jaylike *reeee*. **BEHAVIORS** Hunts from a high perch, or sometimes by hovering; impales prey on thorns or barbed wire, or wedges it into the crook of a tree or bush. Swift and direct flight on fast wingbeats, somewhat undulating on longer flights; typically flies low to the ground and then swoops quickly upward to perch on a shrub. **HABITAT** Sagebrush steppe and desert scrub; pastures and other shortgrass habitats with high perches to hunt from, including roadside utility wires. **STATUS** Uncommon to locally common summer resident; rare year-round resident. **KEY SITES** Idaho: Snake River Birds of Prey National Conservation Area; Minidoka area; Stone Hills and southern Oneida County. Oregon: Malheur NWR; Hart Mountain National Antelope Refuge; SR 140, Adel to Nevada state line. Washington: Toppenish NWR; Quilomene Wildlife Area and Old Vantage Road.

Loggerhead Shrike in flight

Loggerhead Shrike range

Loggerhead Shrike

Northern Shrike (*Lanius excubitor*)

A winter visitor to the Northwest, the Northern Shrike is larger than the very similar Loggerhead Shrike (primarily a summer resident) and prefers a diet of small mammals and birds up to its own size. Populations vary from year to year depending on the availability of prey; both adults and brownish colored juveniles occur in the Northwest, spreading southward from their breeding range in northern Canada and Alaska.

LENGTH 10 inches. Pale gray above, white below, with a distinct black mask (much narrower than on the Loggerhead Shrike); large bill with a hooked tip; black wings with white patches; long, black tail with white outer feathers; appears flashy black and white in flight. **JUVENILE** Light brown to brownish gray overall; mask may be more brown than black; scaled appearance on underparts; dark wings may not show much white in the wing patches. **VOICE** Song is a musical *kee-didi* and a chirped *tur-tur*; calls include a nasal, nuthatch-like *peent-peent-peent*. **BEHAVIORS** Impales prey (small mammals and birds) on thorns or barbed wire, or wedges them into the crook of a tree or bush; often hovers while hunting. **HABITAT** Open habitats with suitable perches for hunting, including pastures, open parks, estuaries, and sagebrush steppe. **STATUS** Uncommon but geographically widespread winter visitor.

Adult Northern Shrike

Northern Shrike range

Juvenile Northern Shrike in flight

Hutton's Vireo *(Vireo huttoni)*

Unique among North American vireos because it is largely nonmigratory, the Hutton's Vireo presents a special identification challenge to Northwest birders because of its strong resemblance to the similar-sized Ruby-crowned Kinglet. The kinglet's secondary flight feathers are black at their bases, forming a rear black border to the lower white wingbar, but the Hutton's Vireo's secondary flight feathers are uniform olive gray. This is generally the easiest field mark to verify species if the identity is in question.

LENGTH 4.25 inches. Grayish overall, usually with a pale olive cast; prominent white oblong eye rings and pale lores; 2 white wingbars, with no black bar bordering rear edge of the lower white wingbar as in the Ruby-crowned Kinglet; solid blue-gray legs and feet (Ruby-crowned Kinglet has light gray legs and yellow feet); fairly thick bill with tiny hook at tip (Ruby-crowned Kinglet has a small, thin bill). **VOICE** Song is a series of ringing, slightly scratchy whistles, *cheweee cheweee cheweee*; sometimes a more hurried, scratchy *cheerep*; call is a high, raspy, buzzing *ziraaaa aack aack*.

BEHAVIORS Generally inconspicuous and easily overlooked; in nonbreeding season, often forages in loose mixed-species flocks, including chickadees and kinglets. **HABITAT** Mixed woodlands, especially where oak is a primary species, but also mixed communities of conifers and other deciduous species, such as maple, ash, madrone, and tall-growing shrub species; also overgrown abandoned orchards, parks, wooded neighborhoods. **STATUS** Fairly common year-round resident.

Hutton's Vireo

Hutton's Vireo range

Hutton's Vireo

Warbling Vireo (*Vireo gilvus*)

One of the great summer songsters of Northwest deciduous woodlands, the aptly named Warbling Vireo, despite its ebullient singing, tends to be inconspicuous, serenading from hidden perches high in the trees, where foliage and the play of light and shadow make it difficult to see. But in prime habitat—especially riverside cottonwood stands and similar environs—an early morning stroll in May or June is sure to be rewarded by indefatigable song from several if not numerous Warbling Vireos.

LENGTH 5 inches. Pale and drab overall; light grayish olive above, darkest on crown; whitish below, with flanks and sides ranging from nearly white to pale yellow; pale lores and eyebrow stripe; dark eyes. **VOICE** Song is a rich, rousing, musical series of trills and whistles, ending with an upward note; calls include a harsh, dry, agitated, buzzy *screeeah* or *peeyear*, often mixed with brief, sharp *chit* chatter notes. **BEHAVIORS** Forages slowly and inconspicuously in tree foliage, gleaning insects. **HABITAT** Breeds in mature deciduous stands, especially near water, including cottonwoods and aspens, as well as poplars and other nonnative ornamentals. In migration, utilizes a wider variety of woodland habitats. **STATUS** Common summer resident (April–September) and migrant west of the Cascades; fairly common to uncommon summer resident and fairly common migrant east of the Cascades.

Warbling Vireo

Warbling Vireo

Warbling Vireo range

Red-eyed Vireo *(Vireo olivaceus)*

Somewhat enigmatic because its breeding range in the southern half of the Pacific Northwest remains difficult to pin down, the Red-eyed Vireo has a seemingly odd distribution. Common in eastern North America, its primary breeding range shifts northward in the West. An incessant songster, the male Red-eyed Vireo may sing at any time of day and often more or less all day, but the constant vocalization does not make him easy to find in the upper levels of foliage he blends with.

LENGTH 5 inches. Grayish olive-green above, darkest on wings and tail; gray crown and distinct light supercilium and dark eye line; white below, sometimes with yellowish wash to sides; red irises (usually) visible at close range in good light. **VOICE** Song is a series of musical, robinlike warbles and whistles, often nearly nonstop for substantial lengths of time; calls include a high, nasal *cheeeah* as well as various dry chatter notes. **BEHAVIORS** Tends to forage high in trees, but often on outer ends of branches, gleaning insects. **HABITAT** Riparian deciduous woodlands; deciduous groves in parks and neighborhoods. **STATUS** Summer resident and migrant (May–September); fairly common to northern Idaho, rare in southern Idaho; rare to uncommon in Oregon; fairly common in Washington (except rare Columbia Plateau and southern Washington); common in interior British Columbia, uncommon in coastal British Columbia. **KEY SITES** Idaho: Kootenai NWR; Mica Bay, Coeur d'Alene Lake. Washington: Upper Skagit Valley; Marymoor Park (Redmond); Three Forks Natural Area (Snoqualmie); Lake Pend Oreille NWR. British Columbia: Creston Valley Wildlife Area; Waldie Island (Castlegar); Cheam Lake Wetlands Regional Park (Rosedale).

Red-eyed Vireo

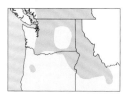

Red-eyed Vireo range

Red-eyed Vireo

Cassin's Vireo (*Vireo cassinii*)

Often gregarious and less secretive than other Northwest vireos, the Cassin's Vireo serenades woodland visitors with its buzzy, slurred song—unless they somehow perturb the little bird and solicit its scolding reprise. Largely a Northwest species (during the breeding season), the Cassin's Vireo has a near lookalike in the region in the Plumbeous Vireo that breeds in southeast and south-central Idaho.

LENGTH 5 inches. Olive-gray above, with bold white spectacles, and 2 yellowish white wingbars; whitish below, with pale yellowish sides of breast and flanks; flight feathers edged with pale olive. **VOICE** Song is a buzzy, up- and down-slurred *cheeriup cheerip chewew* (variable); call is a scolding, rapid, raspy *cheep-cheep-cheep-cheep*. **BEHAVIORS** Gleans insects from foliage and branches, and sometimes hovers briefly and also fly-catches in midair. In migration, often mixes with warblers and other species. **HABITAT** Breeds in dry conifer and mixed forests; brushy riparian areas and riparian deciduous groves in juniper, pine, or Douglas fir forests. In migration, uses wide variety of wooded habitats. **STATUS** Uncommon summer resident and migrant (April–September).

Cassin's Vireo range

Cassin's Vireo

Cassin's Vireo

Plumbeous Vireo
(*Vireo plumbeus*)

The Plumbeous Vireo's breeding range extends northward into southern Idaho and does not overlap with the breeding range of the more northwesterly Cassin's Vireo, with which it was once considered conspecific. The two are quite similar, and their migration paths overlap to some extent, so great care is needed to separate them in areas of likely overlap in late spring and late summer, especially given that they use a wider array of habitats in migration.

LENGTH 5.25 inches. Uniform bright slate-gray above, with bold white spectacles, and 2 white wingbars; rump may be tinged slightly olive; clean white throat; whitish breast with pale gray edges; whitish belly and undertail coverts; ashy gray or whitish flanks, sometimes with a pale olive or yellowish wash; flight feathers edged with white. **VOICE** Song is a robinlike *cheeriup cheerip chewew* (variable); call is a scolding, rapid, raspy *cheep-cheep-cheep-cheep*. **BEHAVIORS** Often forages amid outer limbs

at all levels, gleaning insects from the foliage and branches, and sometimes hovering briefly and also fly-catching in midair. **HABITAT** Open montane juniper and Douglas fir woodlands, often with deciduous component or understory; riparian margins within juniper woodlands. **STATUS** Uncommon summer resident (May–August) of south-central and southeast Idaho; rare vagrant in southwest Idaho and southeast Oregon. **KEY SITES** Idaho: Big Cottonwood Wildlife Area; Castle Rocks State Park and City of Rocks National Reserve; Cherry Spring Nature Area (Pocatello); Stone Hills and Black Pine Mountains.

Plumbeous Vireo

Plumbeous Vireo

Plumbeous Vireo range

361

Jays, Crows, Ravens, Magpies, and Nutcrackers

Gregarious, noisy, and intelligent, members of the Corvidae family include such birds as jays, crows, ravens, and magpies. Some species are very well-known—the American Crow, Common Raven, and California Scrub-Jay, as well as the vibrantly hued Steller's Jay and the comparatively demure Gray Jay, known in many quarters as the camp robber jay for its habit of learning to roam campgrounds and mountain picnic areas for scraps or even handouts from humans. Other species, such as the striking Clark's Nutcracker and flocking Pinyon Jay, are lesser known. Corvids feed on a wide variety of food, and some species cache food for later use. These interesting birds frequently surprise not only birders, but even researchers, with their ability to solve problems.

California Scrub-Jay

Gray Jay (*Perisoreus canadensis*)

Despite an impressive litany of vocalizations, the Gray Jay is perhaps the quietest of the corvids, generally content to move unobtrusively through the forest as it forages. These birds of the high country can be incredibly tame around humans and quickly learn to patrol forest campgrounds and picnic sites looking for handouts, a behavior that has earned them the nickname "camp robber."

LENGTH 11–12 inches. Pale gray overall, palest on breast and belly; white face; dark gray crown and nape in Pacific race; white crown with small, grayish nape patch on birds in Idaho and eastern Oregon and Washington; small, black bill. Juvenile is buffy gray overall, lacking strong facial contrast. **VOICE** Extremely wide range of sounds; commonly a single-note contact call, *whooit*, and musical, whistled *whee-yep*, but many variations. **BEHAVIORS** Typically subdued and quiet, Gray Jays readily and noisily mob predatory birds, such as hawks, accipiters, and owls that they happen upon during foraging. Slow and graceful flight on quiet wings, alternating flapping and gliding. **HABITAT** High-elevation conifer forests up to timberline. **STATUS** Fairly common year-round resident.

Gray Jay

Gray Jay range

Pinyon Jay
(*Gymnorhinus cyanocephalus*)

Highly social and gregarious, Pinyon Jays are communal, forming flocks from a few birds to many dozens and often noisily foraging and traveling, constantly uttering their distinctive flight call. They are pine and pinyon forest specialists, closely tied to ponderosa pine forests in their Oregon range and juniper woodlands in their Idaho range. On rare occasions when pinecone crops fail, Pinyon Jays will disperse outside their normal range.

LENGTH 10.5 inches. Pale blue overall with darker, more lustrous blue head and face; lightly streaked pale bluish white throat; powder-blue wing coverts, grayish blue belly; long, straight, daggerlike bill; shorter tail than other jays. **JUVENILE** Gray overall. **VOICE** Commonly a variety of caws, especially a subdued, somewhat grating, *crawk*, and a loud, high *ack ack ack*. Flight call, a smooth, wavering laughlike cawing, *hahahaha hahahaha*. **BEHAVIORS** Pinyon Jays will harvest and cache pinecone seeds for later use, and will also come to backyard feeders, where a large winter flock can rapidly pillage the available supply of seeds and nuts. Flight is direct and often fairly high, typically in noisy flocks. **HABITAT** Ponderosa pine forests, ponderosa and juniper woodlands, juniper woodlands. **STATUS** Fairly common year-round resident of Oregon; uncommon in Idaho. **KEY SITES** Idaho: City of Rocks National Reserve area; Pocatello area (Pocatello Creek Road, Mink Creek Road). Oregon: Cabin Lake Guard Station; Sisters.

Pinyon Jay

Pinyon Jay

Pinyon Jay range

Steller's Jay *(Cyanocitta stelleri)*

An iconic denizen of campgrounds, parks, and trailheads in the Northwest's forested mountains, the strikingly colorful Steller's Jay habituates well to humans and human activity and learns to steal food from campers, and even beg for handouts. These oft-vocal jays also commonly inhabit forested urban parks and neighborhoods, and can be found from sea level along the Northwest coast and interior valleys to timberline in the Idaho Rockies.

LENGTH 12–13 inches. Black head with prominent ragged black crest; black breast; dark brown nape and back; radiant blue rump, flanks, wings, tail; fairly large black bill. Juvenile identical but duller plumage. **VOICE** Commonly a loud, harsh, rapid-fire, scolding *shack-shack-shack-shack-shack*, and a raspy, drawn-out *chaaack*. **BEHAVIORS** Often feeds on the ground, hopping (not walking), flushing quickly to the cover of trees when disturbed; flight is fairly slow, with deep, soft wingbeats. Readily attracted to bird feeders. **HABITAT** Conifer forests, mixed woodlands, wooded city parks and neighborhoods; from lowlands, including coastal evergreen forests and deciduous river bottoms, to high mountains. **STATUS** Common year-round resident.

Steller's Jay range

Steller's Jay

Blue Jay *(Cyanocitta cristata)*

A rare but fairly regular winter vagrant to the Northwest, the strikingly beautiful Blue Jay is widespread from the Great Plains to the East Coast. The species has undergone a slow westward expansion over the past century, and may have nested within the Pacific Northwest, but all Blue Jays found in the region are considered vagrants.

LENGTH 10–11 inches. Distinctive blue, black, and white pattern; cobalt-blue above, with strongly contrasting flight feathers barred with black and white; black breast band frames white face and extends up along sides of neck to bright blue, crested crown; buffy white below. In flight, note flashy white terminal band on tail, widest on outer feathers, and white trailing edge of flight feathers. Fairly long, bright blue tail with black bars. **VOICE** Typical call is a loud, scolding, repeated, *jayeeer*, with many different inflections; also clicks, grunts, rattles, and many other vocalizations. **BEHAVIORS** Winter vagrants are typically found within urban and suburban areas, including parks and neighborhoods; readily utilizes backyard feeding stations, especially for peanuts and hulled sunflower seeds. **HABITAT** Open woodlots, wooded parks, mixed woodlands, neighborhoods. **STATUS** Rare winter vagrant (October–April), especially northern Idaho, eastern Washington, northeast Oregon.

Blue Jay

366

California Scrub-Jay
(Aphelocoma californica)

The familiar jay of the lowlands and foothills west of the Cascades, the California Scrub-Jay, often raucous and bold, adapts well to human activity and thus abounds in cities, towns, suburbs, and farmlands. This adaptability has allowed the species to steadily expand its range eastward and northward, including into southeast Oregon, the Columbia Basin, and the Puget Sound area. In 2016, the Western Scrub-Jay was divided into two species. The Pacific form is now called California Scrub-Jay, and the paler interior form found in southern Idaho is the Woodhouse's Scrub-Jay (*Aphelocoma woodhouseii*).

LENGTH 11.5 inches. Azure-blue above, brightest on nape, crown, and wing coverts; light grayish brown mantle; long, blue tail; dark gray, wedge-shaped face patch bordered above by prominent white eyebrow stripe; white throat with partial blue necklace; white belly. **JUVENILE** Grayish overall with dull white belly, and dull blue tail and wing coverts. **VOICE** Most commonly a high, throaty, scolding *aayike*, and a rapidly repeated, harsh *aak-aak-aak-aak*; numerous other vocalizations. **BEHAVIORS** California Scrub-Jay caches food, and oak and filbert trees often spring to life where jays have buried nuts just under the ground detritus. Flight is slightly undulating, with alternating gliding and then flapping on soft wingbeats; agile in flight, they can even snatch insects out of the air. **HABITAT** Brushy hillsides, oak woodlots, residential neighborhoods, sparse mixed woodlands, riparian corridors. **STATUS** Common year-round resident.

California Scrub-Jay range

California Scrub-Jay

Clark's Nutcracker
(*Nucifraga columbiana*)

A pine-seed harvesting specialist, the Clark's Nutcracker is well adapted to prying seeds out of cones, and it ranges widely to find food sources. It collects seeds in a specialized throat pouch, and like the Pinyon Jay (which is behaviorally somewhat similar), it then caches the seeds for later use, typically in the ground or in stumps, crevices in tree bark, and other locales.

LENGTH 12 inches. White face and large, black, daggerlike bill; silvery gray nape, crown, neck, breast, back contrasting sharply with jet-black wings; white outer tail feathers, black center tail feathers; in flight, shows white patches on trailing edge of secondaries. **VOICE** Commonly a distinctive high, nasal *aaaaack*, and a similar wavering call. **BEHAVIORS** Often occupies conspicuous perches on treetops or rocks. Among the most agile of corvids; strong, direct flight, but capable of rapid maneuvering; for longer distances, often

flies high above the forest canopy. **HABITAT** Conifer forest and open woodlands, usually with a pine component, especially ponderosa pine and whitebark pine; occurs as high as timberline. **STATUS** Fairly common year-round resident.

Clark's Nutcrackers in flight

Clark's Nutcracker range

Clark's Nutcracker

Black-billed Magpie
(*Pica hudsonia*)

The common and striking Black-billed Magpie is a familiar, gregarious, and loud resident of the interior Northwest. Like ravens and crows, magpies are highly intelligent. They readily feed on carrion, but also eat a wide range of other foods and will even visit backyard bird-feeding stations.

LENGTH 18–20 inches **WINGSPAN** 24–26 inches. Bold black-and-white pattern (shiny and iridescent in good light), with black head, breast, and back, and white belly; long, black, streaming tail is distinctive. Juvenile is identical but with shorter tail. **VOICE** Commonly a harsh, loud, scolding, rolling *ack-ack-ack-ack*; also quieter, less persistent single *aack* notes; various similar calls. **BEHAVIORS** Magpies build huge ball-like nests largely composed of sticks and twigs and placed in trees where they

are especially evident in winter with no foliage to help conceal them. The nest floor is lined with mud. Direct, fairly slow flight on deliberate wingbeats; note streaming tail in flight. **HABITAT** Farmlands and ranch lands, cities, parks, juniper woodlands, brushy desert steppe, riparian corridors, and roadsides. **STATUS** Common year-round resident.

Black-billed Magpie in flight

Black-billed Magpie range

Black-billed Magpie

American Crow
(*Corvus brachyrhynchos*)

Abundant and widespread, the American Crow adapts readily to a variety of habitats and easily integrates into human-dominated ecosystems. Highly intelligent, these familiar birds are both admired and vilified for their ingenuity in finding food and harassing people, pets, and other animals.

LENGTH 16–21 inches. **WINGSPAN** 34–39 inches. All black, smaller than the very similar Common Raven, with less robust bill and lacking the raven's long, shaggy throat plumes; American Crow's tail is fan shaped in flight, or slightly square, whereas the raven's tail is distinctly wedge shaped unless widely spread, when it appears steeply rounded. **VOICE** Commonly a series of persistent caws, in a series; widely varied repertoire of other calls. **BEHAVIORS** Crows adapt readily to urban and suburban environments and are common and typically noisy city residents. Strong, fluid flight; highly agile. **HABITAT** Found in many habitat types, especially near human settlement; generally avoids dense forests, open deserts, and alpine zones. **STATUS** Common year-round resident throughout most of Northwest; withdraws from much of eastern Oregon in winter.

SIMILAR SPECIES Northwestern Crow (*Corvus caurinus*). Despite descriptions of the Northwestern Crow being somewhat smaller than the American Crow, and its call being somewhat hoarser, the 2 species have so much overlap in both voice and size that they are indistinguishable. The Northwestern Crow inhabits the coastal strips of Washington and British Columbia, as well as Puget Sound; crows on Vancouver Island are generally considered to be Northwest Crows, and this species is common on the coast of the Olympic Peninsula. Hybrids appear to be commonplace. For birders wishing to add this species to their life list, Vancouver Island (or along the coast as far as Alaska) is the safe bet.

American Crow

American Crow range

Northwestern Crow range

Common Raven (*Corvus corax*)

Among the most intelligent of wild animals, Common Ravens are capable of problem-solving and learning by experimentation and imitation. Ravens are opportunists, feeding on a wide array of prey and nonprey items; they will graze on grain and vegetable matter, kill rodents, insects, and other birds, and raid nests for eggs and chicks. They are especially fond of carrion.

LENGTH 22–26 inches. **WINGSPAN** 45–48 inches. Very large and all black, with a large, heavy bill; shaggy throat; in flight, tail is long and distinctly wedge shaped except when fanned, when it appears steeply rounded. **VOICE** Commonly a deep, hoarse, throaty cawing and abrupt screamlike caws. Ravens have a deep repertoire of sounds. **BEHAVIORS** Unlike crows, Common Ravens rarely form flocks, but communal roosting and congregating at food sources are both common behaviors. They are frequently seen in pairs, routinely patrolling roadways for carrion, and as such are easily found along many major highways, especially in the Northwest interior. Strong, fluid flight, with deep wingbeats; frequent gliding and soaring (unlike crows); often performs acrobatic midair rolls and half rolls. **HABITAT** Widespread in a variety of habitats, especially sagebrush steppe and open juniper woodlands, forested uplands, ranch lands, open coasts and estuaries, alpine tundra, and highway corridors. **STATUS** Common year-round resident.

Common Raven

Common Raven in flight

Common Raven range

Larks and Swallows

Two lark species (family Alaudidae) inhabit the Northwest, one an import of very limited range (Eurasian Skylark) and the other a widespread and handsome native (Horned Lark); they are ground birds of open country. The swallows (family Hirondinidae) are winged insect assassins, agile and athletic fliers that specialize in snatching insects in midair and even skimming them from the water's surface. Most are cavity nesters, with several species happy to use bird boxes, another species (Cliff Swallow) that builds gourdlike mud nests affixed to hard surfaces, and still another (Bank Swallow) that excavates tunnels in soil bluffs.

Violet-green Swallow

Horned Lark *(Eremophila alpestris)*

Ground-dwelling songbirds perfectly at home in very sparse cover, Horned Larks range from coastal beach to above timberline and from damp shortgrass habitats west of the Cascades to remote grasslands and sagebrush deserts in the interior. The Northwest is home to several of the nearly 24 subspecies; they differ slightly in plumage shades and pattern, but all have the characteristic "horns" for which the bird is named—tufts of black feathers on each side of the head. One subspecies, the Streaked Horned Lark *(Eremophila alpestris strigata)* is a threatened subspecies, with a total population of less than 2000 individuals, most of which breed in Oregon's central Willamette Valley.

Male Horned Lark

Female Horned Lark

LENGTH 7.25 inches. **MALE** Black mask, black forehead stripe and "horns," black bib; whitish to yellow face; brownish above, pale buff to white below; black or blackish tail with white outer feathers. **FEMALE** Paler overall, with less prominent horns and duller facial marks and bib. **VOICE** Song is a high, ringing, slightly raspy twittering *tsit-tsit-tsit-tiwetwetwee*; calls, in flight and on the ground, include an airy *tsiweet* and also a huskier *zeeit*. **BEHAVIORS** Forages almost exclusively on the ground and frequently along gravel and dirt roadways; in winter and migration, large numbers often gather on plowed fields and stubble fields, and along highways and other roads when snow cover is heavy. **HABITAT** Open short-grass prairies, fields, sagebrush steppe, desert scrub, coastal grasslands, dunes; dry, sparsely vegetated subalpine and alpine habitats. **STATUS** Year-round resident and migrant; common interior; locally uncommon west of Cascades.

SIMILAR SPECIES Eurasian Skylark (*Alauda arvensis*) is an Old World species introduced to Vancouver Island, British Columbia, from Great Britain in 1903, the only successful attempt to establish this species in North America. The current population has shrunken considerably and is estimated at fewer than 100 individuals. The most reliable location is Victoria International Airport. Tannish and streaked above, with finely streaked breast and buff belly; in flight, white-tipped secondaries and white outer tail feathers are diagnostic.

Horned Lark range

Similar species: Eurasian Skylark

Purple Martin (*Progne subis*)

Largest of the North American swallows, Purple Martins are cavity nesters that are well-known throughout much of their range in North America for using nesting houses made and erected specifically for them. Popular models include a multi-unit apartment-style birdhouse erected on poles and hollowed-out, dried gourds often strung from wires. One study in Oregon determined that some 75 percent of the state's population of Purple Martins nested in such manmade houses.

LENGTH 8 inches. Largest swallow; long, narrow, sharply pointed wings; forked tail. **MALE** Glossy, dark bluish purple overall, but can appear all black in poor light. **FEMALE** Dark charcoal-gray above with some glossy bluish purple sheen, especially on wing coverts; light gray collar and forehead; light gray below. **JUVENILE** Similar to female, but streaked below. **VOICE** Songs consist of sharp chirps, musical rattling chirps, and insectlike buzzes. **BEHAVIORS** Acrobatically takes insects out of the air, often swooping low over water; frequently nests in hollowed-out gourds and nest boxes, usually placed over or near water. **HABITAT** Open areas, frequently near water, including wetlands, lakes, reservoirs, and estuaries; clear-cuts in Coast Range forests. **STATUS** Uncommon to locally common summer resident (April–August). **KEY SITES** Oregon: Fern Ridge Reservoir; Pacific Oyster Company jetty (Bay City); Sauvie Island; Smith and Bybee Lakes Wetlands (Portland). Washington: East Bay (Olympia); Marymoor Park (Redmond); Nisqually Reach Nature Station (Olympia); Steigerwald NWR. British Columbia: Blackie Spit at Crescent Beach (South Surrey); Cowichan Bay; Ladysmith Community Marina (Ladysmith Harbor).

Male Purple Martin

Female Purple Martin

Female Purple Martin in flight

Purple Martin range

Tree Swallow (*Tachycineta bicolor*)

Although generally seen foraging over open country or water, the starkly beautiful and widespread Tree Swallow does in fact nest in trees. It is a secondary cavity nester, relying in part on woodpecker holes for nest sites, and will also readily utilize artificial nest boxes, as well as a wide array of natural holes and crevices.

LENGTH 5.5 inches. **SEXES SIMILAR** Gleaming iridescent, deep blue-green above, with color extending below eye line (unlike in Violet-green Swallow); snow-white below;

plumage somewhat variable with age, with younger adults showing less extensive shiny iridescence; long, all-dark wings; slightly notched tail. **JUVENILE** Dark gray-brown above; white below, often with pale gray partial breast band (see Bank Swallow) and partial white collar, with distinct contrast between dark face and bright white throat and neck. **VOICE** Song is a high, liquid twittering; flight call is similar but generally faster and more uniform. **BEHAVIORS** Frequently hawks insects over water and open fields; often clings to inside of nest holes with head sticking out. **HABITAT** Wide array of open and semi-open habitats, especially near water, with ample wooded tracts, snags, or scattered trees where nesting holes are available. **STATUS** Common summer resident and migrant (February–September).

Tree Swallow

Tree Swallow

Tree Swallow in flight

Tree Swallow range

376

Violet-green Swallow
(*Tachycineta thalassina*)

More so than the similar-looking Tree Swallow, the elegant little Violet-green Swallow is perfectly at home in urban areas, often perching on overhead wires in residential neighborhoods between bouts of hawking (catching insects on the wing); so too, these widespread, purely western birds occupy the most remote regions of the Northwest where suitable nesting habitat is available.

LENGTH 5 inches. **SEXES SIMILAR** In all plumages, wingtips extend well past tail when bird is perched; starkly dark above and white below; velvety iridescent green mantle, rump, nape, and crown; rump and nape sometimes appear purplish; iridescence somewhat extensive on female; white on face extends above the eye line, unlike Tree Swallow; white extends up onto both sides of rump (visible in flight); female's face slightly dusky rather than pure white as in male; dark, sharply pointed wings and short, slightly notched tail. **JUVENILE** Whitish below, usually somewhat dingy on the breast; gray-brown above; eye usually partially encircled by white (juvenile Tree Swallow generally has distinctly white throat contrasting with dark face). **VOICE** Song is a series of high, clear, whistled chirps; calls are thin, twittering whistles. **BEHAVIORS** Swift and agile flight; commonly nests in crevices and cavities on buildings and other human-built structures; otherwise uses natural crevices and holes. Where they coexist in the same general region, Violet-green Swallows often mix with other swallows and swifts during foraging flights. **HABITAT** Open woodlands, including suburban settings; coast to mountains to arid desert, but generally near water in the arid interior Northwest; often nests on cliffs. **STATUS** Common summer resident and migrant (March–September).

Female Violet-green Swallow

Violet-green Swallow range

Male Violet-green Swallow in flight

Northern Rough-winged Swallow (*Stelgidopteryx serripennis*)

The somber-colored Northern Rough-winged Swallow derives its name from a feature only discernible if the bird is in hand: the leading edge of the outermost primary flight feathers is rough, like a tiny sawblade, because the tiny feather barbs stand out slightly from the feather shaft. The purpose of this feature remains a mystery.

LENGTH 5.25 inches. **SEXES SIMILAR** Soft brown above, with dark gray flight feathers; light buff throat and breast; white belly. **JUVENILE** Similar to adult, but with cinnamon-colored shoulders, wingbars, and flight feather edging. **VOICE** Song and calls composed of bright, bubbling *brrrrrit* notes. **BEHAVIORS** Frequently feeds low over water, and will skim insects off the surface; nests in burrows (pre-existing or its own) and sometimes crevices, usually in steep banks. **HABITAT** Open and semi-open areas, usually near water; almost always near water in arid interior Northwest. **STATUS** Fairly common summer resident and migrant (April–August).

Northern Rough-winged Swallow range

Northern Rough-winged Swallow

Bank Swallow (*Riparia riparia*)

The smallest Northwest swallow, the Bank Swallow is named for its nesting strategy in which the male digs a burrow into vertical or near-vertical, soft-soil inclines. These burrows can reach 5 feet in length. The female builds the nest and then both parents incubate the eggs and feed the young. They are colony nesters, with each pair using its own burrow, and such colonies can number in the hundreds of pairs. During the nesting season, activity is frenetic, as birds busily fly in and out of burrows to deliver food to hatchlings.

LENGTH 4.75 inches. Brown above, with brown extending to below eye line; white below with wide brown breast band; flight feathers distinctly darker than mantle; in flight, sharply angled wings and fairly long, very narrow, slightly forked tail. **JUVENILE** Similar to adult, but breast band ranges from complete and distinct to faint and partial; yellow gape and base of bill in younger juveniles. **VOICE** Calls and songs are harsh, burry chattering notes, *chaap chapchap*, and a mildly screeching, scolding *cheeeeeep cheap cheap*; nest colonies can be alive with these buzzy calls. **BEHAVIORS** Nests in colonies in burrows in steep banks adjacent to or near water, and forms foraging flocks over water or fields. **HABITAT** Open and semi-open areas along lakeshores, rivers, wetlands, wet fields, riparian valleys. **STATUS** Summer resident and migrant (May–September); rare west of Cascades; locally fairly common east of Cascades.

Bank Swallow

Bank Swallow nesting colony

Bank Swallow range

Cliff Swallow
(*Petrochelidon pyrrhonota*)

Working nonstop, toting mud in their bills from water sites to nest sites, smartly patterned Cliff Swallows build their nest "boxes" in the forms of amazing gourd-shaped shelters anchored to cliff faces, overhangs, cement structures, and buildings. Colony nesters, they pack these globular structures tightly together, often on the shady underside of structures, with the holes facing out and down. While some colonies include hundreds of nests, in other places, a single outhouse or picnic shelter at a lonely open-country campground might harbor just 2 or 3 nests.

LENGTH 5 inches. Deep blue mantle and crown, divided by pale gray nape; chestnut face, dark throat, and straw-colored forehead patch; pale rust rump; dark gray dorsal wing and tail surfaces; whitish belly and pale buff breast and flanks. **JUVENILE** Similar in overall pattern to adult, but grayish above and mostly white below; buffy nape; pale cinnamon cheeks; pale forehead patch absent or partially developed. **VOICE** Song is a slow, squeaky, mechanical-sounding combination of creaking and twittering; calls include soft but excited *cheer-cheer-cheer* alarm calls, and soft but husky twittering. **BEHAVIORS** During nesting season, Cliff Swallows swarm at prime mud sites to collect the mud for their nests; at these mud-collection locations, they frequently hold wings straight upward, butterflylike; adults can form a swirling virtual tornado of swallows when alarmed by an intruder coming too close to the colony. **HABITAT** Generally inhabits open country near water, with sufficient cliffs or manmade structures for nesting, and sources of mud for building nests. **STATUS** Common summer resident and migrant (April–August).

Cliff Swallow

Cliff Swallow nesting colony

Cliff Swallow in flight

Cliff Swallow range

Barn Swallow (*Hirundo rustica*)

Perhaps the most familiar swallow to many Northwesterners, and also the most ornate and colorful, the graceful Barn Swallow has adapted well to the trappings of humanity. It readily builds its cup-shaped mud nests under eaves of homes, barns, and other structures. Barn Swallows frequently form large migrating flocks, often mixed with other swallow species, totaling hundreds or even thousands of birds; communal roosting sites during migration can attract tens of thousands of birds.

LENGTH 7 inches. **SEXES SIMILAR** Long, slender, deeply forked tail; dark steel-blue above; pale buffy orange below, with dark rust-orange throat and forehead; female slightly less deeply colored than male. **JUVENILE** Similar to adult, but paler overall and without long tail forks. **VOICE** Song is a raspy twittering mixed with squeaks and ending with a crisp whining note and rapid,

rattling bill clicks; calls include a high, snappy *twip*, often doubled to *twip-twip*; and a musical *peeyear-peeyear*. **BEHAVIORS** Forages over open country and water, frequently close to the ground; often easily seen along roadways; most frequently nests on manmade structures. **HABITAT** Open and semi-open country; farmlands and ranch lands, prairies, shrub steppe, parks, residential areas; open woodland margins. **STATUS** Common to abundant summer resident and migrant (March–September).

Barn Swallow

Barn Swallow range

Barn Swallow in flight

Chickadees, Titmice, Nuthatches, Creepers, Bushtits, Wrens, Dippers, Gnatcatchers, Wrentits, and Kinglets

Representing 9 different taxonomic families, and in the Northwest, 22 different species, most of these active little songbirds share some common traits—they tend to be energetic, even frenetic, and vocal. They may not always be obvious, however—the Brown Creeper is superbly camouflaged, the Wrentit loves heavy cover, and Golden-crowned Kinglets often forage high in the trees. Among this broad conglomeration of species are some of the most persistent and ebullient singers in the bird world, such as the tiny Pacific Wren, whose feverish song goes on and on almost to the point of wonder. Also among these birds are some acrobatic tree-trunk specialists—the aforementioned Brown Creeper, along with the 3 nuthatch species, forage by climbing about on trunks and limbs and even seed cones. The American Dipper meanwhile is an aquatic specialist, adapted to forage in and even under water. Many of these little birds are welcome regular visitors to bird-feeding stations, where their cheery energy is nearly infectious.

Blue-gray Gnatcatcher

Black-capped Chickadee
(*Poecile atricapilla*)

The Northwest's most common chickadee is found in a variety of habitats throughout the region, and is easily attracted to bird feeders where it prefers suet, shelled peanuts, and shelled sunflower seeds. Black-capped Chickadees from the Pacific population are the most colorful, with light cinnamon-colored flanks; Rocky Mountain birds show little to no buff on the flanks, and the Great Basin race is the palest.

LENGTH 5.25 inches. Black cap and bib; large head; wings have white edges; flanks range from light cinnamon to white. **VOICE** A sweet, whistled *fee-bee* or *fee-beehee*; *chicka-dee-dee-dee-dee* or *sicka-dee-dee-dee*. **BEHAVIORS** Hops amongst branches where it often hangs upside down; creeps up vertical surfaces searching for food. Flits between trees with weak, undulating flight; at feeders, often dashes in for a morsel and flits back into the trees. **HABITAT** Deciduous woods and mixed woodlands, river bottoms, parks, and urban areas. **STATUS** Common year-round resident.

Black-capped Chickadee range

Black-capped Chickadee

Chestnut-backed Chickadee
(*Poecile rufescens*)

The Black-capped Chickadee is familiar from coast to coast, but the far West boasts a chickadee all its own. Especially common west of the Cascade Range, the handsome Chestnut-backed Chickadee thrives in conifers, but in many cities and towns, it has adapted to modest-sized parcels of woodlands, including pure deciduous stands in winter. Some birds disperse to higher-elevation conifer forests in summer, and back to lower elevations in winter.

LENGTH 4.75 inches. Distinctive black-and-white head and face pattern, though cap is actually rich, dark brown; chestnut-colored back and flanks; grayish wings and tail. **VOICE** High-pitched and often steady *peep-peep-peep-peep*, and also a rapid and airy *chick-a-dee-dee*, less musical than the namesake call of the Black-capped Chickadee. **BEHAVIORS** Active and acrobatic when foraging for seeds, nuts, and insects; hangs at all angles and hovers momentarily; at feeders, tends to dash in for a nut or seed and fly up into a tree with it, but will linger when feeding at suet stations. Flight tends to be quick and fluttering from point to point; flock members typically fly one at a time across openings. **HABITAT** Coniferous and mixed forests and woodlots; parks and yards even in urban areas. **STATUS** Common year-round resident.

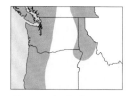

Chestnut-backed
Chickadee range

Chestnut-backed Chickadee

Boreal Chickadee
(*Poecile hudsonica*)

A common bird of northern Canada and
Alaska, the Boreal Chickadee's southern
range just reaches the Northwest, where
the species occurs above 5000 feet in
conifer forests. This tame bird is known
to cache food for the extreme winters of
the north country.

LENGTH 5.5 inches. Brown cap with black
bib; pinkish brown flanks; grayish brown
back, with gray wings. **VOICE** High, raspy
itsa-chee itsa chee-chee whit-whit-chee-chee;
brief, airy *tsi-tsi-tsi*, often followed by raspy,
excited *speechaa-chass*. **BEHAVIORS** Tame;
tends to stay in the densest part of a tree or
group of trees, often making the birds difficult to find. Flight is direct, without undulation; flits and hops when going from branch

to branch. **HABITAT** High-elevation conifer
forests above 5000 feet. **STATUS** Uncom-
mon year-round resident. **KEY SITES** Idaho:
Selkirk Mountains; Smith Creek Road.
Washington: Salmo Pass and Forest Road
2220; Tiffany Springs Campground. British
Columbia: Cathedral Provincial Park; E.C.
Manning Provincial Park.

Boreal Chickadee range

Boreal Chickadee

Mountain Chickadee
(*Poecile gambeli*)

This inquisitive, wide-ranging chicka-dee—a common winter feeder bird—is most frequently found in the mountain-ous regions of the interior Northwest. During irruption years, its range expands and it may show up anywhere in the region. Like all chickadees, the Mountain Chickadee can be quite approachable, though they tend to be less gregarious during nesting season.

LENGTH 5.25 inches. Black cap and bib with bold white eyebrow stripe dividing crown from black eye stripe; upperparts gray, with gray wings; whitish breast and belly. **VOICE** Song is a sweet, whistled *febee-fee-fee*; calls include a raspy, scolding, fast *cheecheechee*. **BEHAVIORS** Highly territorial during breed-ing season; joins mixed-species flocks during winter; often hangs upside down when feeding and probing for food. Flight is weak and undulating, but acrobatic when hawking insects. **HABITAT** During breeding season, conifer forests, including juniper stands; also mixed forests and urban areas during winter. **STATUS** Common year-round resident; rare west of Cascades.

Mountain Chickadee range

Mountain Chickadee

Juniper Titmouse
(*Baeolophus ridgwayi*)

Found in remote sections of south-central Oregon, as well as southern Idaho, the Juniper Titmouse was once considered conspecific with the Oak Titmouse and was aptly named the Plain Titmouse. The 2 species are nearly identical in appearance, and songs are quite variable among both. The best identification keys to separate them in Oregon are habitat and location: the Juniper Titmouse (Idaho's only Titmouse) is found east of the Cascade Range and the Oak Titmouse occurs west of the Cascades, but southern Klamath County appears to be an overlap area where distribution status of one or both species is still not completely understood.

LENGTH 5.25 inches. Distinctive crest; very plain grayish olive above, grayish white below. **VOICE** Song is a warblerlike *tchee-tchee-tchee*; calls include a raspy, scolding *chee-hee-hee-hee*. **BEHAVIORS** Not flocking birds, Juniper Titmice mate for life; they are attracted to bird feeders during winter. **HABITAT** Mature, dense, juniper woodlands. **STATUS** Uncommon to rare, and declining, in scattered locations; generally difficult to find in Oregon. **KEY SITES** Idaho: Castle Rocks State Park and City of Rocks National Reserve; Massacre Rocks State Park; Mink Creek area south of Pocatello (Cherry Springs, Gibson Jack Creek, Kinney Creek, Scout Mountain). Oregon: Deep Creek Canyon area (west of Adel).

Juniper Titmouse

Juniper Titmouse range

Juniper Titmouse

Oak Titmouse
(*Baeolophus inornatus*)

Found only in scattered areas of southwest Oregon, this titmouse was once considered conspecific with the Juniper Titmouse and was aptly named the Plain Titmouse. The two are nearly identical in appearance, and their songs are quite variable. Best identification keys to separate them are habitat and location: the Oak Titmouse does not occur east of the Cascade Range, except in southern Klamath County, where its range appears to overlap that of the Juniper Titmouse. Distribution status of one or both species is still not completely understood in Klamath County. The Juniper Titmouse does not occur west of the Cascades, and is the only titmouse found in Idaho.

LENGTH 5 inches. Distinctive crest; plain grayish olive with a brownish tinge above, grayish white below. **VOICE** Songs and calls include a strong, whistled *dweet-dweet*; a chattering with emphasis on the last part of the call, *dwee-CHUR*; a fast, chickadee-like *see-see-see-chee*. **BEHAVIORS** Not flocking birds, Oak Titmice, which mate for life, may form loose aggregations with chickadees and nuthatches during winter, when they are attracted to bird feeders. **HABITAT** Oak-dominated woodlands and mixed oak-pine woodlands. **STATUS** Fairly common year-round resident. **KEY SITES** Oregon: Agate Lake; Ashland Pond; Emigrant Lake; North Mountain Park; Touvelle State Park and Ken Denman Wildlife Area (Jackson County).

Oak Titmouse range

Oak Titmouse

Bushtit (*Psaltriparus minimus*)

This tiny bird of Northwest woodlands, brushy places, and urban areas is a favorite for many birders with backyard feeders: traveling in gregarious flocks, they descend on suet feeders, totally covering them in as many as dozens of birds. Bushtits west of the Cascades exhibit brown crowns and faces; the interior form's head is entirely gray.

Male Bushtit

Bushtit range

LENGTH 4.5 inches. **SEXES SIMILAR** Gray overall, and long tail; birds west of the Cascades have a brownish cap and face, whereas the interior form is completely pale gray; adult female has light-colored eyes, and the male has dark eyes. **VOICE** Songs and calls include a soft, high-pitched, twittering *seep-see-see-see-see*; and a wavering, airy, high-pitched *see-ee*. **BEHAVIORS** Travels in vocal flocks except during breeding season; readily attracted to suet feeders, which it often visits in swarms; weak, fluttering flight; in flight, seems to drag tail. **HABITAT** Most Northwest habitats except open desert. **STATUS** Year-round resident; common west of Cascades, uncommon interior.

Female Bushtit

Red-breasted Nuthatch
(*Sitta canadensis*)

A common and widespread nuthatch with a preference for fir and spruce forests, the Red-breasted Nuthatch can be highly irruptive and may be seen in any habitat in the Northwest. It has even been found foraging in rocks well above tree line and in nearly treeless desert scrub.

LENGTH 4.5 inches. **SEXES SIMILAR** Black crown (bluish gray on female), white eyebrow stripe, bluish back, cinnamon underparts (paler on female). **VOICE** Calls include a nasal *yank-yank-yank* or *wheep-wheep-wheep*; notes also uttered singly, and in a variety of pitches. **BEHAVIORS** Forages much like a woodpecker, climbing on tree trunks and hanging upside down on branches and cones. Caches seeds for later use, and is a common visitor to bird feeders, where it usually dashes in for a nut or seed and dashes back into the trees. Quick, undulating flight. **HABITAT** Conifer forest, especially with fir and spruce, including neighborhoods and parks. During irruption years, can be found in almost any habitat. **STATUS** Common year-round resident.

Male Red-breasted Nuthatch

Female Red-breasted Nuthatch

Red-breasted Nuthatch range

White-breasted Nuthatch
(*Sitta carolinensis*)

The largest nuthatch in North America, the White-breasted Nuthatch occurs in a variety of habitats. Two subspecies in the Northwest may be given full species status in the future. The Pacific Coast subspecies (*Sitta carolinensis aculeata*) inhabits the valleys of western Oregon and southern Washington and is strongly associated with Oregon white oak. The Interior West subspecies (*S. c. nelsoni*) occupies drier pine forests of the Northwest. The ranges of the 2 subspeciess do not overlap, although status of White-breasted Nuthatches in Oregon's Klamath Basin remains unclear. They are similar in appearance but have different calls.

LENGTH 5.75 inches. **SEXES SIMILAR** Long, thin, sharp bill; male has black cap, and female a more grayish brown cap; white face; bluish gray above and white below; chestnut-colored vent. **VOICE** Calls include a series of nasal, single-pitch *whi-whi-whi-whi-whi-whi* notes; a nasal, rattled descending *eeerrr*; and various other sounds. **BEHAVIORS** Forages much like a woodpecker, climbing up and down tree trunks, and hanging upside down on branches and cones. Caches seeds for later use, and is a common bird feeder species. **HABITAT** *Sitta carolinensis aculeata*: Oregon white oak groves and mixed woodlands, from river bottoms to foothills, including urban areas. *Sitta carolinensis nelsoni*: primarily pine forests; also mixed pine and juniper woodlands. **STATUS** Year-round resident. *Sitta carolinensis aculeata*: locally common in Oregon, rare in Washington; declining in some areas; extirpated from most of western Washington except for Clark County. *Sitta carolinensis nelsoni*: common year-round resident.

White-breasted Nuthatch, Pacific Coast subspecies

White-breasted Nuthatch, Interior West subspecies

White-breasted Nuthatch range

Pygmy Nuthatch (*Sitta pygmaea*)

Smallest and quietest of the 3 Northwest nuthatches, the acrobatic, gregarious, and social Pygmy Nuthatch readily flocks with chickadees, kinglets, and other birds. They are cavity nesters and communal roosters. Within their ponderosa forest habitat, they will visit feeders, especially in winter. Often the easiest ways to find Pygmy Nuthatches is to listen for their oft-continuous airy chatter.

LENGTH 4 inches. Compact; short tail; long, straight, sharp bill; grayish overall; slate-gray above, buff below, with brown head; brown on the head extends below the eye line to form a distinct border with the buff-colored throat; juvenile lacks the brown cap. **VOICE** Airy but often nondescript *tee-tee-tee* or *peep peep*, varying in tone and speed; persistent, but can be difficult to hear on windy days or when louder birds are vocal. **BEHAVIORS** Scales up and down tree trunks and branches; feeds from all angles, including hanging upside down from cones and branches. Usually found in flocks ranging from a few individuals to dozens; bouncy, direct flight, often flitting between trees. Often associates with chickadees, kinglets, woodpeckers, and other birds. **HABITAT** Strongly tied to mature ponderosa pine forest, but also occurs in mixed conifer forest with a ponderosa component; ventures into juniper stands near ponderosa forest. **STATUS** Common year-round resident. **KEY SITES** Idaho: Coeur d'Alene (North Idaho College Beach; Tubbs Hill trails); Moscow area (Idlers Rest Nature Preserve; Phillips Farm County Park). Oregon: Sawyer Park, Tumalo State Park (Bend); U.S. Forest Service station and Best Western Hotel (Sisters). Washington: Little Spokane Natural Area and Riverside State Park (Spokane); Wenas area and Wenas Campground (Ellensburg). British Columbia: Okanagan Valley (Okanagan Mountain Provincial Park; University of British Columbia Kelowna; Hall Road and Mission Creek Regional Park, Kelowna).

Pygmy Nuthatch

Pygmy Nuthatch

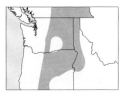

Pygmy Nuthatch range

Brown Creeper
(*Certhia americana*)

Cryptic colored birds that can be hard to spot as they forage on tree trunks, Brown Creepers use their stiff tails as support, like woodpeckers, and usually probe by climbing upward. The down-curved bill is used to probe in bark furrows for insects. The creeper's hammock-style nest is built behind loose bark. In addition to their excellent camouflage, which makes them difficult to find, Brown Creepers also exhibit a maddening habit of flying to the opposite side of tree trunks from would-be observers.

LENGTH 5 inches. Streaked brown above, white below; thin, curved bill; fairly long, pointed tail. **VOICE** Song is a rich, high-pitched *tsee-tsee-tuubut-see*; call is a thin, high, warbled *tseeee*. **BEHAVIORS** Starts on the lower section of a tree and climbs upward, probing bark for insects, often spiraling around the tree; flies to the next tree and repeats process. During winter, associates loosely with flocks of chickadees and kinglets. **HABITAT** Conifer forests and mixed woodlands, especially those with oak; old-growth forests preferred; will use purely deciduous stands as well, especially between fall and early spring. **STATUS** Common year-round resident.

Brown Creeper range

Brown Creeper

Blue-gray Gnatcatcher
(*Polioptila caerulea*)

The energetic little Blue-gray Gnatcatcher lives up to its name by feeding primarily on small insects, which it typically hunts in the outer canopy of trees and large shrubs. This tiny, handsome bird breeds as far north as central Oregon, but is more common in the oak and scrub zone of southwest Oregon, and also occurs in southern Idaho, with these areas forming the northern extent of its range.

LENGTH 4.5 inches. Long tail, often cocked upward, is black above with distinctive white edges, especially showy in flight. **MALE** Gray above, white below; white eye ring; black eyebrow stripes, bolder in breeding season; rich blue-gray crown and back. **FEMALE** Paler gray or soft gray brown above, no black eyebrow stripes. **VOICE** Calls include soft, urgent peeps and *zweet zwee* notes; can be highly varied. Male's song is a rambling series of mixed chirps, whistles, and scratchy notes, and often contains mimicry of other birds. **BEHAVIORS** Highly active, especially when feeding or when males compete for females; forages for insects among tree and shrub foliage, frequently flicking its tail side to side. **HABITAT** Deciduous forests and woodlots, especially oak and mountain mahogany; brushy margins; juniper stands. **STATUS** Locally common summer resident of southwest Oregon, uncommon elsewhere. **KEY SITES** Idaho: Castle Rocks State Park and City of Rocks National Reserve. Oregon: Emigrant Lake; Lower and Upper Table Rock (near White City).

Male breeding Blue-gray Gnatcatcher

Female Blue-gray Gnatcatcher

Blue-gray Gnatcatcher range

Ruby-crowned Kinglet
(*Regulas calendula*)

A tiny, active forest bird, the Ruby-crowned Kinglet is often seen in mixed flocks of other kinglets, chickadees, and warblers, but may also turn up alone. When the male is alarmed or agitated, it raises its ruby-red crown patch, but most often its namesake crown is concealed. The kinglet's loud, ringing voice belies its diminutive stature. Ruby-crowned Kinglets can lay as many as 12 eggs in a single clutch, the most of any songbird.

LENGTH 4 inches. **SEXES SIMILAR** Large head; broken eye ring (Hutton's Vireo has eye ring and complete pale lores, forming a bespectacled appearance); drab olive-brown above; olive wings with a single prominent whitish wingbar contrasting strongly with dark base of flight feathers; pale whitish olive below; fine, thin, tacklike black bill unlike the thicker pale-colored bill of Hutton's Vireo; yellow feet unlike Hutton's Vireo's gray feet. Male has a ruby-red patch on its crown that is generally concealed and difficult to see. **VOICE** Song is a jumbled, warbled *toowee-toowee-toowee*, and a high, thin *tsee-tsee-seeseeseesee*; call a harsh *jetet-jetet-jetet*, much like 2 small metal balls hitting each other. **BEHAVIORS** Very active and vocal, often flicks wings, unlike Hutton's Vireo, which rarely flicks wings. Sometimes hovers near foliage to catch insects. **HABITAT** Breeds in conifer forests, generally above 5000 feet. After leaving breeding grounds, descends to lower elevations and can be found in a variety of habitats throughout the Northwest, down to sea level during winter. **STATUS** Common year-round resident.

Male Ruby-crowned Kinglet

Female Ruby-crowned Kinglet

Ruby-crowned Kinglet range

Golden-crowned Kinglet
(*Regulus satrapa*)

This tiny, compact, grayish olive songbird is most at home in forested mountains, and is often identified by its near-constant calling from the forest canopy. Highly gregarious, these colorfully crowned birds forage and roam at all levels in the forest, often in loose association with other birds, including Ruby-crowned Kinglets, chickadees, warblers, nuthatches, and even woodpeckers.

LENGTH 3.5 inches. **SEXES SIMILAR** Grayish olive overall, with distinctive, colorful crown of bright yellow bordered on each side by bold black stripes wrapping around forehead; male's crown has a flame-orange center; whitish eyebrow stripes bordering darker eye stripes. Light-colored flight feather edges contrast with a central dark spot on the wings, and above that, a whitish wingbar; thin, small, black bill. **VOICE** Airy, high-pitched series of *tsee-tsee-tsee* chirps increasing in cadence to end with a chickadeelike trill; also single or double *tsee* notes and similar chattering. **BEHAVIORS** Forages frenetically for insects in trees, often chattering constantly; fluttering and slightly bounding flight on swift, shallow wingbeats. **HABITAT** Conifer forests during breeding season; conifers, mixed forests, deciduous woods, and brushy habitats with some tree cover during winter. **STATUS** Fairly common year-round resident; wintering birds disperse throughout the region.

Male Golden-crowned Kinglet

Female Golden-crowned Kinglet

Golden-crowned Kinglet range

Wrentit (*Chamaea fasciata*)

Though fairly common, the secretive little Wrentit is difficult to see because it rarely emerges fully from its dense-shrub habitat, making its distinctive song an important identification key. Brownish gray overall, it frequently appears dark sooty brown because it lurks deep in the shadows. By staking out dense coastal shrubbery, including salal, patient observers can often catch glimpses of these active little birds, especially when the Wrentits are foraging or singing on late spring and early summer mornings. The bird's hesitancy to cross open space may explain its absence from Washington—the mighty Columbia River mouth forms a very wide-open space, indeed.

LENGTH 6.5 inches. Gray-brown overall, with gently streaked more reddish brown breast and rump, gray crown and face; long tail often cocked upward; whitish iris. **VOICE** Song (given year-round) is a clear, sharp, resonant, rapidly accelerating *chip chip chip chi-chi-chi-chi-chi-chi*, sometimes slightly rolling in a *chitip chitip chitip chi-chi-chi-chi-chi-chi*. Call is flat, loud, rapid, scolding chatter: *cha cha cha-cha-cha-cha-cha-cha*. **BEHAVIORS** Generally slinks through dense shrubbery, but is quite approachable; weak, fluttering flight, and rarely crosses open space for more than a few feet. **HABITAT** Dense, hard-stem, coastal chaparral-type shrubbery, especially thickets dominated by salal (*Gaultheria shallon*). **STATUS** Fairly common year-round resident, especially along coast. **KEY SITES** Oregon: Bandon Marsh NWR; Bayocean Spit (Tillamook Bay); Cape Arago State Park; Harris Beach State Park (Brookings); Siltcoos River Estuary area, including Waxmyrtle Campground, Waxmyrtle Trail, and Lagoon Campground; Yaquina Bay State Park.

Wrentit

Wrentit range

Wrentit

American Dipper
(*Cinclus mexicanus*)

The American Dipper, sometimes called water ouzel, is uniquely adapted to an aquatic life in swift, tumbling streams where—thanks to adaptations such as oil glands to waterproof its dense plumage and nasal flaps—it wades, swims, and dives to search for underwater insects and other invertebrates. These chubby chatterboxes often forage by wading about with their head underwater, and frequently dive completely under the surface.

LENGTH 7.5 inches. Stocky and dark gray, with stubby tail often cocked upward. **JUVENILE** Light ashy gray below, dark gray above. **VOICE** High, often incessant musical warbling, such as *chirpa-chirpa-chirpa tee-tee-tee turwit-turwit-turwit* (substantial variation); typical call is a scolding, swift, high chirping. **BEHAVIORS** Frequently bobs up and down while walking or perching on rocks. Swift flight on stiff, fast wingbeats low over the water and often tilting and turning, following the curves of the stream. **HABITAT** Streams of all sizes, from small mountain creeks to large lowland rivers. **STATUS** Fairly common year-round resident; some altitudinal migration during winter, with higher-elevation birds wintering at lower altitudes.

American Dipper range

American Dipper

House Wren (*Troglodytes aedon*)

The region's most widespread wren, though not necessarily the most common, the House Wren is adaptable and highly gregarious—and a talented songster. They are cavity nesters that readily use nest boxes, and are known for nesting in common yard objects of many descriptions, from boots to cars.

LENGTH 4.75 inches. Brown above; fairly short brown tail, wings lightly barred with darker brown; buff-white below, with faint barring on rearward portion of pale tannish flanks; minimal, indistinct buff-white eye-brow stripe (compared to pronounced eyebrow stripe of Bewick's Wren); thin, slightly curved bill. **VOICE** Song is a rich, complex, somewhat variable series of very rapid trills and buzzy chatters. Various agitated alarm calls, especially an accelerating dry chatter, and a high, somewhat raspy *peee-peeyiiit* and *whirrrr-wheeiiip*. **BEHAVIORS** Forages frenetically by gleaning insects in dense shrubbery, usually low to ground. **HABITAT** Brushy areas, including riparian corridors, edges of clear-cuts, parks, yards; open conifer woodlots, including juniper; oak, aspen, and mixed-deciduous woodlands. **STATUS** Summer resident. Common interior, fairly common west of Cascades.

House Wren range

House Wren

400

Pacific Wren (*Troglodytes pacificus*)

Its bright, frenetic, conspicuous song belies the Pacific Wren's diminutive stature. These tiny, secretive birds rarely venture far from brushy cover, but within or just above that cover, they sing ebulliently, often from a favorite perch, such as a bare branch, stump, or fallen log. Formerly considered conspecific with the eastern Winter Wren, the Pacific Wren was given full species status in 2010.

LENGTH 4 inches. Rufous-brown overall, darkest above, with lightly barred flanks and belly; strongly barred wings and tail; short, narrow tail often cocked upward; pale eyebrow stripe. **VOICE** Song is a fast, very high-pitched energetic series of varying trills, insectlike buzzes, and chirps. Calls include a high, kissing *chir chir-chir*, and sometimes *chir-chir-chir*. **BEHAVIORS** Frequently forages on or near the forest floor. **HABITAT** Moist, brushy woodlands understory, especially conifer forests and mixed forests, from valley bottoms to mountains; retreats to lower elevations in winter. **STATUS** Common year-round resident.

Pacific Wren range

Pacific Wren

Bewick's Wren
(*Thryomanes bewickii*)

Sometimes an incessant chatterbox and other times content to slink quietly about in low, dense vegetation in its search for insects, the handsome Bewick's Wren is easily identified by its bold white eyebrow stripe. It frequently flicks its long tail side to side, and often cocks it upward. Though interior Northwest populations are scattered, their numbers seem to be increasing in the Columbia Basin region.

LENGTH 5.25 inches. Plain brown above with pale streaking on wings and long, strongly barred tail with black-and-white outer edges; pronounced and striking bold white eyebrow stripe. **VOICE** Song is a complex series of musical whistles and trills, typically with a bright, loud whistle or warble followed by a very rapid high trill, ending in wavering whistles; highly variable. Calls include a sharp *whit-whit*, an annoyed chattering buzz, and numerous variations. **BEHAVIORS** Weaves through heavy cover searching for insects; sometimes sings from high perches in trees. **HABITAT** Dense shrubbery, including blackberry tangles, oak woodlands, clear-cuts; fairly common in urban and suburban settings; from valley floors to lower-elevation mountain slopes. **STATUS** Common year-round resident west of Cascades; fairly common to Columbia Basin and western Idaho; rare in eastern Idaho.

Bewick's Wren range

Bewick's Wren

Rock Wren (*Salpinctes obsoletus*)

Among the most melodious of the wrens, all of which are impressive singers, the Rock Wren is best known as a denizen of rocky habitats in the dry interior Northwest. It tends to be much easier heard than seen, owing to its excellent camouflage, but like most wrens it is fairly approachable.

LENGTH 6 inches. Grayish above, finely speckled, with lightly barred flight feathers; fairly long barred tail with black subterminal band and pale rufous outer edges (most visible in flight); buffy below; pale eyebrow stripe; long, slightly curved bill. **VOICE** Song is a varied, musical series of chirps, trills, and whistles, such as *cheer-cheer-cheer chee-chiiiirrr twirtwirtwirtwir*. **BEHAVIORS** Scampers about on boulders and rock formations; often bobs from side to side, as well as slowly up and down. **HABITAT** Arid, rocky areas, particularly scree slopes, rimrocks, boulder fields, sagebrush and scrub steppe, dry juniper or ponderosa forests; from desert floor to alpine zones. **STATUS** Common summer resident interior, rare winter resident; rare and local west of the crest of the Cascades.

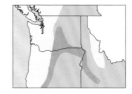

Rock Wren range

Rock Wren

Canyon Wren
(*Catherpes mexicanus*)

The distinctive call of the Canyon Wren, echoing through its canyon habitats, is one of the iconic sounds of the arid realms of the interior West. With long claws, these tiny, handsome wrens crawl effortlessly about on cliffs and rocks, probing crevices with their long bills. They often appear briefly atop a boulder or slab only to disappear from view by climbing down the other side.

LENGTH 5.75 inches. Rich rufous-brown overall, brightest on rump and tail, with contrasting white throat and breast; lightly speckled above; long, barred tail; very long, slender, slightly curved bill. **VOICE** Song is a distinctive, clear, sweetly descending, slowing series of liquidy whistles, *chechee chee chee chee cheer cheer cheer cheer.* Common call is high but raspy buzzing *zeeeeep.* **BEHAVIORS** Crawls and scampers along cliff faces, boulders, and crevices foraging for invertebrates. **HABITAT** Dry and riparian rocky desert canyons with cliffs, especially box canyons and rimrock-bound canyons. **STATUS** Fairly common year-round resident; rare in eastern Idaho. **KEY SITES** Idaho: Celebration Park (Melba); Lucky Peak State Park and Foote Park (Boise). Oregon: Malheur NWR; Sawyer Park (Bend); Succor Creek (Malheur County). Washington: Huntzinger Road (south of Wanapum Recreation Area, Kittitas County); Sun Lakes State Park and Grand Coulee; Yakima River Canyon. British Columbia: Haynes Lease Ecological Reserve (Osoyoos); Vaseux Lake cliffs (Okanagan Falls).

Canyon Wren range

Canyon Wren

Marsh Wren (*Cistothorus palustris*)

An incessant chatterer, especially during breeding season, the Marsh Wren is nonetheless surprisingly shy—it often flits into view to check on an intruder, such as a birdwatcher, then instantly disappears into the cattails or bulrushes. They abound in most marsh habitats, and while many are year-round residents, populations in colder parts of the Northwest are migratory.

LENGTH 5 inches. Warm rufous-brown above, with strongly barred tail, often cocked upward; rufous-tan flanks, white throat, and blackish back patch with fine white streaks; a distinct whitish eye line divides light face from dark crown. Juvenile paler overall, with faint eye line. **VOICE** Song is a chattering combination of buzzes, rattling trills, and scratchy whistling notes, delivered rapidly and repeatedly. Calls include an excited *chur-chur-chur-chur*, and a persistent *chit-chit-chit-chit*. **BEHAVIORS** Generally shy and retiring, although noisy; when approached, often disappears

into marsh vegetation. **HABITAT** Marshes, especially with cattail or bullrush stands; estuary and tidal creek margins; during winter and migration, also occurs in a variety of nonmarsh brushy habitats. **STATUS** Common summer resident; fairly common year-round resident west of Cascades.

Marsh Wren

Marsh Wren range

Marsh Wren

Thrushes and Bluebirds

The thrush family (Turdidae) includes one of our most familiar birds, the American Robin, as well as the startlingly beautiful bluebirds. Thrushes, some of which are melodic and even seemingly melancholic singers, feed largely on insects and other invertebrates, as well as a wide array of fruits. American Robins and Varied Thrushes sometimes mob berry bushes and fruit trees en masse, especially during fall and winter. Some thrushes are largely secretive and difficult to see, while others prefer open spaces, making them easy to observe. In the Northwest, some are year-round residents, moving about seasonally to avoid the harshest winter conditions, while others migrate southward for the winter. Throughout the Northwest (and the nation), concerted efforts have been made to mitigate for loss of bluebird nesting habitat by erecting wooden bluebird nesting boxes, which observant birdwatchers will notice attached to fence posts and poles in many places throughout the region.

American Robin

Western Bluebird
(*Sialia mexicana*)

The brightly colored Western Bluebird is a cavity nester with a lot of friends in the Pacific Northwest. In response to declining numbers of Western Bluebirds, especially in western Oregon and western Washington, erecting bluebird-specific nesting boxes in suitable habitat has becomes a popular tactic for trying to aid the bluebird's recovery. Organizations specifically dedicated to improving conditions for bluebirds include Oregon's Prescott Bluebird Recovery Project and Washington's Puget Sound Bluebird Recovery Project. These nonprofits strategize to address the causes of bluebird declines: loss of habitat, competition from nonnative starlings and English sparrows, and reduction in winter food supplies owing to widespread use of insecticides.

LENGTH 7 inches. Small, thin, black bill. **MALE** Deep blue above, darkest and brightest on head, wings, tail; chestnut-brown flanks and breast, and often chestnut-brown upper back. **FEMALE** Grayish brown above with pale blue tinge to tail and wings; light gray below, with pale brownish wash over flanks and breast. **JUVENILE** Similar to adults, but paler and less distinctly marked. **VOICE** Calls include a slightly raspy, rapid, chattering *cha-cha cha-cha-cha* (or *chir-chir*), often singly or in a series; a rapid *pi-pi-pi-pi*; and a high, airy, *pyeer pyeer* in various cadences. Song, often beginning before dawn, is a series of the *pyeer* calls. **BEHAVIORS** Often occurs in small flocks during

Male Western Bluebird

migration, especially in open, brushy areas. **HABITAT** Open woodlands and woodland edges, including burned forests and clear-cuts; savannah and prairie-type habitats with scattered trees; farmlands and parks. **STATUS** Locally common summer resident and migrant east of Cascade Mountains, uncommon in winter; uncommon summer resident, migrant, and year-round resident west of Cascade Mountains. **KEY SITES** Idaho: Hoodoo Valley and Kelso Lake area (north of Coeur d'Alene); McCroskey State Park (north of Potlatch). Oregon: Finley NWR; Idlewild Campground (north of Burns); Metolius Preserve and Camp Sherman area; Spring Creek Road/FR 21 (I-84 exit 248 west from La Grande); White City area (Jackson County). Washington: Bethel Ridge Road (White Pass); Glacial Heritage Preserve and Mima Mounds Natural Area (south of Tumwater); Little Pend Oreille NWR; Swauk Prairie (Kittitas County); Turnbull NWR; Umptanum Road (south of Ellensburg). British Columbia: Cranbrook area; Okanagan Valley.

Western Bluebird range

Female Western Bluebird

Mountain Bluebird
(*Sialia currucoides*)

Idaho's official state bird, the beautiful Mountain Bluebird certainly lives up to its name, frequenting the mountainous regions of the Northwest, but it is also the common bluebird of the high desert sagebrush steppe and juniper woodlands from southeast Idaho to central Oregon. Bluebirds are secondary cavity nesters, and they have benefited from widespread installment of bluebird-specific nest boxes throughout the Northwest, on both public and private lands. For natural cavities, they use woodpecker holes, natural cavities, and even crevices in buildings.

LENGTH 7.25 inches. Small, thin, black bill; wing primary flight feathers project to lower third of tail (shorter in Western Bluebird). **MALE** Bright azure-blue above, pale sky-blue below; white undertail coverts and belly. **FEMALE** Pale grayish tan overall, lightest on breast, with baby-blue wash over rump, tail, and flight feathers; white undertail coverts; often a slight rusty tan wash over breast and flanks; white eye ring. **JUVENILE** Similar to female, but with light streaking on the breast. **VOICE** Airy, musical, whistled *chew* notes in a drawn-out series varying in cadence; also a rapid *chip-chip-chip*. **BEHAVIORS** Frequently perches on roadside fences and power lines in open areas. Males select cavities to show off to the females they are suiting and in display, sing from atop or near the cavity, often flying back and forth between the singing perch and the nearby perch occupied by a female. **HABITAT** Open forests, including burns and clear-cuts; savannah-type habitats; sage-dominated plains and uplands; juniper woodlands; to at least 12,000 feet in breeding season; high-elevation populations move to lower elevations in winter. **STATUS** Fairly common summer resident and migrant throughout Idaho, rare winter resident in southern Idaho; fairly common year-round resident in central and southeast Oregon; fairly common summer resident and migrant in north-central Oregon and eastern Washington; fairly common summer resident of south-central and southeast British Columbia. **KEY SITES** Idaho: Castle Rock State Park and City of Rocks National Reserve; Challis Golf Course Bluebird Trail (Challis); Craters of the Moon National Monument; Elma Goodman Bluebird Trail (Glenns Ferry); Owyhee Mountains. Oregon: Hart Mountain National Antelope Refuge; Highway 20 Bend to Burns; Steens Mountain. Washington: Badger Mountain area (Waterville); Mount Spokane State Park; Okanogan Highlands; Quilomene Wildlife Area and Old Vantage Highway; Whiskey Dick Wildlife Area Bluebird Trail (east of Ellensburg). British Columbia: Anarchist Mountain; Creston area; Manning Park; Okanagan Valley; Princeton area.

Male Mountain Bluebird

Female Mountain Bluebird

Mountain Bluebird range

Townsend's Solitaire
(*Myadestes townsendi*)

A bird of high-elevation open forests, the Townsend's Solitaire can easily go unnoticed as it perches quietly on the tops and upper limbs of trees or snags, stations from which it often makes insect-catching flights, much like a flycatcher. During winter, solitaires feed heavily on fruit, especially juniper berries, and substantial concentrations of the birds may occupy prime juniper woodlots, sometimes alongside flocks of American Robins.

LENGTH 8–8.5 inches. Uniformly gray overall, slightly darker above; prominent white eye ring; long gray tail with white outer feathers; pale rusty tan inner wing stripe in flight. **JUVENILE** Dark gray with heavy to moderate (depending on age) scalloped or spotted appearance on breast and back; otherwise like adult, but tail often shorter and incomplete. **VOICE** Cheery, sometimes continuous song is finchlike, but with clearer whistles and warbles; call is a single short, bright whistle. **BEHAVIORS** Perches upright, often on high, bare limbs or tops of snags and junipers; fly-catches from perches. Flight is soft, fluttery, irregular. **HABITAT** Open conifer woodlands, including forest burns with standing snags, and juniper woodlots; occasionally mountain mahogany stands; commonly uses more open habitats in winter. **STATUS** Fairly common summer resident; uncommon winter resident, except fairly common winter resident of central and eastern Oregon; uncommon migrant west of Cascades. **KEY SITES** Idaho: Boise foothills; City of Rocks National Preserve; University of Idaho Arboretum and Gardens (Moscow). Oregon: Idlewild Campground (Harney County); Pilot Butte, Sawyer and Shevlin city parks (Bend); Ponderosa Best Western and U.S. Forest Service offices (Sisters). Washington: Okanogan Highlands; Tiffany Mountain and Freezeout Ridge area (Okanogan County); Wenas Campground and Umtanum Ridge area (south of Ellensburg). British Columbia: Okanagan Valley foothills.

Townsend's Solitaire

Townsend's Solitaire
range

Townsend's Solitaire

American Robin
(*Turdus migratorius*)

One of North America's most recognized songbirds, the American Robin is a large thrush that frequents suburban yards and parks, having proven highly adaptable to the trappings of human civilization. They often nest on manmade structures, where the nesting and rearing activities are easy to observe and follow. During winter, American Robins form flocks that can number in the hundreds, and during long-distance and local migrations, they can occur just about anywhere, from the coast to the high mountains to the sagebrush steppe and juniper woodlands of the interior Northwest.

LENGTH 10 inches. **MALE** Rich, rusty red-orange breast; gray above, with black head; black face with prominent broken white eye ring; black-and-white-striped throat; yellow bill. **FEMALE** Similar to male, but paler. **JUVENILE** Prominent dark spots on breast that ranges from cream-colored to pale orange; pale orange flanks; grayish above; face patterned with dark gray and buff. **VOICE** Song is a loud, high, musical *cheerily-cheeryup-cheeryup-cheerily-cheerup*, drawn out and repeated. Calls include a loud, oily *zewit-chuk-chuk-chuk-chuk*, and an urgent, bubbly *pup-pup* and *weeet-pup-pup*. **BEHAVIORS** Habitually hunts for worms and other invertebrates on lawns; in fall and winter, frequents fruit-bearing trees and shrubs to eat ripened or over-ripened fruit, including crabapple, berries, and even wine grapes. **HABITAT** Wooded or semi-wooded areas, including urban housing tracts, city parks, golf courses, farmlands, open forests, marshlands, juniper stands. **STATUS** Abundant and widespread year-round resident and migrant.

Male American Robin

Juvenile American Robin

Female American Robin

American Robin range

412

Varied Thrush (*Ixoreus naevius*)

Haunting the deep, dark, moist conifer woods of the Northwest mountains and foothills, the Varied Thrush ranks among the most strikingly marked songbirds in the region, and also delivers one of the iconic sounds of Northwest forests—eerie, echoing whistles that ring out at dawn and dusk. During periods of heavy snow, Varied Thrushes often invade mountain towns and lowland valleys.

LENGTH 9.5 inches. **SEXES SIMILAR** Robin-like; orange throat and orange breast with bold black necklace band (male) or pale gray necklace band (female); dark gray above, with 2 pronounced orange wingbars; black mask (gray in female) bordered above by orange eyebrow stripe; scalloped grayish flanks; prominent pale orange wing stripe in flight. **JUVENILE** Similar to female, but less distinctly marked. **VOICE** Well-spaced series of drawn-out, high, ringing and/or buzzing whistles, each lasting about 2 seconds, and each usually a different pitch; commonly delivered around dawn and in the evening. **BEHAVIORS** During fall and winter, often congregates in fruit-bearing trees and shrubs, primarily berries, and sometimes mixes with American Robins at prime feeding sites. **HABITAT** Mature and mixed-age conifer forest and mixed woodlands, including conifer-dominated residential woodlands; in winter and migration, occurs in a wider variety of habitats, including suburban areas. **STATUS** Common year-round resident; fairly common to uncommon migrant and winter resident outside of breeding range.

Female Varied Thrush

Male Varied Thrush

Varied Thrush range

Swainson's Thrush
(*Catharus ustulatus*)

The spritely, upbeat song of the Swainson's Thrush belies this common bird's shy and retiring nature. A summer resident of Northwest forests, and quite similar in appearance to the equally widespread Hermit Thrush, the Swainson's Thrush brings the forest to life with its early morning and evening crooning, especially during the courtship period of early summer. But birders wishing for a prolonged view of these birds find them challenging to locate and frequently unwilling to provide more than fleeting glimpses in the understory. The Northwest is home to 2 subspecies: the rusty brown "russet-backed" thrush inhabits the Pacific Slope, while the "olive-backed" subspecies breeds in the interior Northwest and eastward across the continent.

LENGTH 7 inches. Warm, rusty brown above (Pacific Slope) or brownish olive above (interior and Rocky Mountains), with tail similar in shade or slightly redder than back; fairly bold spotting on pale buff breast, fading to indistinct spots on lower breast; distinctive buff-colored eye rings extend forward from eyes to the lores to form "spectacles" (bolder in interior olive-backed subspecies); white throat framed by dark throat stripes. **VOICE** Song is a flutelike, warbling, upward spiraling series of clear, musical whistles. Calls include a high, sharp *whit*. **BEHAVIORS** Shy and retiring; more easily detected by song than sight. More so than Hermit Thrush, Swainson's Thrush will forage above ground in branches, briefly hover to glean insects from foliage, and even snatch flying insects. **HABITAT** In breeding season, mature closed-canopy woodlands, especially moist woods and riparian areas within woodlands; in migration, widespread in forested habitats. **STATUS** Common and widespread summer resident and migrant (May–September).

Swainson's Thrush range

Swainson's Thrush

Hermit Thrush
(*Catharus guttatus*)

Very similar to the Swainson's Thrush, and equally widespread in the Northwest, the Hermit Thrush is a ground-loving troubadour whose song reverberates through the forest on early summer mornings and evenings. Western Oregon and western Washington host wintering populations of Hermit Thrushes, and birders with backyard feeding stations in or near tree cover are especially apt to see these birds when the occasional snowstorm blankets the westside valleys.

LENGTH 6.75 inches. Warm gray-brown to grayish above with brighter cinnamon-brown tail and wing panels (Swainson's Thrush more uniform color above); white below with bold black spots on breast and belly; thin white eye ring; white throat framed by dark throat stripes. **VOICE** Song is a loud, clear whistle followed by a series of beautiful, flutelike warbling whistles, but not definitively rising in pitch like the song of the Swainson's Thrush. Calls include a high, echoing squeal, and a sharp, agitated *chir, chir-chir.* **BEHAVIORS** Habitually and distinctively bobs its tail slowly upward while flicking its wings. **HABITAT** Understory in mature woodlands, including conifer forests, hardwood stands, and mixed forests, ranging from remote isolated aspen and mountain mahogany groves in southeast Oregon, to neighborhoods and city parks within cities as large as Portland, Seattle, and Vancouver. **STATUS** Common and widespread summer resident; fairly common to uncommon winter resident of western Oregon and western Washington; rare winter visitor to southwest British Columbia.

Hermit Thrush

Hermit Thrush range

Hermit Thrush

Veery *(Catharus fuscescens)*

Of the 3 spot-breasted thrushes routinely found in the Northwest—the others being the Swainson's and Hermit Thrushes—the Veery is the most secretive, and like the others, has a unique and mesmerizing song. These ground-dwelling forest birds reach the western extent of their range east of the Cascades in the Northwest and are more common in the northernmost forests of eastern Washington and southern British Columbia than farther south. Veeries arrive in the Northwest by early June, but depart in August, leaving a narrow window of time for birders to find them.

LENGTH 7 inches. Rust-brown above; whitish buff breast with modest spotting compared to the pronounced spotting on the Hermit and Swainson's Thrushes; pale gray flanks and white belly; white or whitish throat usually framed by dark throat stripes. **VOICE** Song is a ringing, bubbly, downward spiraling musical trill, *zeeee-wit veeree-veeree-veeree*. Calls include an echoing downward warbled *cheeyou* and a similar *zoowee*. **BEHAVIORS** Secretive and retiring; spends most of its time lurking in dense understory and frequently feeds on the ground; usually detected by song. **HABITAT** Large tracts of cottonwood, aspen, alder, or willow with intact understory; riparian mixed woodlands; brushy riparian corridors. **STATUS** Summer resident (late May–early September); uncommon in Oregon and Idaho; rare south of Snake River except far eastern Idaho; fairly common in Washington (except rare in Blue Mountains); fairly common in British Columbia. **KEY SITES** Idaho: Grimes Creek Road (10 miles south of Idaho City); Kootenai NWR; McArthur Lake Wildlife Area (south of Bonners Ferry). Oregon: Catherine Creek State Park (Union County); Lookingglass Creek (difficult access); Minam State Recreation Area and vicinity, including Big Canyon Road and waysides along Wallowa River; Ochoco Ranger Station. Washington: Bullfrog Pond (Cle Elum); Number 2 Canyon (Wenatchee); Oak Creek Wildlife Area (west of Yakima); Sinlahekin Wildlife Area; Waterfront Park and Blackbird Island (Leavenworth); Wenas Campground and Wenas Creek (south of Ellensburg). British Columbia: Kokanee Creek Provincial Park (West Arm Kootenay Lake); Okanagan Valley.

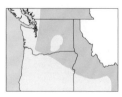

Veery range

Veery

Mimids, Pipits, Starlings, and Waxwings

Mimids (family Mimidae) exhibit incredibly complex vocal arrays, with some species frequently inserting imitations of other birds and even non-avian sounds into their repertoires. The 3 Northwest species are quite divergent: the Sage Thrasher gushes out its complex, rolling song across sagebrush plains, often long before sunrise in late spring; the Gray Catbird seems part ventriloquist—it is difficult to pinpoint the exact location from which its exotic calls emanate amid heavy cover; the Northern Mockingbird, a regular occurring species only in parts of southwest Oregon, ranks among the champion vocalists in the region. Pipits (family Motacillidae), meanwhile, are less conspicuous; they are well-camouflaged open-country ground birds. But starlings (nonnative imports from Europe) and waxwings are loud and gregarious; these birds form large flocks and often swarm favorite food sources.

Sage Thrasher

Gray Catbird
(*Dumetella carolinensis*)

Secretive, and interesting because of its spectacular vocal range and ability, the Gray Catbird is more often heard than seen. It generally inhabits extensive patches of dense shrubbery and rarely emerges to offer birders more than fleeting glances, except during late spring and early summer when males often sing from conspicuous perches. The name Catbird derives from the bird's catlike mewing call, which can be loud and persistent.

Gray Catbird

Gray Catbird range

LENGTH 8.5 inches. Slate-gray overall, with a distinctive combination of black crown and long, black tail, often cocked upward; rust-colored undertail coverts. **VOICE** Common call, often described as a catlike mewing or meow, is a drawn-out, squeaky, slightly descending *weeeaa* or *waaaaa*. Song is a complex, varied, often unhurried mix of melodius whistles, squeaks, and chatters, which can include mimicry of other birds. **BEHAVIORS** Generally secretive; feeds on the ground and in dense shrubbery and understory; males often sing from exposed or semi-exposed perches, such as limbs and shrub crowns extending above dense vegetation. **HABITAT** Brushy riparian corridors, primarily in semi-arid regions. **STATUS** Fairly common summer resident (late May–mid-September); rare winter visitor. **KEY SITES** Idaho: Camas NWR; Chain Lakes and Lane Marsh (Coeur d'Alene River); Foote Park and Boise River Greenway (Lucky Peak Dam to Boise); Mann Lake (near Lewiston); Phillips Farm County Park (north of Moscow); River Front Park (McCall); South Fork Snake River. Oregon: Rhinehart Bridge and Canyon (Union County). Washington: Blackbird Island and Leavenworth Fish Hatchery (Leavenworth); Little Spokane River Natural Area and Riverside State Park; Sinlahekin Valley and Sinlahekin Wildlife Area; Toppenish NWR. British Columbia: Creston Valley Wildlife Area; Haynes Point Provincial Park (Osoyoos Lake); Penticton area (Okanagan River and Vaseux Lake); Road 22, north end of Osoyoos Lake.

Sage Thrasher
(Oreoscoptes montanus)

One of the iconic songbirds of the sage-brush plains of the West, the Sage Thrasher reaches its highest Northwest population densities in the Great Basin of southeast Oregon as well as southwest Idaho (and adjacent Nevada). In these remote areas, an early morning summer drive through any large sagebrush basin is likely to reveal Sage Thrashers flitting quickly across the road, low to the pavement. Farther north, however, they become increasingly scarce and are a species of special concern in British Columbia.

LENGTH 8.5 inches. Gray-brown above, lightly mottled; medium-length tail with white-tipped outer feathers; off-white below with dense speckling of dark spots; thin, black-tipped bill and yellow eyes; juveniles and worn adults in later summer are less boldly speckled below. **VOICE** Song is a loud, melodious, rapidly delivered mixture of a wide array of mostly soft warbles, trills, whistles, and other notes that can go on uninterrupted for minutes. On late spring and early summer mornings, especially with bright moons, singing may begin long before sunrise (or even carry on all night), and routinely begins before dawn. Alarm call is a persistent *chuck*. **BEHAVIORS** Often sings from and perches on top of sage-brush and other desert shrubs, and also while making lively display flights; when disturbed, flushes downward, flying low to the ground to another shrub. On perches, habitually slowly raises and lowers its tail. Frequently runs instead of flies. **HABITAT** Sagebrush plains, greasewood-dominated salt desert; montane scrubland. **STATUS** Summer resident (April–September) and rare winter resident. Common in southeast Oregon and southern Idaho; fairly common in Columbia Plateau, Washington;

Adult Sage Thrasher

uncommon elsewhere in eastern Washington; rare in southeastern British Columbia (and listed as endangered in Canada). **KEY SITES** Idaho: Craters of the Moon National Monument; Duck Valley Indian Reservation; Sage Junction to Hamer Frontage Road (Old Highway 91; Jefferson County); Silver City Road (Owyhee County); Snake River Birds of Prey National Conservation Area. Oregon: Guano Basin (Lake County); Fields-Denio Road; Fort Rock State Park; Hart Mountain National Antelope Refuge; Highway 395 Riley to Lake Abert; Malheur NWR. Washington: Beezley Hills (Baird Springs Road and Monument Hill Road, east of Ephrata); Quilomene Wildlife Area and Old Vantage Road (Kittitas County); Umtanum Road (south of Ellenburg). British Columbia: White Lake area (west of Okanagan Falls).

Sage Thrasher range

Juvenile Sage Thrasher

Brown Thrasher
(*Toxostoma rufum*)

A familiar bird throughout the east half of the United States, the Brown Thrasher is a rare vagrant in the Northwest. These handsome birds—among the continent's most skilled vocalists—are generally unlikely to be confused with other, more common Northwest species, provided the observer gets more than a fleeting glimpse. Over the past few decades, Oregon has about 30 accepted records—enough that the Oregon Birds Record Committee no longer reviews sightings. Idaho has about 30 recorded sightings as well, though not all of them have been reviewed by the Idaho Birds Record Committee. Brown Thrasher records for Washington and British Columbia number more than a dozen each. Most Northwest sightings occur in late spring or fall

(with a few winter records)—migrating Brown Thrashers that are well off course.

LENGTH 11.5 inches. Rich rusty brown above, with a long, rust-colored tail and 2 narrow wingbars; buffy white below with pronounced dark spots forming streaks along the breast, belly, and flanks; slender, slightly decurved bill; yellow eyes. **VOICE** Calls include a husky *chuck* and a high, squeaky, abrupt kissing sound. Rarely found singing in the Northwest, but song is a tremendously musical, exotic combination of chirps, whistles, warbles, and other sounds. **BEHAVIORS** Generally somewhat skulking and secretive, but will come to backyard bird feeders; feeds on the ground, using its bill to pick through debris; typically walks and runs, sometimes hops. **HABITAT** Brushy areas, including hedgerows, parklands, backyards, roadways, and brambles. **STATUS** Very rare vagrant; most sightings are in May, June, and autumn, and are geographically unpredictable.

Brown Thrasher

Northern Mockingbird
(*Mimus polyglottos*)

A common, familiar, vocal bird through-out much of the United States, including California, the Northern Mockingbird's primary range does not reach into the northern tier of states. Nonetheless, mockingbirds turn up regularly and unpredictably throughout the North-west. Most sightings occur in Oregon and Washington, far fewer in British Columbia and Idaho. Both Oregon and Washington have hosted nesting pairs on rare occasions, leading to speculation of possible northward expansion of the species in the far West.

LENGTH 10 inches. Light gray above, whitish below; in flight it shows conspicuous white wing patches; fairly long, dark tail has white outer feathers that are especially conspic-uous in flight; short bill, light colored eye, and dark eye line. **VOICE** Typical call is a rapid, high, scolding *cheep* or *cheeup*; songs are complex and highly varied, and typically include a number of different, often highly divergent, whistles, trills, and other sounds repeated and strung together into a lengthy and startling repertoire; frequently includes mimicry of other birds. **BEHAVIORS** Often feeds by hopping, walking, and running on the ground, including lawns and other landscaped areas, and often makes repeated wing flashes by quickly spreading and raising wings. **HABITAT** Brushy margins, yards, parks, riparian areas in urban, sub-urban, and farm regions; riparian areas in desert-scrub regions of the interior North-west. **STATUS** Rare but regular winter visi-tor, very rare breeder. **KEY SITES** The most consistent area for finding mockingbirds in the Northwest has generally been Jackson County in southern Oregon, especially near the Rogue River west of White City, includ-ing Denman Wildlife Area, The Nature Conservancy's Whetstone Savanna Reserve, and along Kirtland Road.

Northern Mockingbird in flight

Northern Mockingbird

Northern Mockingbird range

European Starling
(*Sturnus vulgaris*)

North America's most notorious and successful feathered invasive species, the European Starling was first introduced—purposefully—to New York City in the 1890s and has since become one of this continent's most abundant birds. Estimates place the continental population at about 200 million. These aggressive, highly adaptable mimics prefer cavities for nesting and compete with native birds, such as bluebirds, for such nesting sites. Starlings, in huge flocks, are also culpable in damaging raids on crops and livestock feed. Despite the disdain with which they are treated, starlings are here to stay, and they are, in fact, attractive and oft-entertaining birds.

LENGTH 8.5 inches. **ADULT BREEDING** Iridescent black-green overall, with iridescent purple head and neck; belly and back feathers tipped with silvery white to varying degrees; bright yellow daggerlike bill; short tail. **ADULT NONBREEDING** Fall and winter birds are blackish overall, each body feather tipped with a silvery white chevron, giving the bird a heavily speckled appearance. **JUVENILE** Plain gray overall; head and neck remains gray for a time after body feathers are molted to adult speckled nonbreeding pattern. **VOICE** Calls include a wide array of whistles, screeches, warbles, and other noises. Skilled mimics, starlings imitate many different birds and other sounds, including human speech and mechanical noises. **BEHAVIORS** In winter, starlings often form massive, dense flocks (called murmurs) that move in long, undulating waves. They nest in cavities and crevices and frequently find suitable nest sites on buildings. **HABITAT** Urban, suburban, and agricultural areas; also wild lands of many descriptions. **STATUS** Abundant year-round resident.

Adult nonbreeding European Starlings

European Starling range

Adult breeding European Starling

American Pipit (*Anthus rubescens*)

In the Far North of Canada and Alaska, American Pipits reach the zenith of their breeding range, but similar tundra habitat in the Northwest is found only in the highest mountains. Hence these birds nest on the volcanic peaks of the Cascades and the high ramparts of the Rockies, along with a few isolated mountain ranges in between. During migration and in winter, however, they occur on a variety of wide-open habitats, from agricultural fields to broad lakeshores and even beaches. Nondescript and easy to miss because they tend to blend in well with their surroundings, American Pipits rarely perch and so are almost always seen on the ground or in flight.

LENGTH 6.25 inches. Uniform pale brownish gray above; warm, light tan below, with dark speckles and streaks on breast and flanks to varying degrees; fairly long brown tail has white outer feathers; unmarked pale tan throat and chin; pale eyebrow stripe; fairly short, thin, mostly dark bill. **VOICE** Calls include a sharp *pipit-pipipit* and a simple *whit*. Song is a persistently repeating, rapid, airy *pseew-pseew-pseew-pseew*. **BEHAVIORS** Habitually bobs tail, and walks upright while searching for food. **HABITAT** Pastures, tilled fields, lakeshores, mudflats during migration and winter. Nests on alpine tundra and high-elevation steppe. **STATUS** Common migrant and fairly common winter visitor; locally common breeder in Washington, British Columbia, and Idaho; locally uncommon breeder in Oregon and southern Idaho. **KEY SITES** Idaho: American Falls Reservoir; Deer Flat NWR. Oregon: Agate Lake and Kirtland Road ponds (White City); Emigrant Lake (Jackson County); Fern Ridge Wildlife Area; Finley NWR. Washington: Gog-Le-Hi-Te Wetlands (Tacoma); Nisqually NWR; Skagit Wildlife Area. British Columbia: Boundary Bay and Iona Island; Martindale Flats (Saanich Peninsula); Okanagan Valley.

American Pipit range

American Pipit

Bohemian Waxwing
(*Bombycilla garrulus*)

A winter visitor to the Northwest from Canada, the colorful and gregarious Bohemian Waxwing feeds heavily on lingering fruit and berry crops, including highbush cranberry, juniper, cedar, rose hips, hawthorn, crabapple, mountain ash, serviceberry, and myriad domestic varieties. Large flocks often feed heavily in one location until the food source is exhausted and then move on, making these birds somewhat nomadic, but also frequently tied to towns and suburbs where planted food sources abound. Numbers vary substantially from year to year. In some years, large flocks of Bohemian Waxwings occur in numerous Northwest locations; other years, the birds are quite scarce.

LENGTH 8 inches. Uniform pale gray below (Cedar Waxwing has pale brownish breast and pale yellowish belly); brownish rust undertail coverts (whitish in Cedar Waxwing); yellow-tipped tail; brownish gray head with long, feathery crest, and handsome black face mask and chin; white wing patches both in flight and when perched. Often appears much stockier than Cedar Waxwing, and primary flight feathers show bright yellow and/or white tips at close range, along with tiny bright red tips on secondaries. **JUVENILE** Similar to adult, but more grayish overall, and softly streaked below (juvenile plumage may be retained into winter). **VOICE** Call is a rapid, high, harsh, bouncing trill. **BEHAVIORS** Often forms large winter flocks at prime feeding sites; occasionally feeds so heavily on fruits that have fermented on the bush that the birds become drunk and bumbling; flocks are typically very noisy. **HABITAT** Often found in residential, suburban, and agricultural areas where fruit-bearing trees and shrubs retain their fruit into winter. **STATUS** Rare but fairly regular winter visitor to northeast Oregon and southeast Washington; rare in southern Idaho; fairly common winter visitor in northeast Washington, northern Idaho, south-central and southeast British Columbia. **KEY SITES** Idaho: Bonners Ferry and Sandpoint; University of Idaho Botanical Gardens and Moscow area. Oregon: Wallowa Valley. Washington: Okanogan Highlands; Spokane Valley. British Columbia: Castlegar and Nelson; Creston area; Okanagan Valley.

Bohemian Waxwing range

Adult Bohemian Waxwing

425

Cedar Waxwing
(*Bombycilla cedrorum*)

One of the Northwest's common suburban birds, and also common in more remote settings, the Cedar Waxwing feeds primarily on fruits, especially berries. It supplements its diet regularly with insects, which it often snatches out of the air. Outside of the breeding season, waxwings form flocks—often large, gregarious flocks—that descend upon prime food sources and feed ravenously (and it's worth searching such flocks for the occasional Bohemian Waxwing mixed in). Waxwings, named for the bright red, waxy tips on their secondary flight feathers, are especially notorious in the Northwest for descending upon fields of berry crops and wine grapes.

LENGTH 7 inches. Soft brown above, with grayish flight feathers and rump; pale brownish breast transitioning to pale yellowish belly and white undertail coverts; distinctive feathery crest and black face mask; yellow-tipped tail; small bright red tips on secondaries. In flight, shows plain unmarked grayish wings. Typically appears more slender than stocky Bohemian Waxwing (a winter visitor only). **JUVENILE** Similar to adult, but streaked breast and less black in face and chin. **VOICE** Calls include an airy, high-pitched *zeeee*, and high, rapid, insectlike trills. **BEHAVIORS** Feeds heavily on fruits and berries, and in fall and winter may swarm prime feeding sites in large, noisy flocks. Often hawks insects out of midair like a flycatcher. **HABITAT** Found in myriad habitats where fruits and berries are prominent; generally nests in young, open, deciduous or mixed woods with brushy understory and edges; common in residential areas. **STATUS** Widespread and generally common year-round resident.

Cedar Waxwing range

Adult Cedar Waxwing

Warblers

The warbler family (or more correctly, the wood-warbler family, Parulidae) includes some of our most colorful songbirds. Warblers are small, energetic gleaners, feeding largely on insects and other invertebrates found in foliage. Some species also routinely hawk insects in midair, flycatcher style, and many will readily take various small fruits, nectar, and even meaty seeds, such as those of hulled sunflowers.

During migration, warblers often form mixed-species flocks, making such congregations especially popular with birdwatchers who hope to find the occasional vagrant species among the Northwest regulars. All warblers in the Northwest are migratory, but individuals of a few species, especially Yellow-rumped and Townsend's Warblers, routinely overwinter in moderate-climate areas west of the Cascades.

MacGillivray's Warbler

Yellow-rumped Warbler
(*Setophaga coronata*)

Perhaps no other warbler is more recognizable to Northwest birders than the widespread Yellow-rumped Warbler. From its striking spring plumage to its penchant for visiting bird feeders, this warbler is always a welcome sight. The Yellow-rumped Warbler includes 2 subspecies that were formerly considered separate species: the Audubon's Warbler is the Northwest's resident breeding subspecies, while the "Myrtle" Warbler is an uncommon migrant and winter visitor, which, like other species of warbler, often occurs with flocks of Yellow-rumps.

LENGTH 5.5 inches. **MALE BREEDING** Bluish gray above with black streaks on mantle; bright yellow cap, throat, and rump; black breast band bordering white belly; white undertail coverts; yellow flanks bordered below by heavy black streaks; bold, partial white eye ring and black lores; large white wing patches; white patches on each side of tail. **FEMALE BREEDING** Similar to male, but with gray face; yellow cap, white wingbars, and a dark-streaked breast. **"MYRTLE" SUBSPECIES** Breeding male is similar to Audubon's, but throat is white instead of yellow, and wings have 2 white bars instead of a large white patch; also white supercilium. Breeding female similar to male, but mask not as dark and supercilium not as evident. **ADULT NONBREEDING** (both subspecies) Light gray-brown above; streaks on a buff-colored breast and belly; throat ranges from yellowish to white; yellow flank patches greatly reduced from spring or absent; yellow rump; note that Audubon's subspecies lacks white supercilium, which is always evident to some extent in "Myrtle"

First-fall female Yellow-rumped Warbler

Adult nonbreeding Yellow-rumped Warbler

Warbler subspecies. **JUVENILE** (both subspecies) Brown above with fine streaks, including on crown; buffy below with brown streaks; lacks yellow rump. **VOICE** Song is a whistled, warbled *siti-siti-siti-siti-siti-surisur*; call is a loud *chep*. **BEHAVIORS** Often seen in loose flocks as they persistently and thoroughly glean insects from foliage at all levels, from ground to tree canopies. Adept fly catchers, they often make long, acrobatic flights chasing insects. Common visitor to bird feeders, especially suet feeders. **HABITAT** Breeds in conifer forest (and Pacific madrone in the San Juan Islands). Outside of breeding season they inhabit a variety of habitats. **STATUS** Widespread and common summer resident; abundant migrant; uncommon winter resident.

Male breeding "Myrtle" Warbler

Yellow-rumped Warbler range

Female breeding Yellow-rumped Warbler

Male breeding Yellow-rumped Warbler

Hermit Warbler
(*Setophaga occidentalis*)

This beautiful warbler of the mountains of western Oregon and Washington is heard more than seen, as it forages in the forest canopy. Hermit Warblers and Townsend's Warblers interbreed where their nesting ranges overlap in the Cascades, and hybrids often have the Hermit Warbler's head colors with the Townsend's Warbler's underparts, with wide variation.

LENGTH 5 inches. **MALE** Lemon-yellow face and crown contrasting vividly with black throat; black nape; gray back with black streaks; white below; white wingbars; white outer tail feathers edged in dark gray. **FEMALE** Similar to male but slightly paler, with a throat that is less solid black. **JUVENILE** Similar to female, but duller color overall, with unmarked throat and brown back. **VOICE** Song is a series of high, buzzing, rising *zee* notes, with much variation, such as *zeezeezee-zeeeup-zeeeup-zeeeup* or *cheecheechee-zeeeup-zeeeup-zee-zee*; similar to song of Townsend's Warbler, but more drawn out and less buzzy. **BEHAVIORS** Generally remains in canopy, though briefly common at lower levels upon arrival in late spring; feeds both by gleaning insects from trees and by fly-catching. **HABITAT** Mature conifer forests, especially Douglas fir-, true fir-, hemlock-, or spruce-dominated stands. **STATUS** Common summer resident and uncommon migrant (April–September); very rare winter visitor. **KEY SITES** Oregon: Gold Lake (Lane County); Little Crater Lake (Clackamas County); Maxwell Butte Sno-park (Highway 22); Mount Ashland. Washington: Capital State Forest (Olympia); Mount Rainier National Park.

Hermit Warbler

Hermit Warbler

Hermit Warbler range

Townsend's Warbler
(*Setophaga townsendi*)

The striking Townsend's Warbler is primarily a denizen of coniferous forests, but many remain year-round in the lowlands west of the Cascades. During the drab Northwest winters, these little splashes of bright color are welcome surprises at backyard feeders and will often visit daily into early spring.

LENGTH 5 inches. **MALE** Overall, contrasting pattern of bright yellow and black; black cheek patch (with yellow lower eyelash patch) surrounded by bright yellow; black throat and crown (crown can be dusky olive black in some birds); olive back with dark streaks; bright yellow breast with black-streaked flanks; white belly; blackish wings with 2 white wingbars; black tail with white outer feathers. **FEMALE AND JUVENILE** Similar in overall pattern to adult male, but more muted in color, with olive markings instead of black, and yellow throat. **VOICE** Song is a series of high-pitched buzzing, rising *zeee* notes, with much variation, such as *zee-zee-zee, zeeeup-zeeeup, zeeeup* or *chee-chee-chee-zeeeup-zeeeup-zee-zee.* **BEHAVIORS** Forages for insects at all levels, from near the ground to the forest canopy.

Wintering birds will eat suet and seeds, especially shelled sunflower chips. **HABITAT** Nests in coniferous forests from the foothills into the high mountains; winters in lowlands, in a variety of habitats, usually within or near conifers. **STATUS** Fairly common summer resident (May–September); uncommon winter resident.

First-fall juvenile Townsend's Warbler

Female Townsend's Warbler

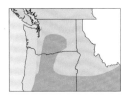

Townsend's Warbler range

Male Townsend's Warbler

431

Black-throated Gray Warbler
(*Setophaga nigrescens*)

The striking Black-throated Gray Warbler, to an astute observer, is actually nearly identical in pattern to the Townsend's Warbler, but the latter has yellow highlights. In fact, the 2 species—both of which nest in the Northwest—are closely related. Often, the best way to steal a glimpse of the Black-throated Gray Warbler in its forest habitat is by following its buzzy song.

LENGTH 5 inches. In all plumages, note small yellow spot at the base of the bill, and dark-edged white outer tail feathers. **MALE** Black-and-white head with bold white supercilium and moustache stripes; yellow supraloral spot; dark gray back with black streaks; white wingbars; black throat and breast, and black streaks along white flanks. **FEMALE** Similar to male, except no streaks on the back, and throat is white. **JUVENILE** Similar to female, but grayer overall; incomplete breast band. **VOICE** Song is a buzzy, whistled *zeesu-zeesu-zeesu see-see*; call note is a flat *chut*. **BEHAVIORS** Forages in mid-level foliage for insects, with males typically foraging higher than females. **HABITAT** Juniper woodlands, open conifer forest, mixed woodlands; partial to oak woodlands west of Cascades. **STATUS** Common summer resident and migrant (May–October); rare winter resident. **KEY SITES** Idaho: Castle Rocks State Park; Cress Creek Nature Trail (near Idaho Falls). Oregon: Finley NWR; Page Springs Campground (Harney County); Tualatin River NWR. Washington: Marymoor Park (Redmond); Olympic National Park. British Columbia: E.C. Manning Park; Maplewood Conservation Area (Vancouver).

SIMILAR SPECIES The Black-and-white Warbler (see next), a vagrant, has a black-and-white streaked mantle rather than a gray mantle, and lacks yellow lores; its broad white eyebrow stripe envelopes the top of the eye, whereas the Black-throated Gray Warbler's eye is entirely within the broad black eye stripe.

Black-throated Gray Warbler range

Black-throated Gray Warbler

Black-and-white Warbler
(*Mniotilta varia*)

This striking eastern warbler, a rare but regular visitor to the Northwest, is typically seen during May and early June in migrant traps of the interior Northwest (oasislike wooded enclaves in otherwise open country); occasionally a Black-and-white Warbler shows up west of the Cascades. Nuthatchlike in its habits, the Black-and-white Warbler's strong legs and long bill are adaptations for probing bark crevices for prey.

LENGTH 5.25 inches. **MALE** Distinctively streaked black and white; black striped crown with white median; black cheek and throat; long, thin, slightly decurved bill. Fall males may have white throat and smaller black cheek patch. Immature male similar to adult, but with white cheek patch. **FEMALE** Narrower flank streaks than male; pale gray to buff cheeks; black stripe extends rearward from eye; white chin and throat. Fall females may have buff-colored underparts. **VOICE** Song is a dry, high-pitched *see-it see-it see-it see-it*; call is a high *chink*, like stones struck together. **BEHAVIORS** Forages on tree trunks and branches like a nuthatch; can climb up and down trunks; also gleans insects from foliage. Often found with other warblers during migration. **HABITAT** Deciduous trees, especially cottonwoods and willows. **STATUS** Rare but annual vagrant, primarily in late spring.

Male Black-and-white Warbler

Blackpoll Warbler
(Setophaga striata)

Most Northwest sightings of Blackpoll Warblers occur during fall migration, after the males have molted into a plumage dramatically different than their striking spring dress. A vagrant in the region, the Blackpoll Warbler is a breeder of the Far North, its summer range extending from eastern Canada all the way to Alaska, although it is rare in the western half of the continent.

LENGTH 5.5 inches. Short tail and long undertail coverts. **ADULT NONBREEDING** Light olive above with dark streaks on mantle; pale yellowish or dusky whitish below with dark streaks on flanks; pale legs, and pale feet with yellow pads; 2 pale yellow or whitish wingbars. **MALE BREEDING** Black cap extending down to eyes; white cheeks; white throat framed by black malar stripes; white below, with black streaks along flanks; gray above with black streaks; 2 white wingbars; yellow legs and feet. **FEMALE BREEDING** Pale gray above, slightly cast in olive, with darker streaks; crown often finely streaked; white below; white undertail coverts; bright yellow legs and feet. **JUVENILE** Similar to adult nonbreeding, but more yellowish below and less streaking on flanks. **VOICE** Song is a series of high, rapid, insectlike *tsit* notes; call is a very high *tsip*. **BEHAVIORS** Forages more slowly than most warblers, somewhat like a vireo. **HABITAT** Varies, depending on food supplies. **STATUS** Rare fall vagrant, very rare spring vagrant.

Male breeding Blackpoll Warbler

First-winter female Blackpoll Warbler

Female breeding Blackpoll Warbler

Magnolia Warbler
(*Setophaga magnolia*)

With a breeding range extending from Canada's Maritime Provinces and the New England states westward and northward across Canada to northeastern British Columbia, it is little wonder that vagrant Magnolia Warblers occasionally show up in the Northwest. Most sightings occur in desert-oasis enclaves in otherwise open country. In his spring breeding plumage, the male is a striking mix of black, yellow, and white; in both sexes and all plumages, the Magnolia Warbler's black-and-white tail pattern is unique among warblers.

Male breeding Magnolia Warbler

LENGTH 5 inches. Broad white band across middle of black tail. **MALE BREEDING** Gray crown, black mask, and bright yellow throat; bold white supercilium; yellow throat with black necklace; yellow breast and flanks with bold black streaks; black back; yellow rump; white undertail coverts. **FEMALE BREEDING** Similar to male, but with gray mask and grayish olive back. **ADULT NON-BREEDING** Gray instead of black mask; white eye ring, but lacks white supercilium; greenish olive back; reduced black necklace; narrow white wingbars. **JUVENILE** Similar to adult nonbreeding, but with no streaking on the breast and a gray neckband. **VOICE** Song is a short, high-pitched *weeta-weeta-wetah*; call is a dry *jeep*. **BEHAVIORS** Forages in brush and tangles to feed on insects; sometimes fans its tail like an American Redstart. **HABITAT** In the Northwest, most sightings occur in deciduous trees near water. **STATUS** Rare spring and fall vagrant.

Female breeding Magnolia Warbler

Orange-crowned Warbler
(Oreothlypis celata)

The widespread Orange-crowned Warbler nests in the Northwest, and some
also overwinter in the region. Despite the
bird's name, its orange crown is rarely
seen. The Pacific subspecies (*Oreothlypis
celata lutescens*), found west of the Cascades, is brightest. Interior birds (*O.c.
orestra*) are duller, often with a gray head
slightly contrasting with the drab, pale
yellowish body. The Tennessee Warbler,
which is a rare vagrant in the Northwest, is similar to the Orange-crowned
Warbler, but the latter always has yellow
undertail coverts, while the Tennessee,
which appears short-tailed, has white
undertail coverts.

LENGTH 5 inches. Greenish yellow overall,
brightest in coastal birds; yellow below, with
diffused streaking; pale eyebrow stripe, eye
line, and partial eye ring; some birds have
grayish heads which may indicate juveniles,

Orange-crowned Warbler, Pacific subspecies
(spring)

Orange-crowned Warbler, interior subspecies

but drab interior birds (*Oreothlypis celata orestra*) are also gray headed, especially in fall; sharp, slightly decurved bill; 2 faint wingbars. **VOICE** Song is a trill that rises and/or descends in pitch. **BEHAVIORS** Often hovers when gleaning insects from foliage; visits sap wells created by sapsuckers for insects caught in the sap, and the sap itself. **HABITAT** Riparian thickets, older clear-cuts; conifer and mixed woodlands with shrub understory, deciduous woodlands. **STATUS** Common summer resident; common spring and fall migrant; uncommon winter resident (primarily west of Cascades).

Orange-crowned Warbler range

Pale Orange-crowned Warbler, Pacific subspecies

Orange-crowned Warbler, Pacific subspecies (fall)

Tennessee Warbler
(Oreothlypis peregrina)

A plain-looking warbler that is easily confused with the far-more-common Orange-crowned Warbler as well as the Warbling Vireo, the Tennessee Warbler is a vagrant to the Northwest and turns up annually. The Warbling Vireo has a larger bill than the Tennessee Warbler; the Orange-crowned Warbler has yellowish rather than white or whitish undertail coverts, and faint streaks on the underparts.

LENGTH 5 inches. Plump and short-tailed. **MALE** In spring, greenish above, white below; grayish nape and cape with white supercilium and dark eye line; in fall, buff-yellow below, with white undertail coverts. **FEMALE** Similar to male, except tinged with yellow below, and duller overall. Juvenile similar to female, but more yellowish below, and undertail coverts may be tinged yellow. **VOICE** Song is a high, loud, forced *chi chi chi chi chi* zwitzwitzwit-*cheecheecheecheecheecheecheechee* (rarely sings during migration); call is a high *chip*. **BEHAVIORS** Gleans insects while flitting amongst foliage; more frenetic than vireos. **HABITAT** Deciduous and mixed woodlands. **STATUS** Rare vagrant.

Tennessee Warbler

Yellow Warbler
(*Setophaga petechia*)

The striking little Yellow Warbler, which is especially fond of willow thickets, is one of the Northwest's more common warblers, and a favorite of many bird-watchers. Brown-headed Cowbirds, which lay their eggs in the nests of other birds, frequently parasitize the nests of Yellow Warblers, but these tiny lemon-yellow songsters have developed at least a partial defense: when a cowbird lays an egg in the warbler's nest, the warbler often builds a new nest on top of the original.

LENGTH 5 inches. Dark eyes; short light olive tail with elongated yellow spots. **MALE** Bright yellow overall, with varying-intensity chestnut-colored streaks on breast and flanks; plumage becomes duller after breeding season. **FEMALE** Paler yellow than male, with faint chestnut streaks on breast and flanks. **JUVENILE** Similar to female, but with no streaks. **VOICE** Song is a clear,

high-pitched sweet-sweet-sweet-sweete r-than-sweet; call is a single sharp tseep. **BEHAVIORS** Flits through foliage, gleaning insects; makes short fly-catching flights. **HABITAT** Deciduous riparian habitats dominated by willows and other dense brush; wet meadows with willow groves. **STATUS** Common summer resident and migrant (late April–late September).

Female Yellow Warbler

Yellow Warbler range

Male Yellow Warbler

Palm Warbler
(*Setophaga palmarum*)

With most of its breeding range in the boreal forest belt across central and eastern Canada, Palm Warbler might seem like an odd name for this rare but regular fall and winter visitor to the Northwest. However, the species was first described for science in the 18th century from a wintering specimen found in the Caribbean—the bird's primary wintering grounds—by the German naturalist Johann Friedrich Gmelin. In the Northwest, Palm Warblers show up annually in fall and winter, mostly along the coast.

LENGTH 5.5 inches. **ADULT NONBREEDING** Drab grayish brown overall, with faint streaks on flanks; yellowish rump and yellow undertail coverts; dark eye line; white corners of tail visible in flight. **ADULT BREEDING** (rare in Northwest) Rufous crown; dark eye line and malar stripe; grayish buff above with faint streaks on mantle;

unmarked yellow throat; pale buff below with faint streaks along flanks; olive-yellow undertail coverts and rump; white corners of tail visible in flight. **VOICE** Song is a series of musical *zeeki* notes; call is a sharp, high-pitched *chik*. **BEHAVIORS** Constantly wags tail up and down; often chases insects on the ground. **HABITAT** Weedy fencerows, low-growing shrub borders, vegetated sand dunes; primarily coastal, but occasionally found inland. **STATUS** Rare but regular fall and winter visitor; very rare spring migrant. **KEY SITES** The most consistent locations for Palm Warblers in the Northwest include the nature trail and grounds at Oregon's Mark Hatfield Marine Science Center in Newport and the Ocean Shores area in Washington.

Palm Warbler range

Adult nonbreeding Palm Warbler

Wilson's Warbler
(*Cardellina pusilla*)

Male Wilson's Warbler

The beautiful little Wilson's Warbler tends to present itself as a frenetic bright yellow blur, constantly in motion, flitting rapidly about in shrubbery, searching for insects. It shows up in a variety of habitats—almost anywhere during migration. Individuals spotted west of the Cascades, where Wilson's Warblers are most numerous, tend to be brighter than those in the interior Northwest. The large, dark eye stands out against its yellow face.

LENGTH 4.75 inches. **SEXES SIMILAR** Bright olive-green above, male with black crown; bright yellow below; dark eye stands out prominently on yellow face. **VOICE** Song is a series of rapid, high *tsip* notes, dropping in pitch and ending with an emphatic *chu*; call is a sharp, gruff *chip*. **BEHAVIORS** Forages in heavy shrubbery and small trees with ample cover; very active, hopping and flitting nonstop from branch to branch. **HABITAT** Tall shrubbery and willow thickets in moist areas; brushy coastal clear-cuts; blackberry patches; open mixed woodlands with brush understory. **STATUS** Common summer resident and migrant (April–October).

Female Wilson's Warbler

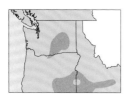

Wilson's Warbler range

441

MacGillivray's Warbler
(*Geothlypis tolmiei*)

The beautiful MacGillivray's Warbler is surprisingly common but often difficult to observe because it tends to skulk in heavy cover, rarely emerging. However, its metallic *chip* calls often reveal its presence. Widespread, MacGillivray's Warbler seems to thrive in areas that have been disturbed by clear-cuts or fires. Because birders often get no more than a fleeting glimpse of a MacGillvray's Warbler, the bird's broken white eye ring is the primary diagnostic field mark.

LENGTH 5.25 inches. **SEXES SIMILAR** Dark gray head, breast, nape (lighter gray on female); distinctive broken eye ring forming white arcs above and below eyes; yellow belly and undertail coverts; yellowish olive above. **VOICE** Song is a rolling, sweet *chee-chee-chee-chee-chee-chuwee*; call is a sharp, toneless *chip*. **BEHAVIORS** Skulking behavior in thick cover makes it difficult

to detect, but males are very vocal during breeding season. **HABITAT** Low shrub habitat in mixed forests; conifer forests with clear-cuts or fire damage; arid mountain drainages with deciduous trees and understory. **STATUS** Fairly common summer resident and migrant (April–September).

Female MacGillivray's Warbler

MacGillivray's Warbler range

Male MacGillivray's Warbler

Nashville Warbler
(*Oreothlypis ruficapilla*)

This handsome little ground-nesting warbler breeds throughout the Northwest and is a welcome sight to many because of its bold colors, but is typically secretive, sticking to heavy cover. Unlike the region's other gray-headed warblers—the widespread MacGillivary's Warbler and the rare Virginia's Warbler—the Nashville Warbler has a yellow throat and complete, bold, white eye ring.

LENGTH 5 inches. **SEXES SIMILAR** Gray head with bold, white eye ring; olive upperparts and tail; yellow throat and breast; white belly and yellow undertail coverts; rufous cap is rarely visible. **VOICE** Song is a high, sweet, rich *tesee-tesee-tesee-tesee-sweeswee-sweeswee*; call is a sharp, dry *chic*. **BEHAVIORS** Frequently pumps tail; gleans insects from foliage. Males sing from forest canopy and often forage there. **HABITAT** Nests on or near the ground in deciduous stands in lowland valleys, and in shrub and deciduous openings in conifer forests; more widespread in migration. **STATUS** Uncommon but widespread summer resident and migrant (April–September). **KEY SITES** Idaho: Intermountain Bird Observatory (Ada County); Tubbs Hill Park (Coeur D'Alene). Oregon: Suttle Lake; TouVelle State Park and Ken Denman Wildlife Area (Jackson County). Washington: Calispell Lake (Pend Oreille County); Mount Rainier National Park. British Columbia: E.C. Manning Provincial Park; Okanagan Valley.

Nashville Warbler

Nashville Warbler range

Nashville Warbler

Virginia's Warbler
(*Oreothlypis virginiae*)

The range of the Virginia's Warbler barely reaches the Northwest, where it breeds in just a few high-elevation locations in extreme southern Idaho; suspected nesting in southern Malheur County, Oregon, has not been confirmed. Its specialized nesting habitat, especially steep slopes with stands of mountain mahogany, tends to be remote and rugged, making precise delineation of its very limited Northwest range difficult.

Virginia's Warbler

Virginia's Warbler range

LENGTH 4.75 inches. **SEXES SIMILAR** Grayish overall, with varying amounts of yellow on the breast (more extensive on the male); yellow rump and undertail coverts; white eye ring; rufous patch on crown often hidden. **VOICE** Song is a melodious series of up to 10 *twee* notes followed by 5 lower-pitched *chuwee* notes; call is a sharp *chink*. **BEHAVIORS** Frequently bobs tail; flight is fast and direct. Gleans insects at various levels, from the ground to up in trees; prefers feeding amid leafy vegetation, but will also forage in junipers and sagebrush. **HABITAT** Nests in stands of mountain mahogany, other deciduous species, and juniper on steep slopes between 6000 and 9000 feet; brushy riparian zones at lower elevations used during migration and post-breeding dispersal. **STATUS** Very rare and local summer visitor to southern Cassia County, Idaho (May–August); vagrant elsewhere; Trout Creek Mountain and Oregon Canyon Mountains in southeast Oregon have suitable habitat and sightings have been recorded in those areas, but nesting has not been documented. **KEY SITE** Idaho: Castle Rocks State Park (Cassia County).

American Redstart
(*Setophaga ruticilla*)

The male American Redstart's black, orange, and white pattern is unique among warblers of the Northwest. Like that of a flycatcher, the redstart's broad bill is flanked by fine bristles, and indeed this warbler actively fly-catches more than most warblers. As it flits through foliage, it often fans its tail and partially spreads and droops its wings in an apparent attempt to flush out insects.

LENGTH 5.25 inches. **MALE** Silky black above with bright orange patches alongside breast, and on wings and tail; white belly and undertail coverts. **FEMALE** Grayish olive above, with gray head; pale gray below; yellow patches alongside breast, and on wings and tail; broken eye ring. **JUVENILE** Similar to female. **VOICE** Song is a variable, high-pitched, rapid, rising or falling *sweetasweetasweeta* or *tue-tue-tue-chuchuteetee*; call is a dry *chip*. **BEHAVIORS** Actively fly-catches in acrobatic sallying flights; forages in foliage, where it will spread its tail for seconds at a time and spread and droop its wings,

Male American Redstart

445

apparently to flush insects into flight where it pursues them. No other Northwest warbler displays this type of fanning and wing spreading. **HABITAT** Near water in deciduous woods with an understory of younger trees or willows; aspen stands at higher elevations; riparian areas and willow stands in open conifer forests. **STATUS** Uncommon summer resident (May–September) of northeast Washington, central and northern Idaho, and southeast British Columbia;

rare summer resident of northeast Oregon. Rare migrant throughout Northwest. **KEY SITES** Idaho: Heyburn State Park; Hoodoo Valley (near Clagstone). Oregon: no reliable sites, but most frequently found in Wallowa and Union counties. Washington: Amazon Creek Marsh (east of Colville); Skagit and Whatcom County line ponds (Newhalem). British Columbia: E.C. Manning Provincial Park; Grohman Narrows Provincial Park.

Female American Redstart

American Redstart range

Northern Parula
(*Setophaga americana*)

This smallest North American warbler, the Northern Parula is a regular vagrant to parts of the Northwest, especially Oregon's well-known migrant traps, such as Malheur National Wildlife Refuge headquarters, the oasis at the tiny town of Fields, and along the south coast. Most Northwest sightings of this eastern North American species occur in spring, but they occasionally show up in autumn.

LENGTH 4.5 inches. **MALE** Bluish gray above with a green mantle; bright yellow throat and breast divided by a black or grayish band, which is bordered below by a rust-orange lower band; yellow lower mandible blends with yellow throat; white arcs above and below eyes; white belly; 2 conspicuous white wingbars; white spots on each side of tail underside. **FEMALE** Similar overall to male, but paler and lacks breast band. **VOICE** Song is a flat, buzzy trill, rising in pitch, then dropping to a single concluding note: *zeezeezeezee-zup*; call is a sharp, musical *chip*. **BEHAVIORS** Frenetic when feeding, often high in tree canopy; erratic flight with rapid wingbeats. **HABITAT** Mature riparian woods; coniferous forests with a variety of tree species. **STATUS** Rare spring vagrant; very rare fall vagrant.

Female Northern Parula

Male Northern Parula

Chestnut-sided Warbler
(*Setophaga pensylvanica*)

The handsome Chestnut-sided Warbler, an eastern species that breeds as far west as central Alberta, is a vagrant to the Northwest, seen almost annually in the region. Closely related to the common Yellow Warbler, its song is quite similar, but easily recognized with a phrase that sounds something like *please-please-pleased-to-meetcha*. Note that the spring and fall plumages are divergent.

LENGTH 5 inches. **MALE BREEDING** Bright white breast, belly, undertail coverts, throat, and cheeks; chestnut stripes along the flanks and neck; black mask through eyes and around to nape, also extending down to form mustache stripes; bright yellow crown; pale yellow wingbars; when tail is spread, outer feathers are white centered. **FEMALE BREEDING** Similar to male, but black mask and chestnut flanks not as extensive. **ADULT NONBREEDING** Lime-green above, pearly gray below; white eye ring; pale yellow wingbars; less chestnut on flanks. **JUVENILE** Similar to adult nonbreeding, but lacks chestnut flanks. **VOICE** Song is a musical *please-please-pleased-to-meetcha*; call is a loud, sweet *chew*. **BEHAVIORS** Tail is often held half-cocked and wings often droop; forages below forest canopy and at ground level. **HABITAT** Deciduous groves, mixed woodlands, brush. **STATUS** Rare spring and fall vagrant.

Male breeding Chestnut-sided Warbler

Black-throated Blue Warbler
(*Setophaga caerulescens*)

The Black-throated Blue Warbler occasionally strays widely from its normal range—it breeds in the northeast United States and adjacent Canada. The species is extremely sexually dimorphic, with the male being remarkable handsome and the female quite drab. These warblers show up in the Northwest annually and among the best places to find them are migrant traps—oasislike wooded enclaves in otherwise open country.

LENGTH 5.25 inches: **MALE** Black face and chin extending down the flanks; deep blue above; white below, with white patches on the wing and tail. **FEMALE** Drab olive-gray above, pale buff below, with off-white undertail coverts; whitish supercilium and partial eye ring; dark cheek; white spot on wings. **JUVENILE** Similar to female but may lack white spot on wings. **VOICE** Song is a buzzy, rising *zuu-zuuzee-zwee*; calls include a juncolike *chewp* and a high-pitched *tseep*. **BEHAVIORS** Sometimes droops wings when foraging for insects in foliage. **HABITAT** Various habitats during migration, especially low shrubbery and willow groves. **STATUS** Rare vagrant, with most sightings in autumn.

Female Black-throated Blue Warbler

Male Black-throated Blue Warbler

Common Yellowthroat
(*Geothlypis trichas*)

The male Common Yellowthroat is a familiar black-masked warbler found in a variety of damp and brushy habitats throughout the Northwest. It is easily detected by its recognizable *wichety wichety wichety* song. Female yellowthroats are highly reclusive, especially during the breeding season, and not often seen; comparatively drab, they can be confused with Orange-crowned Warblers or female Yellow Warblers.

LENGTH 5 inches. **MALE** Bright yellow throat borders broad black face mask; mask bordered above by white; greenish olive above; grayish or yellowish belly; yellow undertail coverts. **FEMALE** No facial mask; soft olive-brown above, pale grayish below with yellow throat and undertail coverts. **JUVENILE** Similar to female but more brownish and may lack yellow on throat; juvenile males may have formative mask. **VOICE** Song is a rich, rapidly whistled *wichety wichety wichety wit*; call is a sharp *chit*. **BEHAVIORS** Skulks in dense brush or marsh vegetation, bobbing its tail as it moves. Weak, fluttering flight close to cover, with tail seeming to drag. **HABITAT** Low, dense vegetation, especially marshes, bogs, ditches, and deciduous thickets. **STATUS** Common summer resident and migrant (April–October) rare winter resident west of Cascades.

Female Common Yellowthroat

First-fall male Common Yellowthroat

Male breeding Common Yellowthroat

Common Yellowthroat range

Yellow-breasted Chat
(*Icteria virens*)

The most unusual of Northwest warblers—if indeed it is a warbler at all, for its taxonomy is puzzling—the Yellow-breasted Chat's beautiful colors and exotic song make is especially popular with birders, although not necessarily with streamside campers being serenaded all through the night. Chats inhabit the densest brambles and thickets, usually near water, and sing from anything that protrudes above the tangle of vegetation.

LENGTH 7.5 inches. **SEXES SIMILAR** Yellow-breasted Chat is the largest warbler; it has a stout bill and long tail; bright yellow throat and breast; olive above; broken white eye ring; black lores bordered by white stripes. **JUVENILE** Much more plain than adult; dusky below, sometimes with olive wash; some white around the eye may be visible. **VOICE** Song is a study in individuality and mimicry. Croaks, raspy rattles, whistles, along with imitations of other bird songs, frogs, or even local machinery can be strung together to make a song;

call is a descending *chew*. **BEHAVIORS** Elusive and secretive; rarely emerges from dense cover except to sing or investigate a disturbance. Sings from a high perch and while in display flight, which begins from a high perch and consists of fluttering, tail spreading, and singing. Diet is varied and includes insects and fruit. **HABITAT** Dense brush interspersed with small trees, usually near water; also drier habitats with appropriate cover. **STATUS** Uncommon summer resident and migrant (April–September). **KEY SITES** Idaho: Hulls Gulch Reserve (Ada County); Niagara Springs Wildlife Area. Oregon: Page Springs Campground (Harney County); Rhinehart Canyon (Union County); E.E. Wilson Wildlife Area (Benton County). Washington: Douglas Creek area (Douglas County); Hawk Creek Canyon (Lincoln County). British Columbia: Inkaneep Provincial Park.

Yellow-breasted Chat

Yellow-breasted Chat range

Yellow-breasted Chat

Northern Waterthrush
(Parkesia noveboracensis)

A large ground warbler, the Northern Waterthrush, during breeding season, is a denizen of bogs, swamps, and dense deciduous riparian zones. Even where fairly common, these secretive birds are difficult to observe owing to their dense habitat. A very similar species, the Louisiana Waterthrush, is a very rare vagrant to the Northwest; it has brighter, pinkish legs, a white eye line, and white underparts with no streaking on the throat.

LENGTH 6 inches. Plump, large-bodied warbler with a short tail; dark brown above; white to pale yellowish buff eyebrow line and dark eye stripe; white to pale yellowish buff below, with dense black streaks from throat to breast to belly and flanks; flesh-colored legs. **VOICE** Song is a bright, whistled *swee-swee-swee-swee-chur-chu r-chur-ch-ch-churew.* Call is a sharp *cheet.* **BEHAVIORS** Almost constantly bobs its rear end. Primarily ground dwelling, but sings

from a low perch and gleans insects from low foliage as well as ground. **HABITAT** Dense deciduous cover bordering streams, lakes, wooded swamps, and bogs. **STATUS** Fairly common summer resident to eastern British Columbia, northeast Washington, and northern Idaho; locally uncommon summer resident of Cascade Range in northernmost Klamath County, Oregon; possibly nests in northeast Oregon; rare migrant. **KEY SITES** Idaho: Coeur D'Alene River Road (Kingston); Pend Oreille Wildlife Area. Oregon: Gilchrist Crossing (Klamath County). Washington: Amazon Creek Marsh (east of Colville); Sanpoil Road (Ferry County). British Columbia: E. C. Manning Park; Vaseux Lake Bird Observatory.

Northern Waterthrush range

Northern Waterthrush

Ovenbird (*Seiurus aurocapilla*)

A vagrant to the Northwest, the ground-dwelling Ovenbird, named for its domed, ovenlike nest, is often first detected by its unique song. The majority of Ovenbird sightings in the Northwest have been in oasislike wooded enclaves in otherwise open country (migrant traps), but like many vagrant songbirds, this species can occur nearly anywhere with appropriate habitat. These plump, rather large warblers breed and normally migrate east of the Rockies.

LENGTH 6 inches. Olive-brown above; white breast boldly streaked or spotted with black; orange crown bordered by bold, blackish crown stripes; prominent eye ring and large eye; white throat bordered by dark mustache stripes; white belly; short tail. **VOICE** Song, diagnostic, is a loud, whistled *teacher teacher teacher*. **BEHAVIORS** Walks with a bobbing head gait, like a chicken, and flicks tail. "Pishing" (making swishing noises by mouth to draw songbirds out of hiding) may tempt the bird to jump from the ground to a higher perch. **HABITAT** Deciduous stands with brushy understories. **STATUS** Rare vagrant; some reports of singing males during summer, but no confirmed nesting.

Ovenbird

Tanagers, Buntings, and Grosbeaks

The family Cardinalidae includes a variety of colorful Northwest songbirds, ranging from the exquisite and aptly named Blue Grosbeak to the kaleidoscopic Western Tanager. The males of these types of birds are far more radiant than the females, whose more muted tones help them remain camouflaged while sitting on nests. Most are seedeaters, as evidenced by their thick, conical bills adapted to easily cracking open many types of seeds, though they also eat invertebrates and fruits; the tanagers, with their longer, less powerful bills, rely primarily on insects and fruits. One species, the Evening Grosbeak, is the lone North American representative of its genus, and while actually a member of the finch family (Fringillidae), we include it with this group for easy comparison with the grosbeaks of the Cardinalidae family.

Black-headed Grosbeak

Western Tanager
(*Piranga ludoviciana*)

The male Western Tanager is among the most mesmerizingly colorful birds in the Northwest. Happily for admirers, this bejeweled species is quite common throughout the forested expanses of the region. While partial to coniferous forests during the breeding season and much associated with forested mountains, Western Tanagers routinely nest in stands of conifers at low elevation, including large forested suburban parks near major Northwest cities. In migration, Western Tanager can show up virtually anywhere in the Northwest, from the most remote, treeless desert steppe to urban backyards.

LENGTH 7 inches. **MALE** Bright yellow breast, belly, rump, and nape; black back and tail; at rest, black wings have a white wingbar and yellow shoulder stripe; bright flame-orange face or entire head (especially bright in breeding season, but fading to yellow, often with a wash of orange, from late summer through midspring). **FEMALE** Pale to bright yellowish below, with bright yellow undertail coverts; yellowish olive head; olive-gray back; narrow white wingbar and yellowish shoulder bar; larger size and thick bill separate female from superficially similar warblers. **JUVENILE** Similar to female. **VOICE** Common call is a fast, rising, ringing *pit-er-ick*; song is a robin-like musical series of warbling whistles: *cheereep-cheeryup-cherrup-cheerip-cheereer*. **BEHAVIORS** During breeding season, generally remains fairly high in trees, where it hunts insects. In migration, often occurs in small flocks of a few to a dozen or so birds. **HABITAT** Breeds in coniferous and mixed forests; especially prevalent in drier ponderosa

Male breeding Western Tanager

Female Western Tanager

forests and Douglas fir forests; widespread in migration. **STATUS** Common summer resident and migrant (May–September). **KEY SITES** Idaho: Bogus Basin area and Boise National Forest; Island Park area; Ponderosa State Park and McCall area; Saint Joe River. Oregon: Black Butte and Green Ridge area; Idlewild Campground (Harney County); Westside Road (Klamath County). Washington: Leavenworth area; Mount Spokane State Park; Okanogan Highlands; Winthrop area. British Columbia: Castlegar and Nelson area; Okanagan Valley and surrounding highlands.

Western Tanager range

Male nonbreeding Western Tanager

Black-headed Grosbeak
(*Pheucticus melanocephalus*)

During spring and summer, the brightly spangled male Black-headed Grosbeak announces his presence with his resonant, musical song. Finding him can be a different matter, however, as he frequently sings from cover, often fairly high in a large shrub or leafy tree, though at times from well out in the open. In contrast to the male, the female Black-headed Grosbeak does not have a black head and is well cloaked in camouflaging shades that help her stay hidden on the nest.

LENGTH 8.25 inches. Plump, with very thick bill; yellowish wing linings in flight. **MALE** All-black head bordering light orange breast and flanks; orange collar and rump; yellowish sternum streak extending up from belly; blackish mantle; 2 bold white wingbars forming flashy white wing patches in flight; bold white patches on corners of tail especially noticeable in flight. Nonbreeding adult male is less boldly patterned overall. **FEMALE** Mottled brown and tan overall, with buffy breast; flanks lightly speckled; bold white eyebrow lines and crown line divided by dark brown stripes; grayish brown cheeks bordered below by white; narrow white wingbars; bicolored bill has dark upper mandible unlike uniformly pale bill of similar female Rose-breasted Grosbeak (a rare but regular vagrant in the Northwest). **JUVENILE MALE** Similar to female, but with more black in the face and little or no streaking on buffy orange breast (variable). **VOICE** Song is a rich robinlike warbling; call is a persistent high *pink*, reminiscent of a Downy Woodpecker. Females also sing, generally a simplified version of the male's song. **BEHAVIORS** Easily attracted to feeding stations with sunflower placed in or near appropriate habitat. **HABITAT** Deciduous and mixed woodlands with extensive understory; brushy woodland openings; brushy riparian zones; from near sea level to well up into mountainous regions in appropriate habitat, and from city neighborhoods to remote canyons. **STATUS** Common summer resident and migrant (May–August).

Male Black-headed Grosbeak

Female Black-headed Grosbeak

Juvenile male Black-headed Grosbeak

Black-headed Grosbeak
range

Rose-breasted Grosbeak
(Pheucticus ludovicianus)

Although common in northeast British Columbia, where it reaches the western extent of its breeding range, the beautiful Rose-breasted Grosbeak is a rare casual visitor to the remainder of the Northwest. Individuals tend to show up annually somewhere in the region, with most sightings occurring in May and June.

LENGTH 8 inches. **MALE** Breeding male unmistakable, with all-black head and mantle, and bright rosy red, wedge-shaped breast patch yielding to snow-white belly; at rest, shows white shoulder patch and wingbar; in flight, white outer wing patches and pinkish wing linings; white outer tail feathers; white rump. Winter and first-spring male paler overall than breeding male and not as jet-black above. **FEMALE** Very similar to female Black-headed Grosbeak, but more whitish than buffy below, with bolder streaks on breast; bill is uniformly pale (Black-headed Grosbeak's bill has pale lower mandible and dark upper mandible). **VOICE** Common call is a high-pitched squeaky *cheech* or *chich*. Song is a clear, liquid, musical warbling, similar to song of Black-headed Grosbeak. **BEHAVIORS** A common backyard feeder bird in its normal range, the Rose-breasted Grosbeak likewise shows up at feeders well outside its range in the Northwest. **HABITAT** Deciduous forests; vagrants may occur in deciduous forests or tree stands in open areas, brushy fields, mixed woodlands. **STATUS** Rare but fairly regular spring and summer vagrant; very rare fall and winter vagrant.

Female Rose-breasted Grosbeak

Male Rose-breasted Grosbeak

Blue Grosbeak
(*Passerina caerulea*)

The Blue Grosbeak's miniscule range within the Northwest is limited to south-central Idaho, where it has nested, though reports are forthcoming only very rarely. Named for the male's conspicuously beautiful vivid blue plumage, the Blue Grosbeak is widespread across the southern tier of states, the northern limit of its range in the eastern half of the United States aligning surprisingly well with the Mason–Dixon Line.

LENGTH 6.25 inches. Large finch with very robust bill. **MALE** Gleaming, royal purplish blue overall, with black face mask, rust-brown shoulders, and brown wingbars; black flight feathers and tail. First-spring male brown with blue mottling. **FEMALE** Soft brown above, with 2 cinnamon wingbars; tan below; varying amounts of blue on shoulders, upper body, and head. **JUVENILE** Similar to adult female. **VOICE** Song is composed of swift musical warbles and whistles; calls include a sharp, high, repeated *whit*. **BEHAVIORS** Often wags tail side to side, and often fans tail briefly; forages on ground and gleans from shrubs and grasses. **HABITAT** According to information compiled by the Idaho Department of Fish and Game, the Blue Grosbeak in Idaho "...may nest in hayfields or chicory, Russian olive, willow, and wild rose thickets adjacent to sagebrush foothills." **STATUS** Very rare summer resident of southern Idaho (primarily Cassia and Oneida counties); rare vagrant elsewhere in the Northwest.

Male breeding Blue Grosbeak

First-spring male Blue Grosbeak

Female Blue Grosbeak

Blue Grosbeak range

Evening Grosbeak
(*Coccothraustes vespertinus*)

A large, striking, colorful member of the finch family, the Evening Grosbeak is typically seen flying in noisy flocks, or perched high overhead while giving its high-pitched trill. It tends to be highly irruptive and in some years occurs well south of its normal range, including anywhere in the Pacific Northwest where adequate food is available (such as feeders). Its name derives from the huge conical bill used to crack open seeds.

LENGTH 8 inches. Massive greenish bill that changes to a dull ivory in the winter. **MALE** Yellowish overall; dark gray head and face highlighted by bright yellow forehead and eyebrows; upper back and breast dusky yellowish olive; bright yellow rump, flanks, and belly; black wings contrast sharply with bold white secondaries. **FEMALE** Gray with a golden-yellow nape and flanks (often hidden by the wings); white highlights in otherwise black wings and tail. **VOICE** Song is a simple series of musical whistles and

trills. Flight call is a whistled series of trills. **BEHAVIORS** Readily eats sunflower seeds and often stays close to a favorite feeder throughout the winter months, when it typically occurs in flocks; eats minerals from the ground, including from campfire pits. **HABITAT** Conifer forests in breeding season; conifers, mixed conifers and hardwoods, juniper woodlots for balance of year; these high-elevation birds typically move to lower elevations in winter. **STATUS** Uncommon year-round resident.

Female Evening Grosbeak

Evening Grosbeak range

Male Evening Grosbeak

Lazuli Bunting (*Passerina amoena*)

A flashbulblike glint of bright electric blue and a bubbly cheerful song readily identify the male Lazuli Bunting when he sings on breeding grounds—typically dry, brushy, riparian slopes. Behaviorally, this little bunting recalls Lesser and American Goldfinches, and where 2 or all 3 of these species occur in proximity, birdwatchers couldn't ask for a more startling array of colors.

LENGTH 5.5 inches. **MALE** Bright, electric blue head, back, rump; blue throat borders broad orange-rust breast band; white belly; broad white wingbars; black flight feathers with blue edges; bluish tail; first-year breeding males duller than older birds; post-breeding males paler than breeding-season males; some males exhibit only pale orange breast or barely a hint of orange. **FEMALE** Soft gray-tan above, light buff below with pale grayish throat; buff wingbars; sometimes a slight bluish sheen on rump. **VOICE** Song is a high, buzzy, warbled series of excited notes. Alarm call is a sharp, persistent *tsip*, sometimes in rapid succession. **BEHAVIORS** Fond of thistles, cone flowers, wild sunflowers, and other seeds, Lazuli Buntings perch athletically on stalks of these food plants. When agitated, they nervously flick tails from side to side. Males sing from exposed perches, often during the heat of midday. **HABITAT** Brushy areas, especially riparian slopes in arid regions; from residential neighborhoods and suburbs to remote wilderness settings. **STATUS** Widespread and fairly common summer visitor (early May–September) east of the Cascades and in Rogue River Valley, Oregon; uncommon northward west of Cascades. Fairly common migrant throughout range. **KEY SITES** Idaho: Boise area (Foote Park; Boise River Greenway); Camas NWR and Market Lake Wildlife Area; Salmon River Road. Oregon: Deschutes River State Park; Ken Denman Wildlife Area; Malheur NWR; Mount Ashland. Washington: Biscuit Ridge Road (Walla Walla); Mount Spokane State Park and Mount Spokane foothills; Sinlahekin Wildlife Area and Sinlahekin Valley; Umtanum Road and Wenas area (Wenatchee); Wenatchee area foothills. British Columbia: Okanagan Valley.

Male breeding Lazuli Bunting

Female Lazuli Bunting

First-spring male Lazuli Bunting

Lazuli Bunting range

Indigo Bunting *(Passerina cyanea)*

The eastern counterpart to the West's Lazuli Bunting, the Indigo Bunting occasionally strays into the Northwest. Most sightings occur during summer, from May through July, and most are of males; sightings of females are extremely few in the Northwest, possibly because the females are less prone to vagrancy or possibly because they are not as noticeable as the males and difficult to distinguish from female Lazuli Buntings.

LENGTH 5.5 inches. **MALE** Bright blue overall, with blackish blue wings and tail. **FEMALE** Warm soft brown above, buffy below with faint streaks on breast and flanks; pale throat. **VOICE** Songs similar to Lazuli Bunting but less hurried, with a varied range of paired notes; typical call a sharp *whit* or *whit-whit*. **BEHAVIORS** Forages for seeds and insects in shrubs and on ground.

Hybridizes with Lazuli Bunting where the ranges overlap, and Indigo-Lazuli pairings have been documented in the Northwest. **HABITAT** Brushy habitats, including fields, draws, riparian corridors. **STATUS** Rare summer vagrant (May–September); very rare fall and winter vagrant.

Male Indigo Bunting

Female (left) and male Indigo Bunting

Sparrows

Represented by more than 2 dozen species in the Northwest, the sparrow family (Emberizidae) includes numerous melodious superstars whose springtime songs burst enthusiastically from nearly every type of habitat in the Northwest. Some of these birds are drab, blending easily with their surroundings, while others bear rather striking patterns. Some species are difficult to tell apart, especially in juvenile and nonbreeding plumages. Some members of the family—Dark-eyed Juncos, Song Sparrows, Spotted Towhees, the 3 different crowned sparrows of the genus *Zonotrichia*—are frequent and conspicuous visitors to bird feeders, especially during winter. These birds primarily eat seeds and grains, but some species feed heavily on insects and fruit as well.

White-crowned Sparrow

Spotted Towhee
(Pipilo maculatus)

Often heard scratching through leaves and debris in the densest brush, the Spotted Towhee also frequents bird feeders, typically taking seeds from the ground under hanging feeders, but sometimes boldly dining on hanging feeders and even suet cakes. This widespread towhee winters in many areas of the Northwest, while some populations migrate to more temperate regions.

LENGTH 8 inches. **MALE** Black head, back, throat, and breast; long black tail with white corners; white spots on the black wings; rich, cinnamon-red flanks; white belly; bright red iris. **FEMALE** Similar to male, but dark brown instead of black. **JUVENILE** (May–August) Dusky brown above; tannish below, with darker indistinct streaks; brownish red iris. **VOICE** Song, highly variable, is a whistled, descending *tew tew* followed by a trilled *cheee*; common call is a forceful, harsh, inquisitive *meeeeew*. **BEHAVIORS** Usually forages on or near the ground; skulks in dense, brushy tangles, but is easily drawn out of cover by pishing or squeaking. Like other towhees, the Spotted Towhee double scratches by hopping forward and backward to displace ground litter and expose potential food. **HABITAT** Thickets, tangles, heavy brush, brambles, brushy road margins. **STATUS** Common year-round resident west of the Cascades; common summer resident and rare winter resident of the interior.

Female Spotted Towhee

Juvenile Spotted Towhee

Male Spotted Towhee

Spotted Towhee range

Green-tailed Towhee
(*Pipilo chlorurus*)

A denizen of drier, brushy areas, this large, handsome sparrow can be difficult to find because of its dense habitat, although during the breeding season, singing males are conspicuous. During migration, Green-tailed Towhees often show up with other sparrows at feeders.

LENGTH 6.5 inches. Rufous crown and white throat distinctive; dark gray nape and greenish olive back; bright greenish yellow wings and tail; dark gray underparts. **JUVENILE** (June–August) Brown-tan-streaked back, crown, breast and flanks; whitish throat; greenish tinge to wings. **VOICE** Song is similar to that of the Fox Sparrow (which often occurs in the same ponderosa pine or manzanita habitat east of the Cascades)—a series of cheerful trills and short whistles. Call is a questioning *meew*. **BEHAVIORS** Typically flies low to the ground, below brush-top level; pumps tail in flight. Reclusive in preferred habitat and often only

detected by *meew* calls, unless males are singing. **HABITAT** Dry, brushy areas, often on hillsides; especially groves of mountain mahogany, chokecherry, bitterbrush, and sagebrush; juniper woodlots with heavy shrub understory; old burns; open ponderosa pine forests with brushy understory. **STATUS** Fairly common summer resident of Oregon and Idaho, locally uncommon summer resident of Blue Mountains, Washington. **KEY SITES** Idaho: Boise foothills; City Of Rocks National Reserve; Pocatello-area foothills; Silver City. Oregon: Cold Springs and Indian Ford Campgrounds (Deschutes County); Mount Ashland; Steens Mountain. Washington: Biscuit Ridge Road (east of Walla Walla).

First-fall Green-tailed Towhee

Green-tailed Towhee range

Adult Green-tailed Towhee

California Towhee
(*Melozone crissalus*)

Southwest Oregon is the northern extent of the California's Towhee's range. Otherwise, this aptly named large sparrow is found only in California and Baja California. The California Towhee sometimes builds its nest in poison oak and readily eats the berries of this caustic plant. Like the Spotted Towhee, this sparrow is also fond of bird feeders. It has a penchant for car mirrors and shiny hubcaps, where it is engaged by its own reflection.

LENGTH 9 inches. Brown overall, with long tail; sparse pattern of darker spots on cinnamon-brown throat; light brown belly; rich cinnamon undertail coverts. **VOICE**

Song is a series of loud, clear *tsee* notes starting slowly and accelerating, then ending with rapid, squeaked *chew* notes, fading in volume; call is a loud, sharp *tsee*. **BEHAVIORS** This ground forager usually hops or runs; its jerky flight appears encumbered by a long tail and short wings. Aggressive toward fellow California Towhees, it engages in feet-to-feet battles with others. Sometimes displays with wing flutters, occasionally holding a twig in its bill. **STATUS** Locally common year-round resident in Rogue Valley, Oregon; uncommon northward to Roseburg area; fairly common in the Klamath Falls area. **KEY SITES** Oregon: Denman Wildlife Area, Roxy Anne Butte, Upper and Lower Table Rock, Whetstone Savanna Preserve (Jackson County); Link River Trail (Klamath Falls).

California Towhee range

California Towhee

Dark-eyed Junco (*Junco hyemalis*)

This beloved "snowbird" is a year-round resident in the Northwest, but most people don't see it until it escapes the mountain snows for more temperate lower elevations, where it ranks among the most common species seen at bird feeders. The species is actually composed of a number of subspecies that were considered separate species until the 1970s. Four of these subspecies occur in the Northwest: the Oregon, Slate-colored, Gray-headed, and Pink-sided Juncos. To further complicate juncos, interbreeding between subspecies leads to many plumage variations.

LENGTH 6 inches (geographically variable). **SEXES SIMILAR** Hood and breast varies from rich, dark brown (almost black) to dark gray; wings, flanks, and back can be pink, rufous, or gray; white belly; white outer tail feathers. **OREGON** Black hood (male), dark gray-brown hood (female); pale rust flanks. **PINK-SIDED** Similar to Oregon, but pale gray hood and black lores. **SLATE-COLORED** Uniform slate gray to light gray-brown overall; white belly and undertail coverts. **GRAY-HEADED** Pale gray overall, with distinctive rust-brown mantle; black lores. **JUVENILE** (May–August) heavily streaked brown back; dark gray head; streaked flanks and white belly; white outer tail feathers. **VOICE** Song is an even, high-pitched trill; call is similar to the *chink* of stones being

Male Oregon Junco

Female Oregon Junco

Pink-sided Junco

Slate-colored Junco

struck together. **BEHAVIORS** Forages on the ground for seeds, often in flocks during winter, when it readily mixes with other species of birds. During breeding season, males sing from high perches in conifer forests. **HABITAT** Breeds in conifer forests. During winter, prefers open, shrubby habitats; parks; mixed deciduous woodlands; residential areas. **STATUS** Oregon Junco: common year-round resident throughout the Northwest (widespread in winter, primarily higher elevations in summer). Pink-sided Junco: uncommon summer resident of southeast Idaho, rare winter resident of southern Idaho. Gray-headed Junco: rare and local in mountains of southern Idaho and southeast Oregon. Slate-colored Junco: rare to uncommon winter visitor.

Gray-headed Junco

Dark-eyed Junco range

Sagebrush Sparrow
(*Artemisiospiza nevadensis*)

Formerly called the Sage Sparrow, the handsome Sagebrush Sparrow earned its new name after taxonomists divided regional populations into 2 species (a California population is now called Bell's Sparrow). As its name implies, the Sagebrush Sparrow lives in the expansive sagebrush steppe of the interior Northwest, as well as other arid shrub steppe habitats. During the breeding season, this sparrow is quite vocal, singing exuberantly in the morning, until the heat of the day prompts it to seek shelter in the shade of desert brush.

Adult Sagebrush Sparrow

LENGTH 6 inches. Grayish head, with bold, white supraloral spot and white eye ring; white mustache stripe and black malar stripe; gray, lightly streaked back; brown wings; white below with buffy streaking on flanks, and a black breast spot. **JUVENILE** Gray head; white eye ring; brown back and wings; brown streaking on breast and flanks. **VOICE** Song composed of cheerful, buzzy notes that rise and then drop in pitch, then rise back to the original tone; call is a thin, high-pitched *tsee*. **BEHAVIORS** Walks and hops on the ground with tail cocked upward. Flight is short and weak; prefers to run when alarmed. After breeding, some birds disperse to higher elevations. **HABITAT** Sagebrush steppe on or near broken or uneven ground, hillsides, or ancient dunes. **STATUS** Uncommon spring and summer resident (begins arriving as early as February; a few birds linger into September). **KEY SITES** Idaho: Craters of the Moon National Monument; Snake River Birds of Prey National Conservation Area. Oregon: Hart Mountain National Antelope Refuge; Malheur NWR; Saddle Butte. Washington: Quilomene Wildlife Area; Yakima Training Center.

Juvenile Sagebrush Sparrow

Sagebrush Sparrow range

471

Black-throated Sparrow
(*Amphispiza bilineata*)

This handsome sparrow breeds in arid desert habitats in the Great Basin and a few scattered nearby locales. Preferring sparsely vegetated slopes with boulder fields, the Black-throated Sparrow can gather adequate moisture from insects and plants when no water is available, thanks to its special physiological adaptations. The relative abundance of Black-throated Sparrows may be underestimated because it lives in remote regions seldom visited by birders.

LENGTH 5.5 inches. Gray-brown above, white below; black bib with bold white eyebrow stripe and cheek stripe; fairly long tail, with white outer tail feathers visible in flight. **JUVENILE** Bold whitish eyebrow and throat; lightly streaked breast; gray back with light streaks. **VOICE** Two sharp whistles followed by a crisp musical trill. **BEHAVIORS** Secretive; feeds on the ground, gleaning plant material and insects; often sings from a perch lower than the surrounding brush. **HABITAT** Sagebrush steppe; especially slopes with boulders and rock outcroppings. **STATUS** Fairly common summer resident (May–August); very rare breeder outside normal range, including Washington, where a few have nested along the Columbia River near Vantage, and in the Boardman area of Oregon. **KEY SITES** Oregon: Hart Mountain National Antelope Refuge. Idaho: Mud Flat Road near Grand View.

Black-throated Sparrow range

Black-throated Sparrow

Brewer's Sparrow
(*Spizella breweri*)

The most common sparrow of the Northwest's vast sagebrush steppe, this little songster is also the region's smallest sparrow. Drab and well camouflaged to match its equally drab-colored surroundings, the Brewer's Sparrow is one of the handful of songbirds perfectly adapted to eke out an existence in the seas of sagebrush covering the interior West.

LENGTH 5 inches. Grayish brown overall; small, thin bill; long, notched tail; white eye ring; brown, finely striped crown; thin brown eye line; streaked back and nape; overall grayish brown. **VOICE** Song is a bubbly, musical series of buzzes and trills; call is a high-pitched *chip*. **BEHAVIORS** Feeds on or near the ground, and generally sings from a low perch, below a shrub's crown. Brewer's Sparrows often form vocal flocks away from the breeding grounds. **HABITAT** Sagebrush steppe up to and above timberline, desert scrub, juniper and mountain mahogany woodlands. During migration, uses a wider variety of habitats. **STATUS** Abundant summer resident and migrant (April–September) interior; uncommon migrant west of Cascades.

Adult Brewer's Sparrow

Brewer's Sparrow range

Juvenile Brewer's Sparrow

Chipping Sparrow
(*Spizella passerina*)

The sweet trill of the Chipping Sparrow is a common sound in dry coniferous forests and oak woodlands of the Northwest during spring and summer. Because this sparrow's plumage changes seasonally, it can easily be misidentified during autumn, when it bears a strong resemblance to other *Spizella* sparrows.

LENGTH 5.25 inches. **BREEDING** (April–August) Gray head with bold rufous crown, dark eye line, and white supercilium; pale gray spot just above all-dark bill; clean gray breast and belly; 2 pale beige wingbars; tan mantle with dark streaks; gray rump; long notched tail. **NONBREEDING** (August–March) Thin, light brown to partly rufous crown stripes; pinkish bill; dark eye line with buffy supercilium. First-fall adult has brown-striped crown and grayish rump. Juvenile (summer) has streaked breast. **VOICE** Song is a simple, rapid, single-pitch trill, similar to the song of a Dark-eyed Junco, with which it often shares forest habitat; call is a sharp *chip*. **BEHAVIORS** Males sing often during spring and summer, usually from higher perches; often flocks with other sparrows during migration. **HABITAT** Breeds in drier conifer forest edges or open mixed woodlands and orchards; found in a variety of habitats during migration. **STATUS** Fairly common summer resident and migrant (April–September). **KEY SITES** Idaho: Farragut State Park (Kootenai County); Foote Park (Boise). Oregon: Camp Sherman area; Idlewild Campground (Harney County). Washington: Oak Creek Wildlife Area (Yakima County); Turnbull NWR. British Columbia: Creston Valley Wildlife Area; E.C. Manning Provincial Park.

Breeding Chipping Sparrow

Juvenile Chipping Sparrow

Nonbreeding Chipping Sparrow

Chipping Sparrow range

Clay-colored Sparrow
(*Spizella pallida*)

The Clay-colored is one of 3 sparrows that are easy to confuse with one another in their nonbreeding plumages; others are the Chipping Sparrow and Brewer's Sparrow, both of which are far more common and widespread in the Northwest.

LENGTH 5.25 inches. **BREEDING** (April–August) Brown-striped crown with a pale central stripe; wide buffy to gray-white supercilium and lores; brownish ear patch bordered by darker brown eye line and moustache stripe; dark throat stripes framing white to buffy white throat; clean grayish breast and belly; 2 thin, white wingbars; long, notched tail. **ADULT NONBREEDING** (August–March) Facial stripes more buffy (rather than whitish as in breeding plumage); gray nape; buffy breast and belly. **VOICE** Song is a low, insectlike buzz, *zee-zee-zee*; call is a high, sharp *tsip*. **BEHAVIORS** Males sing constantly on breeding grounds, even at night. When flushed, will fly a short distance and drop to the ground. Often mixes with other sparrows during migration. **HABITAT** Breeds in drier mixed grassland and shrub steppe habitats. During migration, found in a variety of brushy habitats with other sparrows. **STATUS** Rare but regular migrant; rare summer resident near Spokane and Okanogan Valley, Washington; uncommon summer resident of south-central and southeast British Columbia.

Breeding Clay-colored Sparrow

Nonbreeding Clay-colored Sparrow

Clay-colored Sparrow range

American Tree Sparrow
(*Spizella arborea*)

A winter visitor to the Northwest, the American Tree Sparrow occurs singly and sometimes in flocks (primarily east of the Cascades). The name is misleading because the American Tree Sparrow has no particularly noteworthy affinity for trees; rather, it was named for its resemblance to the Eurasian Tree Sparrow. In fact, this bird is more comfortable in weedy fencerows and thicket edges.

LENGTH 5.75 inches. Gray face, with rufous crown and eye line; bicolored bill, with yellow lower mandible and dark upper mandible; streaked back, rufous shoulder, and white wingbars; light gray breast and belly; black breast spot. **VOICE** Call is a high-pitched *taseet*, with the first part of the call a higher pitch than the last. **BEHAVIORS** Often mixes with other ground-feeding species. **HABITAT** Fencerows, weed patches, weedy fields with brushy swales and small trees; often seen at bird feeders. **STATUS** Uncommon to locally common winter visitor (November–March) east of the Cascades, rare west of the Cascades except extreme northwest Washington and southwest British Columbia, where it is more regular. **KEY SITES** Idaho: LQ Drain Wildlife Habitat Area (Twin Falls); Market Lake Wildlife Area (Jefferson County). Oregon: Golf Course Road (Wallowa County). Washington: Sprague Lake (Adams County); Wells Wildlife Area (Colville Reservation). British Columbia: Kootenay River Roads (Creston); Osoyoos Roads.

American Tree Sparrow range

American Tree Sparrow

Vesper Sparrow
(Poocetes gramineus)

This bird of grasslands and sagebrush steppe is known more for its musical song, often delivered from a fence post, than for its unassuming plumage. Two subspecies occur in the Northwest, one found west of the Cascades, and the other from east of the Cascades into the Rocky Mountains. The western subspecies has been in a notable decline because of habitat loss, and extirpation remains a possible outcome.

LENGTH 6 inches. Streaked brown above; thin white eye ring; brown throat stripes framing off-white throat; streaked breast and flanks; clean cream-colored belly; in flight, white outer tail feathers and rusty lesser coverts are distinctive. **VOICE** Song can vary, but generally starts with 2 to 4

musical notes, *hee-hee-whee-whee*, followed by a variety of sweet trills and warbles. **BEHAVIORS** Generally stays on the ground except for singing males. **HABITAT** In western part of range, grass-dominated areas, such as pastures, weedy Christmas tree farms, prairies, and meadows. In the drier interior Northwest, shrub steppe, natural grasslands, open grassy montane areas, pastures, and sparse juniper woodlands. **STATUS** Uncommon to rare summer resident and rare winter resident west of the Cascades. Fairly common summer resident and rare winter resident east of Cascades. **KEY SITES** Idaho: Intermountain Bird Observatory at Lucky Peak; Ponderosa State Park (McCall). Oregon: Bald Hill (Benton County); Steens Mountain North Loop Road. Washington: Scatter Creek Wildlife Area; Turnbull NWR. British Columbia: Kilpoola Lake (Osoyoos); Princeton area.

Vesper Sparrow range

Vesper Sparrow

Lark Sparrow
(*Chondestes grammacus*)

The striking facial pattern of the adult Lark Sparrow is unmistakable, a study in rich chestnut, black, and white. This beautiful sparrow is found primarily east of the Cascades, but southwest Oregon hosts breeding populations in the Rogue and Umpqua Valleys, and a few birds even winter in those areas. Outside of the breeding season, Lark Sparrows tend to occur in flocks.

LENGTH 6.5 inches. Large, slender sparrow with a rounded tail similar to a towhee; white corners of tail most obvious in flight or display. Bold facial pattern, with chestnut crown separated by white median crown stripe and chestnut ear patches; black eye stripes; black throat stripes framing white throat; clean whitish breast and belly with dark breast spot; buffy above, with dark streaks. First-year birds paler than adults. **VOICE** Song is a musical series of *twee* notes followed by a variety of whistles, buzzes, and trills; call is a high *tsip*. **BEHAVIORS** In an unusual courtship display, the male droops its wings, and cocks its fanned tail up at a 45-degree angle; just before copulation and sometimes during, the male will pass a small twig to the female. Generally walks rather than hops, except during the courtship display. **HABITAT** Open grassland and scrub edges, sagebrush steppe, open juniper woodlands, open conifer forests. **STATUS** Fairly common summer resident (March–September) over most of range. Sometimes winters in Oregon's Rogue and Umpqua Valleys. **KEY SITES** Idaho: Castle Rocks State Park; Intermountain Bird Observatory at Lucky Peak. Oregon: Double O Road (Harney County); Kirtland Road ponds (Jackson County). Washington: Quilomene Unit, L.T. Murray Wildlife Area; Riverside State Park (Nine Mile Falls). British Columbia: Haynes Lease Ecological Reserve (Osoyoos).

Juvenile Lark Sparrow

Lark Sparrow range

Adult Lark Sparrow

Savannah Sparrow
(*Passerculus sandwichensis*)

The handsome Savannah Sparrow, a common denizen of open grasslands, is often seen singing from fences or clumps of tall grass. Perfectly at home on or near the ground, it often flits or climbs briefly to a vantage point to check out intruders before dropping back into concealment.

LENGTH 5.5 inches. Brown-striped crown with pale median stripe; dark eye line and moustache stripe; yellow supraloral spot; yellowish to white eyebrow; pale tannish above, with brown streaks; clear white throat forming distinct border with brown-streaked white breast; usually a dark central spot; streaked flanks; white belly and undertail coverts; some individuals have off-white outer tail feathers. **VOICE** Song is a sweet, musical *tzip-tzip-tzip* followed by a trilled *treeeeeee*; call is a high, sharp *tsip*. **BEHAVIORS** Walks, runs, or hops through weeds, not unlike a scurrying rodent. Sings from fences, low brush, or grass tufts. **HABITAT** Open grasslands or grassy margins, overgrown pastures, weedy roadside edges, hay and alfalfa fields, grassy coastal dunes. **STATUS** Abundant summer resident; fairly common winter resident.

Savannah Sparrow

Savannah Sparrow range

Savannah Sparrow

Grasshopper Sparrow
(*Ammodramus savannarum*)

A bird of wide-open spaces, the Grass-hopper Sparrow breeds erratically in the Northwest because it nests in loose colonies that often change locations annually. Unlike most sparrows, the Grasshopper Sparrow's body shape—flat head, large bill, short tail—provides a solid clue to its identity. Although largely a ground-dwelling species, male Grasshopper Sparrows sing from elevated perches, and the availability of such perches seems to be a habitat requirement.

LENGTH 5 inches. Large, flat head; large bill; short tail; pale overall; heavily streaked back; pale buff below; streaked crown, with pale median stripe; eyebrow may be orangish near the bill, but gray farther back on the face. **VOICE** Song is a high, insectlike *tip-tip tip zheeeeee*; call is a weak *tidic*. **BEHAVIORS**

Walks, runs, or hops through weeds like a scurrying rodent. Male sings from perches that stand above the surrounding vegetation, but rarely sings after nesting; nests in loose colonies. **HABITAT** Dry, open grasslands; bunchgrass-dominated slopes with patches of bare ground and low brush. **STATUS** Locally fairly common summer resident (April–July); rare migrant and winter resident. **KEY SITES** Idaho: Avimor (Ada County); Camas NWR. Oregon: Fern Ridge Wildlife Area; Umatilla NWR. Washington: Kahlotus Ridgetop Preserve; Turnbull NWR. British Columbia: Okanagan Valley.

Grasshopper Sparrow range

Grasshopper Sparrow

Fox Sparrow (*Passerella iliaca*)

The Northwest is home to each of the 4 subspecies groups of Fox Sparrow, with each group composed of several subspecies. Each of the groups may gain full species status in the future. The Sooty Fox Sparrow (*Passerella iliaca unalaschensis*) breeds in northwest Washington and along the British Columbia coast and winters south through western Oregon; the Slate-colored Fox Sparrow (*P.i. schistacea*) breeds in the interior Northwest, including parts of central and eastern Washington and eastern Oregon; the Thick-billed Fox Sparrow (*P.i. megarhyncha*) breeds in southeast and central Oregon, and down to about 3,000 feet on the west slope of the Cascades; and the Red Fox Sparrow (the nominate subspecies group) breeds in the Far North and is an occasional migrant and winter visitor to the Northwest.

LENGTH 6–7 inches. Large and robust, with variable plumage; dark above; white belly; white breast heavily streaked with wedge-shaped spots, densest in the middle of the breast to form dark central breast spot; 2-tone bill with dark upper mandible and pale lower mandible. **SOOTY** Uniform dark brown above, with dark spots on the breast

Thick-billed Fox Sparrow

Slate-colored Fox Sparrow

and flanks. **SLATE-COLORED** Gray above, with reddish tones in the wings and tail; heavily spotted breast and flanks. **THICK-BILLED** Similar to Slate-colored, but with larger bill and finer spotting on breast and flanks. **RED** Distinct strongly 2-tone, rust-and-gray facial pattern; back and wings streaked with rust; rust-colored tail; thin whitish wingbars; rust-colored spots on breast and flanks. **VOICE** Variable depending on subspecies.

Song features loud, musical whistles, sometimes ending in a trill. **BEHAVIORS** Does not stray far from cover. Flights are short and jerky. **HABITAT** Dense shrubbery in woodland edges, including chaparral, blackberry thickets, dense streamside scrub, and dense backyard shrubbery. **STATUS** Common summer resident; common winter resident west of Cascades, uncommon winter resident east of Cascades.

Sooty Fox Sparrow

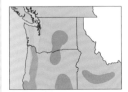

Fox Sparrow range

Song Sparrow
(Ammodramus savannarum)

The Song Sparrow, the most common and widespread sparrow in the Northwest, is also among the most melodious. The male sings throughout the year, and most exuberantly during spring and early summer, when large tracts of prime habitat can provide a symphony of numerous individuals. So effusive is their singing that juveniles burst into song as early as 2 months of age. The Northwest is home to about a dozen subspecies, with plumages varying from the darkest birds west of the Cascades to the lightest-shade birds in the interior.

Song Sparrow

LENGTH 5.5–6.25 inches. Long tail; stout bill; gray supercilium; brown to rufous crown with gray median stripe, eye line, and mustache stripe; streaked dark brown above, with brown to rust-colored wings and tail; heavily streaked below, with dark central breast spot; breast and belly background varies from smoky gray brown to nearly white. **VOICE** Both male and female sing a variable mix of loud, penetrating, musical whistles and trills, often *whit whit whit whit werder churee-ree-ree*; also a distinctive loud *chee*. **BEHAVIORS** Fairly weak, jerky flight, usually close to ground, with tail pumping. Males sing incessantly during breeding season, and a bright day in winter may also induce them to sing. **HABITAT** Open brushy habitats near water; marshes; blackberry patches; hedgerows and gardens, including residential areas. **STATUS** Abundant year-round resident.

Song Sparrow

Song Sparrow range

Lincoln's Sparrow
(*Melospiza lincolnii*)

Named by John James Audubon after a traveling companion, Thomas Lincoln, this small, secretive sparrow buries itself in the thick cover of brushy bogs and willow groves, often refusing to allow birders a good look. Though similar to the larger, longer-tailed Song Sparrow, the Lincoln's Sparrow has distinctive fine, delicate lines on its breast, and frequently shows a slightly peaked crown.

LENGTH 5 inches. Short tail, peaked crown, small bill; rusty brown crown with gray median stripe; gray supercilium, and thin buffy eye ring against gray background; dark eye line and mustache stripe; finely streaked back; rich, light buff breast and flanks with fine dark streaks, and often a central breast spot; white belly and buffy undertail coverts. **VOICE** Song is a sweet, bubbly *wer wer wewewewe teeteeteetee*; call is a high, dry *chip*. **BEHAVIORS** Secretive but can be drawn out of cover by pishing. Males sing from lower branches in thick

cover. Escapes and draws intruders away from the nest by slinking quickly through low cover much like a rodent. **HABITAT** Bogs with thick brush or dense willows, aspens, or cottonwoods; densely vegetated clear-cuts and forest edges; brushy riparian areas. **STATUS** Fairly common summer resident above 3500 feet; fairly common winter resident in valleys west of the Cascades.

Lincoln's Sparrow

Lincoln's Sparrow range

Lincoln's Sparrow

484

Swamp Sparrow
(*Melospiza georgiana*)

The Swamp Sparrow's normal range is
primarily east of the Rocky Mountains,
but its breeding range extends westward
into the Canadian Rockies of British
Columbia, and consequently, these secretive, cover-loving sparrows regularly show
up in the Northwest during fall and winter, primarily west of the Cascades. They
are so reclusive, that often a brief flash
of the rust-colored wings is the first and
maybe only clue to the bird's identity.

Swamp Sparrow

Swamp Sparrow range

LENGTH 5.75 inches. Rufous-brown
crown with gray median crown stripe;
rich rufous wings and tail; streaked back;
clean white throat; breast and flanks have
pale streaks against a gray background;
white belly; buffy undertail coverts.
VOICE Song is a slow, single-pitched trill,
weet-weet-weet-weet-weet; call is a metallic *chip*. **BEHAVIORS** Secretive, but can be
drawn out of heavy cover by pishing. Flits
quickly about within marsh grasses. Infre-
quently sings during the winter, but more
often recognized by its *chip* call as it forages.
HABITAT Densely vegetated freshwater and
brackish marshes; dense riparian tangles
and swamps. **STATUS** Rare but regular fall
and winter visitor; most sightings occur in
valleys west of the Cascades, but recorded at
many locations in the inland Northwest.

White-throated Sparrow
(*Zonotrichia albicollis*)

The striking White-throated Sparrow, an uncommon but regular visitor to the Northwest from fall through early spring, readily visits bird feeders with other sparrows. In years when they are more common, White-throated Sparrows can occur in small flocks in the region, but individuals are typical. Both color variations, white-striped and tan-striped, are found in the Northwest.

LENGTH 6.5 inches. White-striped: black crown with white median stripe; broad white supercilium and black eye line; bright yellow supraloral patch; broad white throat; clean gray breast and belly; mantle is light brownish overall, with darker streaks; dark rufous wings with 2 thin wingbars; grayish tan rump and tail. Tan-striped: similar to white-striped, except head stripes are brown and pale buff, instead of black and white; white throat does not stand out as starkly against buffy breast and neck; supraloral patch is dark yellow; faint streaks on breast and flanks. **JUVENILE** Resembles tan-striped, but with more fine streaking on breast and flanks. **VOICE** Song is a series of 3 or 4 wavering, whistled notes, with first note lower in pitch; call is a sharp, but somewhat muted *seep*. **BEHAVIORS** Often feeds in the open, but not far from cover; when disturbed, often flies up to a tree branch or shrub canopy. Commonly seen at feeders. **HABITAT** Woodlands with heavy thickets or brush piles; blackberry tangles; bird feeders with nearby trees or brush. **STATUS** Rare but regular migrant and winter visitor.

White-stripe White-throated Sparrow

Tan-stripe White-throated Sparrow

White-throated Sparrow range

Golden-crowned Sparrow
(*Zonotrichia atricapilla*)

Occurring in flocks throughout valleys west of the Cascades, this wintertime resident attains its striking breeding plumage while here in the Northwest. When Golden-crowned Sparrows first arrive in the fall, the golden crowns for which they are named are dingy yellow bordered by brown. Beginning in February, their crowns begin to brighten and by April, they are fully adorned with showy bright yellow and jet black. These sparrows respond well to pishing, and a dozen or more curious birds may suddenly appear in a bush where there were none evident before.

LENGTH 7 inches. Large and robust, with a long tail. **NONBREEDING** (August–March) Brown crown with varying amount of yellow in the median crown stripe (some birds have little or none); gray head, breast, and belly; white undertail coverts; pale brownish mantle with dark stripes; dark flight feathers and wing coverts have light buff edges and whitish tips that form 2 faint wingbars; plain tannish gray rump and tail; **BREEDING** (March–August) Black crown with broad, bright yellow median stripe. **VOICE** Song is a series of descending whistles; call is a sharp *chink*. **BEHAVIORS** Often occurs in fairly large flocks that will spend the entire winter in a single location if food supplies last. **HABITAT** Blackberry tangles, woodlands with heavy thickets or brush piles, brushy fence lines and hedges, brushy road margins, bird feeders with nearby trees or brush. **STATUS** Abundant migrant and winter resident west of the Cascades; less common heading eastward from the Cascades; uncommon to interior Oregon, Washington, and British Columbia; rare vagrant in the Rockies. Breeding range extends south to south-central British Columbia, with one confirmed breeding record for the North Cascades of Washington.

Breeding Golden-crowned Sparrow

Nonbreeding Golden-crowned Sparrow

Golden-crowned Sparrow range

White-crowned Sparrow
(*Zonotrichia leucophrys*)

The showy White-crowned Sparrow, with its distinctive zebra-striped head pattern, is among the most common and widespread sparrows in the Northwest, able to utilize a variety of habitats. Several subspecies exhibit minimal differences: westside breeder *Zonotrichia leucophrys pugetensis* (Puget Sound White-crowned Sparrow) is smaller and dingier than the rather local *Z. l. oriantha* (Mountain White-crowned Sparrow), found in eastside mountains and willow thickets, which is more colorful and has a pink bill and black lores; migrant *Z. l. gambelii* (Gamble's or Western Taiga

White-crowned Sparrow) is larger than *pugetensis*, with an orange bill and pale lores. This large sparrow frequently visits bird feeders during fall and winter, where it readily coexists with other species.

LENGTH 7 inches. Large and robust, with long tail. **ADULT** Black crown with broad white median stripe; broad white supercilium, and black eye line; bill color ranges from pink to yellow; clean gray nape, throat, breast, and belly; grayish to tannish mantle with dark streaks; brownish gray rump and tail; wings edged with rufous; 2 faint wingbars. First-year adult (August–April) overall grayish brown; brownish instead of black head stripes; grayish brown median crown stripe. **JUVENILE** Heavily streaked overall, with dark crown stripes. **VOICE** Song (highly variable) is 2 whistled notes, the second higher in pitch, followed by 3 *chew* notes and then a trilled *ghee*. Call is a sharp *pik*. **BEHAVIORS** Often feeds in the open, frequently on the ground, but usually near cover for quick escapes. White-crowned Sparrows can lay up to 3 clutches per year. **HABITAT** Woodland thickets, brush piles, blackberry tangles, chaparral, desert scrub and shrub steppe, mountainsides with dense shrubbery, riparian corridors, brushy road margins, residential shrubbery. **STATUS** Common summer resident; abundant migrant and common winter resident west of the Cascades; abundant migrant and uncommon winter resident east of the Cascades.

Adult White-crowned Sparrow

Juvenile White-crowned Sparrow

White-crowned Sparrow range

Harris's Sparrow
(*Zonotrichia querula*)

A rare winter visitor to the Northwest, the large, handsome Harris's Sparrow is a Far North breeder, and in fact the only songbird that breeds exclusively in Canada; its normal winter range spans most of the American Great Plains and much of Texas. In the Northwest, it is usually seen in the company of other sparrows, and when a Harris's Sparrow mixes in with a flock of other species, it often remains with them all winter.

LENGTH 7.5 inches. **ADULT NONBREEDING** (August–March) Large and robust; pink bill; black face, chin, and throat, sometimes mottled and variable; dark crown and buffy brown ear coverts; striped back and flanks; white wingbars; clean white belly. **ADULT BREEDING** (April–August) Similar to adult nonbreeding, but more complete black bib, face, and crown; sides of head gray instead of buffy as in adult nonbreeding **FIRST-WINTER** white or partly white throat framed by black throat stripes and bordered below by partial black necklace. **VOICE** Rarely sings in winter, but song is a series of 2 to 4 whistles on one pitch, often followed by another series of lower-pitch whistles; call is a loud *wink*. **BEHAVIORS** In the Northwest, nearly always seen with other sparrows; generally forages above ground, in shrubs and trees. **HABITAT** Brush piles, fencerows, woodland tangles, bird feeders. **STATUS** Vagrant; most records October–May.

Adult breeding Harris's Sparrow

Adult nonbreeding Harris's Sparrow

Lark Bunting
(*Calamospiza melanocorys*)

A relatively recent arrival to Idaho, Lark Buntings appear to have expanded their range westward, across the Continental Divide, and have bred in very limited numbers as far east as the Arco Desert and Minidoka areas.

LENGTH 7 inches. Thick gray bill; white-tipped tail. **MALE BREEDING** Black overall, with prominent white shoulders. **MALE NONBREEDING** White breast and belly heavily patterned with black streaks; prominent white shoulders; black on front of face surrounds thick gray bill; grayish mantle and crown. **FEMALE** Pale gray-brown above; white below, with dark streaks over breast and flanks; extensive white on greater wing coverts; pale tan face with faintly lighter supercilium; black throat stripes frame white throat. **VOICE** Song is a sequential combination of a variety trills and buzzy, musical whistles, rolling from one to another, to another. **BEHAVIORS** Nests on the ground in loose colonies. **HABITAT** Sagebrush steppe, agricultural areas, grasslands. **STATUS** Vagrant (a few breeding records from eastern Idaho).

Male breeding Lark Bunting

Female Lark Bunting

490

Snow Bunting
(*Plectrophenax nivalis*)

A breeder of the Far North that regularly winters as far south as northern Washington and southern British Columbia, the Snow Bunting often strays much farther, and has been found throughout much of Oregon. As a general rule, Snow Buntings occur in flocks—sometimes containing many hundreds of individuals—within their normal wintering range, but outside of that range, small flocks and individuals are more typical. Birders in the Northwest won't see Snow Buntings in the stark and showy black-and-white breeding plumage, but even in their less ostentatious winter garb, they are striking birds, especially when they take flight and reveal their black-tipped white wings.

LENGTH 6.75 inches. In flight, shows black tips on white wings. **MALE NONBREEDING** White belly, breast, and throat; whitish face, with pale buff ear patch and crown; full or partial cinnamon breast band; pale grayish mantle with darker streaks; white inner wings form white stripes down the side when folded, bordered below by black. **FEMALE NONBREEDING** Similar to winter male, but slightly darker above. **VOICE** Calls include a very rapid, trilled *chir-wit-wit-wit* and a rapid *pipipipipip*. **BEHAVIORS** Forages on the ground; when flushed, flocks may circle around and land near the same area. In the interior Northwest, Snow Buntings sometimes occur within large flocks of Horned Larks. **HABITAT** Shortgrass prairies, tilled fields, grain lands, playas, dunes, beaches, jetties. **STATUS** Rare but regular winter visitor. **KEY SITES** Idaho: Ashton area; Highway 20 through Camas County. Washington: Okanogan Highlands. British Columbia: Boundary Bay and Iona Island; Okanagan Valley.

Nonbreeding Snow Bunting

Snow Bunting in flight

Snow Bunting range

SPARROWS

Lapland Longspur
(*Calcarius lapponicus*)

Between its Far North breeding territory and its wintering expanse in the Lower 48, the Lapland Longspur's range nearly covers the width and breadth of North America. Primarily a fall migrant and winter visitor in the Northwest, these ground birds of open country are likely somewhat underreported: east of the Cascades, they often mix with large winter flocks of Horned Larks, and such flocks are common on the high plains and wheat country of north-central Oregon and south-central Washington, but a paucity of winter birdwatchers in these sparsely settled regions probably means some Lapland Longspurs escape the winter unobserved.

LENGTH 6.25 inches. White outer tail feathers in all plumages. **MALE NONBREEDING** Streaked dark brown and warm tan above and on flanks; pale black-specked partial breast band and chin; warm buff-colored face, with cheeks framed above, below, and behind by black outline; rufous-colored greater wing coverts; dark crown; bronze-buff nape. **FEMALE NONBREEDING** Similar overall to winter male, but nape is lightly streaked and breast and throat are paler. **MALE BREEDING** (rare) Black breast, face, and crown, with bold white stripes extending from behind the eyes down the sides of the neck and breast; bright rusty red nape. **FEMALE BREEDING** (rare) Similar to winter female, but brighter overall. **VOICE** Calls include a high, sharp, musical *ticheeuw*; a clear whistled *tcheer*; a very rapid, chattering *pi-pit*; the various calls are often strung

Male breeding Lapland Longspur

492

together. **BEHAVIORS** Forages on the ground; often mixes with flocks of Horned Larks, Snow Buntings, and/or American Pipits. **HABITAT** Shortgrass prairies, tilled fields, grain lands, dunes, beaches, jetties. **STATUS** Rare fall and winter and very rare spring and summer visitor to Idaho and Oregon; rare to uncommon winter visitor and rare spring and summer visitor to Washington and British Columbia. **KEY SITES** Oregon: Columbia River south jetty and Fort Stevens State Park. Washington: Ocean Shores and Westport. British Columbia: Boundary Bay; Cattle Point and Clover Point (Victoria); Iona Island and Iona jetties (Vancouver).

Female nonbreeding Lapland Longspur

Lapland Longspur range

Male nonbreeding Lapland Longspur

Chestnut-collared Longspur
(*Calcarius ornatus*)

A small sparrow of the prairie states and provinces, the Chestnut-collard Longspur is a short-range migrant, wintering in the southern Great Plains, the desert Southwest, and northern Mexico. Annually a few birds stray west. They occur annually in California, but become far rarer northward: Oregon has about 20 confirmed sighting records and the Oregon Bird Records Committee no longer reviews sightings of this species. Washington has about 10 records, British Columbia about 20, and Idaho very few.

LENGTH 6 inches. White tail with black wedge tapering from tip toward rump is diagnostic. **MALE BREEDING** Black belly and breast; mottled black-brown back; black cap and black ear mark over white face with pale tannish yellow cheeks and throat; bright chestnut nape. **MALE NONBREEDING** Warm tan overall with streaked back, fall birds with a wash of residual black feathers over breast and belly, lessening in winter; pale buffy face and throat, dark brown cap, whitish eyebrow; pale chestnut nape. **FEMALE AND JUVENILE** Similar to winter male, but lack pale chestnut nape. **VOICE** Calls include a rapid, rattled *pipipipip* and a slightly harsher *per-pi-pit*. **BEHAVIORS** Typically forages on the ground. **HABITAT** Most Northwest records are coastal, generally on or near jetties and surrounding areas; also shortgrass prairie, expansive mudflats, airport grounds; tilled fields. **STATUS** Rare vagrant; most Oregon and Washington records are for late summer and autumn, whereas most British Columbia records are for May–July along the southwest coast.

Male breeding Chestnut-collared Longspur

Male nonbreeding Chestnut-collared Longspur

Female Chestnut-collared Longspur

Blackbirds, Cowbirds, Grackles, Bobolink, Meadowlarks, and Orioles

Gregarious, often colorful, and frequently loud, the icterids (family Icteridae) include such familiar birds as the Brewer's and Red-winged Blackbirds, among the most abundant species in the Northwest. This family also includes the colorful Orioles, represented by only 2 breeding residents in the Northwest, the relatively widespread Bullock's Oriole, and the Scott's Oriole, largely a Southwest species that reaches the northern extent of its nesting range in one small, isolated mountain range in southern Idaho. Many icterids are social, forming nesting colonies to varying degrees, and forming flocks outside of breeding season that often number in the hundreds.

Yellow-headed Blackbird

Bobolink *(Dolichonyx oryzivorus)*

Bobolinks generally arrive at a handful of well-known Northwest breeding sites—hayfields, grasslands, and tallgrass prairies—by late May. The breeding-plumage males are striking, and their enthusiastic display flights are always fun to watch.

LENGTH 7 inches. Compact; short neck, flat head; short, sharp-pointed tail feathers; thick, short, sparrowlike bill. **MALE BREEDING** Unmistakable, with straw-colored nape, large white wing and rump patches, black

Female Bobolink

breast, belly, and flanks. **MALE NONBREEDING** Similar to female but more yellowish below. **FEMALE AND JUVENILE** Light tan below, streaked dark tan above, with dark crown stripes and buffy eyeline; female has light streaking on the breast. **VOICE** Male's song is loud and bubbly—rambling, rich with twittering highs and buzzy lows; each male's songs is composed of 25 to 50 notes and last about 3.5 seconds. Both sexes use a variety of one-note calls, including a metallic *pink* or *quink*. **BEHAVIORS** Males perch on willows, fence posts, and any plant that can support their weight; they make fluttering display flights (while singing) over the nesting territory. Females are inconspicuous on nesting grounds, preferring to forage on the ground when not sitting on the nest. Highly social, often seen in groups. **HABITAT** Open, moist, tallgrass prairies and hayfields. **STATUS** Locally common at breeding colonies; rare elsewhere. **KEY SITES** Idaho: Garden Gulch Road (8 miles east of Potlatch); Landruff Lane (Middleton). Oregon: Blitzen River Valley (south end); Diamond Valley; Malheur NWR (south end Center Patrol Road). Washington: Aeneas Valley; Toppenish NWR. British Columbia: Road 22 just north of Osoyoos Lake (Bobolink Meadow).

Bobolink range

Male breeding Bobolink

Western Meadowlark
(*Sturnella neglecta*)

Oregon's official state bird abounds throughout much of the interior Northwest, but, perhaps surprisingly, is increasingly scarce west of the Cascades in Oregon and Washington, where it once thrived, largely owing to loss of native bunchgrass habitat. Once replete with large tracts of the open prairie that forms the Western Meadowlark's prime habitat, the valleys west of the Cascades are now occupied by human settlements and large farms, with no year-round grassland cover, and substantial acreage of heavy blackberry-riddled fallow—the kinds of places this colorful bird avoids.

LENGTH 9.5 inches. Soft mottled brown and tan above; bright yellow below with black necklace (reduced and paler in fall and winter); brown crown stripes and pale supercilium; long, straight bill; short tail with white outer feathers. Juvenile is similar, but paler overall and lacks black necklace. **VOICE** Song is one to 3 plaintive whistles followed by a rapid, descending, whistled warbling; calls include single-note whistles, a rapid high rattle, an urgent *chup*; flight calls include scratchy jumbled whistles. **BEHAVIORS** Although generally a ground-dwelling species, the Western Meadowlark often sings or surveys its surroundings from shrubs, rocks, fence posts, and other such objects. **HABITAT** Grasslands, pastures, grain field margins, arid shrub steppe, vegetated

dunes. **STATUS** Common summer resident of interior; uncommon year-round resident of interior, but birds generally retreat from Rocky Mountains slopes, as well as much of eastern Oregon, southwest Idaho, and northeast Washington; uncommon year-round resident west of Cascades.

Western Meadowlark

Western Meadowlark

Western Meadowlark range

Brewer's Blackbird
(*Euphagus cyanocephalus*)

One of the most widespread and familiar songbirds in the Northwest, the Brewer's Blackbird occupies a broad array of habitats, from city sidewalks—where it may gather for the handout crumbs of outdoor diners—to remote, wide-open spaces.

LENGTH 9 inches. **MALE** Lustrous black overall, in good light flashing iridescent blue, green, and bronze; bright white eyes. **FEMALE** Unmarked grayish brown overall, darkest on wings and tail. **VOICE** Song is soft slur followed by a loud, shrill, metallic, *sooop-shureeee*; calls include a rapid *chuk*, and a sharp *whit*. **BEHAVIORS** Primarily forages on the ground, eating a variety of foods, including insects, seeds, grains, and small fruit. During migration and winter, forms flocks, often mixing with Red-winged Blackbirds, Brown-headed Cowbirds, and Starlings. **HABITAT** Widespread across myriad open or relatively open habitats, from city streets to remote deserts. **STATUS** Abundant year-round resident.

Male Brewer's Blackbird

Brewer's Blackbird range

Female Brewer's Blackbird

Rusty Blackbird
(*Euphagus carolinus*)

Breeding from coast to coast across most of Canada and Alaska, the Rusty Blackbird migrates primarily east of the Rocky Mountains. It is a rare winter visitor to most of the Northwest, and also a rare summer resident and migrant in south-central British Columbia.

LENGTH 9 inches. **MALE** Similar to Brewer's Blackbird, but with rusty crown and mantle; silvery scaling on throat and breast, becoming a rustier scaling in winter; white eye. **FEMALE** Similar to female Brewer's Blackbird, but rustier overall and with pale iris. **VOICE** Calls include an abrupt, sharp *chuck*; typical song (rarely heard in Northwest) is a piercing metallic whistle followed by a hurried, garbled *kyurdl-uurup*. **BEHAVIORS** In the Northwest, usually found foraging on the ground, and usually in the company of Brewer's Blackbirds, Red-winged Blackbirds, or other species; Rusty Blackbirds often hold their tails cocked upward when foraging. **HABITAT** Generally open to semi-open spaces when found in the Northwest, including agricultural fields, shorelines, mudflats, parks; breeding habitat in British Columbia is typically moist areas in boreal forests. **STATUS** Vagrant. Most sightings occur in winter.

Male nonbreeding Rusty Blackbird

Female nonbreeding Rusty Blackbird

Red-winged Blackbird
(*Agelaius phoeniceus*)

The abundant and widespread Red-winged Blackbird is found in virtually every Northwest wetland habitat during breeding season, from urban and suburban parks and waterways to the most remote desert springs and seasonal wetlands. During his breeding-season courtship displays, the male puffs up his glowing red shoulders in a bedazzling attempt to woo nearby females and intimidate rival males.

LENGTH 8.5 inches. **MALE** Black overall, with showy crimson shoulders bordered by yellowish stripe across median coverts (shoulder patches can be concealed); sharply pointed bill. During winter, black feathers become tipped with brown. First-year male similar to adult male, but browner, with less developed red and yellow shoulder patches. **FEMALE** Rusty brown to brownish overall, heavily streaked with dark brown over pale tan; sometimes a rufous wash on mantle and wings; pale tan supercilium, face, and throat. **JUVENILE** Similar to adult female. **VOICE** Song is a loud, distinctive, gurgling *konk-a-reee*. Calls include a sharp, metallic *chink*, and a high-pitched *seeeeee*. **BEHAVIORS** In migratory parts of its range, males are the first to arrive in the spring, sometimes in large flocks. Winters in large flocks, and retreats from regions experiencing hard freezes. **HABITAT** Freshwater and saltwater marshes, ditches, willow groves, ponds, lakes, and rivers, seasonal wetlands, agricultural areas, suburban areas, bird feeders. **STATUS** Abundant year-round resident.

Male Red-winged Blackbird displaying

Female Red-winged Blackbird

First-spring male Red-winged Blackbird

Red-winged Blackbird range

Tricolored Blackbird
(*Agelaius tricolor*)

Fairly recent arrivals in the Northwest from their traditional range in California, Tricolored Blackbirds nest colonially in cattail and bulrush marshes. Such nesting colonies have sprung up in a handful of locations in Central Oregon and on the Columbia Plateau in Washington, but these gregarious birds sometimes abandon colony sites and choose new nesting grounds elsewhere.

LENGTH 8.75 inches. **MALE** All black, with fire engine–red shoulders bordered by white stripes across the median coverts; red shoulders often concealed, but white median coverts usually visible. **FEMALE** Dark sooty gray overall, with only minimal streaking on the upper breast, neck, and nape (unlike heavily streaked female Red-winged Blackbird), and lacking the light buff-tan facial shades and rufous wash of the female Red-winged Blackbird. **VOICE** Calls are similar to those of Red-winged Blackbird; song is a reedy, oinking *wool-weeeank*. **BEHAVIORS** Nests colonially within marsh hardstem vegetation; outside of breeding season, often flocks with other blackbirds. **HABITAT** Nests in marshes with dense stands of cattails or tule bulrush; outside of nesting season, often found in agricultural areas. **STATUS** Locally uncommon summer resident; uncommon year-round resident. **KEY SITES** Oregon: Juniper Flat Road (Wasco County; check cattail stands, especially 0.75 mile north of Highway 216); Lamonta Road, south of Grimes Road (north of Prineville); Woodward Marsh on Woodward Road, 0.5 mile south of Gerke Road (north of Prineville); Lower Klamath Lake Road, State Line Road, Township Road (Klamath County); Washington: marshes just south of Kanniwai Creek Road, south of Wilson Creek (Grant County); Para Ponds, McManamon Road (north of Othello); Texas Lake (Whitman County).

Female Tricolored Blackbird

Male Tricolored Blackbird

Tricolored Blackbird range

Yellow-headed Blackbird
(*Xanthocephalus xanthocephalus*)

The stark contrast between black body and bright yellow head makes the male Yellow-headed Blackbird one of the must-see species of the Northwest's inland marsh ecosystems. A nesting colony of these gregarious blackbirds creates a cacophony as the males compete for the attentions of females.

LENGTH 9.5 inches. **MALE** Black body and bright yellow head, neck, and breast; white primary coverts. First-year male similar to female. **FEMALE** Dark brown above, with yellow breast, and yellowish face and vent. **VOICE** Song begins with a series of reedy, metallic *chink* notes, then ends with a dry buzzing; calls include a rapid, clucking *chu-chuk* and a quick, repeated *chuk*; the female offers a chattering *chee-chee-chee-chee*. **BEHAVIORS** In his spring courtship display, the male, perching on a reed, puffs up his plumage, fans his tail, spreads his wings, throws his head back, and utters his metallic song. **HABITAT** Freshwater marshes with dense emergent vegetation, especially those with tule bulrush and cattail; agricultural areas adjacent to marshlands. **STATUS** Locally common summer resident and uncommon year-round resident of the interior; uncommon and very local summer resident and migrant west of the Cascades. **KEY SITES** Idaho: Camas NWR and Market Lake Wildlife Area; Hyatt Hidden Lakes Reserve (Boise). Oregon: Finley NWR; Ladd Marsh Wildlife Area; Malheur NWR; Upper Klamath Lake. Washington: Columbia NWR and Potholes Reservoir; Toppenish NWR; Turnbull NWR. British Columbia: Creston Valley Wildlife Area; Road 22 (Osoyoos).

Female Yellow-headed Blackbird

Yellow-headed Blackbird range

Male Yellow-headed Blackbird

Brown-headed Cowbird
(*Molothrus ater*)

Originally, Brown-headed Cowbirds ranged across the North American shortgrass plains, where they tagged along with bison herds, eating insects. But human settlement allowed them to expand their range—problematic for myriad other songbirds because cowbirds are brood parasites. Rather than build a nest and incubate eggs, the female cowbird lays her eggs in the nest of another species and lets the host species do all the work. Cowbird eggs hatch sooner than host-bird eggs and the young cowbirds outcompete the host hatchlings for food.

LENGTH 7.5 inches. Dark eyes (Rusty Blackbird and male Brewer's Blackbird have white eyes); when seen together with blackbirds, Brown-headed Cowbirds are noticeably smaller, with shorter tails, giving them a stubbier appearance; short conical bill. **MALE** Glossy black overall, iridescent in good light, with chocolate-brown hood (difficult to see in bad light). **FEMALE** Pale grayish brown overall, lightest on breast and throat. **JUVENILE** Similar to female, but paler and speckled overall, heaviest on breast; pale throat. **VOICE** Song is a high, liquid, bubbling *gling-goink-eeee*, the first 2 short notes sounding like dripping water; calls include various chatters and whistles. **BEHAVIORS** Often flocks with other blackbirds, feeding in crop fields and around livestock. **HABITAT** Open and semi-open areas, especially near farms, ranches, suburbs; city parks, large parking lots, brushy fallow, open woodlands, and drawn-down reservoirs. **STATUS** Common summer resident and migrant of interior; common summer resident and uncommon year-round resident west of Cascades.

Male Brown-headed Cowbird

Female Brown-headed Cowbird

Juvenile Brown-headed Cowbird

Brown-headed Cowbird range

Great-tailed Grackle
(*Quiscalus mexicanus*)

A fairly recent arrival in the Northwest, and now breeding in southern Idaho, the Great-tailed Grackle has been expanding its range northward from its traditional stronghold in the Southwest. It is an annual vagrant in Oregon, where it has bred, primarily east of the Cascades and especially in southeast Oregon. It remains very rare in Washington and British Columbia.

LENGTH 15–19 inches (male is larger). Long tail; heavy, slightly decurved bill. **MALE** Unmistakable; glossy black (purplish iridescent in good light), with very long, wedge-shaped, bannerlike tail. **FEMALE** Dark sooty brown above, pale cinnamon-brown below, with dark brown cap and pale supercilium; white iris. **VOICE** Highly variable, with a substantial repertoire. Song ranges from a very high, sirenlike *tawee-tawee-tawee churee-churee-churee*, usually with added rattles and whistles, to a more complex, rolling *pipipipipip* mixed with various squeals, tinny

croaks, and spiraling whistles; calls include various whistles, low grunts, and metallic whines. **BEHAVIORS** Readily adapts to human habitation and often forages in parking lots, highway rest areas, ranches, and farms. **HABITAT** Generally open areas with scattered trees, including towns. **STATUS** Rare summer resident (April–August); very rare winter visitor. **KEY SITES** Idaho: Market Lake Wildlife Area and Roberts area.

Female Great-tailed Grackle

Great-tailed Grackle range

Male Great-tailed Grackle

Common Grackle
(*Quiscalus quiscula*)

Like the larger, showier Great-tailed Grackle, the Common Grackle has been expanding its range, in this case, from the east, across the Rocky Mountains into Idaho and Oregon. It breeds in eastern Idaho, and shows up annually in eastern Oregon, though it remains very rare in Washington and southern British Columbia.

LENGTH 12 inches. **SEXES SIMILAR** Long, keel-shaped tail; black overall, but in good light, iridescent bronze with purplish sheen on head (female less lustrous than male); bright, light yellow eyes; slightly decurved bill more robust than blackbird bills; wing-tips often held slightly drooping. **VOICE** Song is a sharp, unmusical, rising, whiny *zeereeep* or *zoo-are-eeep*, something like a rusty gate hinge; calls include clucks, croaks, and a buzzy *cheeeeer*. **BEHAVIORS** Mixes with other blackbirds, especially in nonbreeding seasons, and readily adapts to foraging in urban and suburban settings. **HABITAT** Generally open areas with scattered trees, including towns, farmlands, ranches, marshes, river corridors, roadways. **STATUS** Rare but regular summer resident and fall migrant (May–September) in Idaho; rare vagrant elsewhere. **KEY SITES** Idaho: Market Lake Wildlife Area and Roberts; Rexburg area; Salmon area.

Common Grackle

Common Grackle range

Bullock's Oriole (*Icterus bullockii*)

The western counterpart of the famous Baltimore Oriole of the east half of the continent, the Bullock's Oriole is the most colorful of the blackbird family in the Pacific Northwest. At one point, these superficially similar and widespread orioles were lumped into one species and unceremoniously renamed Northern Oriole, but subsequent genetic research reaffirmed that they were indeed separate species.

LENGTH 9 inches. **MALE** Flashy combination of orange, black, and white; black crown, eye line, nape, mantle, and bib; black wings with white shoulders; bright orange face, neck, breast, belly, and rump; orange tail with black center and terminal band. First-year male much paler than adult male and similar to female; more yellowish than orange, and dusky gray-olive rather than black above; partial black bib. **FEMALE** Bright soft yellow to lemon-yellow face, with darker eye line; light yellow throat, breast, rump, and tail; pale gray mantle; gray wings with white wingbars; white belly. **VOICE** Song (variable) is a mixture of rich warbled whistles and scratchy chatter; calls include a raucous, dry chattering; a sharp, piping single-note whistle; abrupt *chuk* calls. **BEHAVIORS** Like other orioles, Bullock's Oriole builds a woven hanging basket–style nest, which is easy to notice when trees are defoliated during fall and winter, providing evidence that orioles inhabit the area in summer. **HABITAT** Open deciduous woodlots near water (including rivers), brushy canyon-bound streams, brushy springs with

Male Bullock's Oriole

Female Bullock's Oriole

tree cover, wetlands, farmlands; suburban deciduous trees. **STATUS** Summer resident (May–September); uncommon west of Cascades; fairly common to interior.

SIMILAR SPECIES Baltimore Oriole (*Icterus galbula*) is a rare vagrant to the Northwest. Male is similar overall to Bullock's Oriole, but with all-black hood, orange shoulder, and white wingbar. Female is very similar to female Bullock's Oriole, but yellowish orange instead of yellow throat and breast, with color extending onto the belly (generally a more extensive white belly in Bullock's), crisp white wingbars; dusky olive-brown face, crown, and nape; grayish brown mantle; sometimes varying degrees of black or gray scaling on head and neck.

First-spring male Bullock's Oriole

Bullock's Oriole range

Similar species: Baltimore Oriole (male)

Scott's Oriole (*Icterus parisorum*)

The Scott's Oriole is a desert specialist of the American Southwest and Mexico, and strongly associated with yucca and pinion pine. So perhaps surprisingly, it reaches the northernmost limit of its range in what appears to be a disjunctive and tiny breeding population in a small, remote range of low mountains in southeast Idaho, not far from the Utah Border.

LENGTH 9 inches. **MALE** Striking combination of bright yellow and black; black hood, breast, and mantle; black wings with 2 white wingbars and yellow shoulders; yellow belly, flanks, rump, and undertail coverts; first-spring males have varying degrees of black or sooty gray on hood, breast, mantle. **FEMALE** Yellowish below, yellowish olive hood and back with varying amounts of gray or black specks or streaks; gray wings with 2 white wingbars. **VOICE** Song is series of musical whistles that warble slightly, more akin to a Western Meadowlark than an oriole; call is a reedy *chap*, often repeated persistently. **BEHAVIORS** Scott's Orioles will sing any time of day, from predawn to the heat of midday. **HABITAT** In Idaho, slopes in sagebrush steppe and associated juniper woodlands. **STATUS** Very rare and local summer resident of south-central Idaho; very rare vagrant elsewhere. **KEY SITES** Idaho: Black Pine Road area, Oneida County (west by southwest of Stone Reservoir); Stone Hills.

Female Scott's Oriole

Scott's Oriole range

Male Scott's Oriole

Hooded Oriole (*Icterus cucullatus*)

A California and desert Southwest species, the Hooded Oriole wanders into the Northwest just often enough that birders who see an oriole in winter should immediately suspect they may have found this rarity.

LENGTH 7.5 inches. Fairly long, noticeably slightly decurved bill. **MALE** Bright orange, with black bib and face; black mantle; black wings with prominent white shoulders. First-year male more yellowish than orange, with partial to nearly full black bib; grayish back. **FEMALE** Soft yellowish head, throat, and breast, fading to light gray belly, then yellowish undertail coverts; light gray mantle; gray wings with 2 white wingbars. **VOICE** Calls include a high, slightly rising *weeit*, and rapid, snappy chattering. **BEHAVIORS** Attracted to feeders, including hummingbird feeders and suet. **HABITAT** Most sightings occur in residential areas. **STATUS** Very rare but fairly regular vagrant, primarily west of the Cascades; most sightings are in late fall and winter.

Female Hooded Oriole

Male Hooded Oriole

Finches and House Sparrows

Sparrowlike in general appearance, members of the finch family (Fringillidae) tend toward the colorful, and include intriguing species, such as the 2 aptly named crossbills, itinerant forest birds whose bills are specially adapted to extract seeds from the cones of conifer trees, as well as the rosy-finches, alpine specialists perfectly at home atop snowfields and scree on the region's highest peaks. Some finches—the House Finch, Purple Finch, and the 2 goldfinches—frequent bird feeders, where they prefer sunflower and thistle, and one, the Pine Siskin, can arrive at bird feeders en masse, hanging around in noisy approachable flocks all winter. The related House Sparrow (Passeridae family) is an import from Europe that has enjoyed explosive success in populating North America; it is the common songbird of city sidewalks.

Red Crossbills

Gray-crowned Rosy-Finch
(*Leucosticte tephrocotis*)

Observing the relatively little-known Gray-crowned Rosy-Finch in its breeding habitat often requires a substantial investment in boot leather, for these attractive finches are alpine specialists. They nest above timberline, usually on rocky slopes near snow fields or glaciers; to avoid climbing into such inhospitable habitats on foot, intrepid birders often visit high-elevation ski resorts during summer and fall to search for rosy-finches. Luckily, these finches are more widely distributed in winter. Two prominent subspecies show minor plumage variation: in the Hepburn's subspecies, the head is mostly gray; in the interior subspecies, the cheeks are brown and the hind-crown is gray.

LENGTH 6 inches. **SEXES SIMILAR** Black forehead contrasts with pearl-gray hind-crown (interior subspeices) or hind-crown, nape, and cheeks (Hepburn's subspecies); warm brown breast, neck, and mantle; black throat; belly, flanks, and wing coverts washed with pale pink. **VOICE** Calls include high, bubbly, twittering chirps and *cheer* notes. **BEHAVIORS** Frequently feeds on and along snowbanks; in winter, feeds on open ground, including fields, barren hilltops, road margins, alpine areas blown free of snow; wintering flocks tend to take flight in leapfrog fashion. **HABITAT** Breeds above timberline, typically on rocky slopes near glaciers and snowfields; winters at both high and lower elevations in open and semi-open country. **STATUS** Uncommon year-round resident. **KEY SITES** Idaho: Discovery Unit, Lucky Peak State Park (winter). Oregon: Crater Lake National Park; Marys Peak in Benton County (fall and winter); South Sister (hike-in only); Wallowa Valley (winter). Washington: Mount Rainier National Park; Mount Spokane (fall).

Adult Gray-crowned Rosy-Finch

Juvenile Gray-crowned Rosy-Finch

Gray-crowned Rosy-Finch range

Black Rosy-Finch
(*Leucosticte atrata*)

A high-elevation specialist, and in fact, one of the likeliest ground-feeding songbird to be found on the highest ramparts of central Idaho, the Black Rosy-Finch is enigmatic throughout its range: population estimates are nearly impossible owing to the imposing, often inaccessible alpine habitat of this species. Further, a disjunctive Northwest breeding population occupies Steens Mountain in southeast Oregon.

LENGTH 6 inches. **MALE** Blackish overall, with fine silvery scaling on breast and back; distinctive pale gray hind-crown extending down sides of face to eyes; rosy wing coverts and flight feather edges; rosy wash to belly. **FEMALE** Similar overall to male, but largely grayish instead of black, and less colorful. **JUVENILE** Grayish overall, with lighter wing coverts; yellowish bill. **VOICE** Calls include high, bubbly, twittering *cheer* notes; a drier, slower *cheeuw*; a high, slightly raspy, barking *puurt*. **BEHAVIORS** Frequently feeds on and along snowbanks; nests on steep, rocky slopes and cliffs; in winter, feeds on open ground, including fields and road margins, as well as alpine areas blown free of snow; wintering flocks tend to take flight in leapfrog fashion. **HABITAT** Nests above timberline; generally disperses to lower elevations in winter. **STATUS** Uncommon and local year-round resident. **KEY SITES** Oregon: Steens Mountain (summer). Idaho: Discovery Unit, Lucky Peak State Park (winter); Mount Borah, (summer; hike-in only).

First-fall Black Rosy-Finch

Adult Black Rosy-Finch

Black Rosy-Finch range

Pine Grosbeak
(*Pinicola enucleator*)

A large, beautiful finch that nests in high-elevation, open montane forests in its limited Northwest breeding range, the Pine Grosbeak varies substantially in plumage, from bright red males to grayer individuals faintly washed with red. During winter, they are more widespread, but always a rare and welcome sight. Wintering flocks—from a few individuals to large flocks—expand their range to take advantage of fruit-bearing trees and shrubs.

LENGTH 9 inches. Large, plump finch; short, conical bill with slightly curved upper mandible; fairly long, notched, black tail; dark wings with 2 white wingbars. **MALE** Ranges from bright rosy red overall, with gray flanks and belly, to grayish overall with pale pinkish red head, nape, and rump, and reddish wash on upper breast and mantle. **FEMALE** Gray overall with yellow to olive-yellow to bronze-orange crown, nape, and rump. **JUVENILE** Similar to adult female, juvenile males often with rust-red highlights. **VOICE** Calls include a surprisingly musical, soft, high, piping; a high *tew-tew-tew*. Song is a fairly brief but regularly repeated musical warble, reminiscent of an American Robin, but not as rich. **BEHAVIORS** Can be very tame and approachable, but also tends to be quiet and easily overlooked; in breeding season, males often sing from prominent perches, such as the tip-top of a conifer. Pine Grosbeaks can be irruptive in winter, expanding southward well beyond their normal range. **HABITAT** Nests in open coniferous forest, especially in moist areas at high elevation; wintering habitat more varied and based on available food supplies, including fruit and seeds. **STATUS** Rare year-round resident and winter visitor. **KEY SITES** (winter) Idaho: Moscow area. Oregon: Joseph area. Washington: Mount Rainier National Park; Okanogan Highlands; Salmo Pass (Pend Oreille County). British Columbia: E.C. Manning Provincial Park; Okanagan Valley.

Male Pine Grosbeak

Female Pine Grosbeak

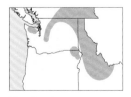
Pine Grosbeak range

House Finch
(*Haemerhous mexicanus*)

One of the Northwest's most familiar backyard feeder birds, the House Finch embraces a wider variety of habitats than the similar and closely related Cassin's and Purple Finches. East of the Cascades, where Cassin's Finches are fairly common, and west of the Cascades, where Purple Finches are common, House Finches can be differentiated by their lack of a pronounced facial pattern as well as more subtle clues.

LENGTH 6 inches. Pale tan overall with darker brownish streaks, including streaked flanks; thick, stubby bill with slightly curved culmen; smoothly rounded head; 2 pale wingbars. **MALE** Red (or sometimes yellow or orange) forecrown, breast, throat, and rump. **FEMALE** Similar to male, but lacks red tones; face is plain grayish brown with fine streaks, lacking the defined dark and light marks of female Purple and Cassin's Finches. **JUVENILE** Similar to female but more grayish and more finely streaked. **VOICE** Song is a rich, musical mix of chirping whistles usually including a buzzy whistled trill at or near the end; calls include a short *chip* and more drawn-out *cheeeup*. **BEHAVIORS** Frequents bird feeders year-round. **HABITAT** Brushy areas, open woodlands and woodland edges, city parks, residential areas. **STATUS** Common year-round resident.

House Finch range

Male House Finch

Female House Finch

Purple Finch
(*Carpodacus purpureus*)

The colorful and musical male Purple Finch is always a welcome sight. Careful study is required to differentiate Purple Finches from Cassin's Finches where their ranges overlap (primarily along the lower east slope of the Cascades, and in Oregon's Rogue River Valley). The Purple Finch's wingtips extend only to the tail coverts, whereas the Cassin's Finch's wingtips extend nearly to the middle of the tail; the Cassin's has a longer bill with a straight culmen; the Purple Finch's bill is stubbier and the culmen is curved.

LENGTH 6 inches. Very thick, short bill with curved culmen; wingtips extend to tail coverts. **MALE** Raspberry-red head and throat; pinkish brown back, pinkish rump; pink-tinged breast appears scaled, not streaked; a few very light pale brown streaks on flanks; pale gray and white belly and undertail coverts. **FEMALE** Streaked, soft grayish brown above, sometimes with a slight olive cast; buffy below with streaks almost forming chevrons (unlike streaks on female House Finch); pale and well-defined eye stripe and cheek stripe. **JUVENILE** Similar to female. **VOICE** Melodious, rich mixture of chirpy warbles, sometimes including short, high-pitched trills; flight call a sharp, high-pitched *chip chip chip*. **BEHAVIORS** Forages at various levels in trees, including canopy; eats a wide variety of seeds and fruits; readily comes to feeders, especially for sunflower seeds. Swift, direct, undulating flight, often with flight call. **HABITAT** Prefers mixed forests of conifers, oaks, ash, and other hardwoods. **STATUS** Fairly common year-round resident.

Female Purple Finch

Purple Finch range

Male Purple Finch

Cassin's Finch
(Haemerhous cassinii)

A forest species of the interior North-west, the Cassin's Finch is quite similar in appearance to the Purple Finch, and their ranges overlap along the east slope of the Cascades and in Oregon's Rogue River Valley. Wingtips of the Cassin's Finch extend nearly to the middle of the tail, whereas the Purple Finch's wingtips extend only to the tail coverts; the Cassin's has a longer bill with a straight culmen, the Purple Finch's bill is stubbier and the culmen is curved. The adult male Cassin's Finch is less extensively red than the male Purple Finch.

Male Cassin's Finch

Female Cassin's Finch

LENGTH 6.25 inches. Chunkier than the House Finch; triangular-shaped bill with straight culmen; crown often peaked with feathers slightly raised rather than rounded; slightly notched tail. **MALE** Bright red crown; rosy red face and breast; rosy hued mantle with distinct streaks; white belly; whitish to pink-tinged flanks, usually with fine streaks. **FEMALE** Light brownish above with distinct streaking; clean white below with pronounced thin streaking; broad brownish eye and cheek stripe framed above and below by whitish stripes. **JUVENILE** Similar to adult but markings not as crisp. **VOICE** Song is a lively musical combination of warbles and whistles, somewhat softer and more hurried than Purple Finch; calls include a high, musical *pity-ip*, and a rather harsh, high *peeryip*. **BEHAVIORS** Within appropriate habitat, often visits bird feeders, especially in winter. **HABITAT** Conifer forests, especially open pine and juniper woodlands; towns and residential areas in appropriate habitat. **STATUS** Uncommon year-round resident. **KEY SITES** Idaho: City of Rocks National Reserve; Craig Mountain Wildlife Area; Lucky Peak (Boise area); Ponderosa State Park (McCall). Oregon: Bend; Calliope Crossing; Cold Springs Campground; U.S. Forest Service offices in Sisters; Crater Lake National Park; Idlewild Campground (Malheur National Forest); Mount Ashland. Washington: Chinook Pass; Leavenworth area; Mount Rainer National Park; Mount Spokane. British Columbia: Kimberley and Cranbrook area; Okanagan Valley and surrounding uplands.

Cassin's Finch range

Red Crossbill (*Loxia curvirostra*)

Named for their distinctive bills (in which the tips of the upper and lower mandibles cross over each other, enabling the birds to pry open conifer cones to get at the seeds inside), Red Crossbills are nomadic denizens of forested Northwest mountains. In most areas, their numbers vary from year to year and season to season based on cone crops. Bill size and calls vary considerably among Red Crossbill populations, leading to frequent speculation that the species could at some point be split into several distinct species throughout its circumpolar range.

LENGTH 6.25 inches. Distinctive crossed mandible tips; long, dark, pointed wings lacking any contrasting markings. **MALE** Reddish overall, ranging from orange to brick red, and occasionally bright yellow to yellow orange; sometimes scalloped or scaled appearance in one or more shades of red; dark wings. **FEMALE** Muted yellowish green and gray overall, brightest on crown, breast, mantle, and rump; dark wings. **JUVENILE** Pale tan overall and heavily patterned with fine brown streaks, often with washes of color ranging from yellow to red. **VOICE** Distinctive call, often in flight, is a musical *jip-jip*; song, quite variable, is a series of thin warbles mixed with airy whistles, such as *pi-pit-pit* or *thi-pit, thi-pit, thi-pit*; also *pur-it pur-it*. **BEHAVIORS** Usually seen in flocks; feeds in pine trees, but often visits the ground for water and to obtain minerals; can nest at any time of year based on available cone crops for food. **HABITAT** Conifer forests, especially ponderosa pine, lodgepole pine, Douglas fir, and western hemlock. **STATUS** Fairly common year-round resident. **KEY SITES** Idaho: Cassia Mountains (also known as South Hills); Lucky Peak Intermountain Bird Observatory; Ponderosa State Park (McCall); Tubbs Hill Park (Coeur d'Alene). Oregon: Ponderosa Best Western and U.S. Forest Service offices (Sisters); Sawyer and Shevlin city parks (Bend). Washington: Discovery Park (Seattle); Mount Spokane; Theler Wetlands (Kitsap Peninsula); Wenas Campground and Wenas area. British Columbia: Okanagan Valley, Swan Lake Nature Sanctuary (Victoria).

Male Red Crossbill

Male Red Crossbill

Female Red Crossbill

Juvenile Red Crossbills

Red Crossbill range

White-winged Crossbill
(*Loxia leucoptera*)

Precise delineation of the range and relative abundance of the highly nomadic White-winged Crossbill is as elusive as the birds themselves. These enigmatic wanderers can nest at any time of year and numbers can vary enough that they can be locally fairly common at times and then absent for years on end, with their movements largely dependent on spruce cone crops, primarily Engelmann spruce in the Northwest (as well as a few other conifers). The Crossbill's unique bill is adapted to prying open cones to get at the seeds.

LENGTH 6.25 inches. Distinctive crossed mandible tips; long, black, pointed wings with 2 prominent wingbars. **MALE** Rosy red overall; black tail; black wings with white wingbars. **FEMALE** Greenish yellow breast, crown, face, and mantle, ranging from bright and crisp to only a yellowish wash over tan background. **JUVENILE** Similar to female, but generally lacking yellow shades and more heavily streaked; dark wings with thin white wingbars. **VOICE** Song is an energetic series of rapid musical chatters, changing pitch with each series. Calls include a harsh, slurred *chef-chef-chef* typically given in flight; also a sharp *peet-peet-peet*. **BEHAVIORS** Typically found in flocks; feeds on cones (especially spruce), and flocks feeding in the same tree produce a shower of cone fragments; licks minerals and salts from the ground, especially roadways. **HABITAT** Conifer forests with substantial Engelmann spruce component. **STATUS** Uncommon year-round resident of British Columbia, becoming increasingly rare and irregular southward; very rare in Oregon and central and southern Idaho.

Male White-winged Crossbill

Female White-winged Crossbill

White-winged Crossbill range

Common Redpoll
(*Acanthis flammea*)

Breeding in the Arctic and subarctic, the pretty little Common Redpoll expands its range as far south as central Idaho and northeast Washington during winter. In some years it strays even farther south. Small groups are typical, but in the species' normal winter range within the Northwest, flocks can total dozens and even hundreds of birds, especially in irruption years.

LENGTH 5 inches. **MALE** White below, with bold, dark flank streaks and rosy wash on breast and throat; red cap; whitish face with black chin; light gray above with darker streaks; black wings with 2 white wing-bars and white edging on flight feathers; stubby, conical, yellowish bill. **FEMALE** Similar to adult male, but more heavily streaked and lacks pink wash on breast. **VOICE** Calls include a ringing, slightly nasal *cheeeUP*; high *cheep* calls; flight call is a series of single and double *chip* notes. **BEHAVIORS** Acrobatic when feeding in shrubbery, often hanging upside down; ground-feeding flocks typically take flight in leapfrogging waves; sometimes mixes with Pine Siskins, goldfinches, and others wintering seed eaters. **HABITAT** Winters primarily in open country, including farmlands, grasslands, roadsides, shrub steppe, fallow fields, and willow groves. **STATUS** Uncommon winter visitor (November–March) to northeast and north-central Washington; rare winter visitor southward to northeast Oregon and central Idaho; very rare winter visitor elsewhere. **KEY SITES** Washington: Okanogan Highlands. British Columbia: Okanagan Valley.

SIMILAR SPECIES Hoary Redpoll (*Acanthis hornemanni*) is a more northerly species that is a very rare winter visitor to the Northwest. Paler overall than Common Redpoll, with white rump and little to no streaking on flanks and breast. The two may prove to be conspecific.

Female Common Redpoll

Male Common Redpoll

Common Redpoll range

Pine Siskin (*Spinus pinus*)

Northwesterners who enjoy maintaining backyard bird feeders eagerly anticipate the winter invasion of Pine Siskins from the mountains and more northerly latitudes. These gregarious little finches typically arrive in flocks and frenetically mob feeding stations, remaining at favorite locales all winter until suddenly departing come spring. Siskins are endlessly entertaining, and observers may be left wondering why the little birds don't spend more time eating and less time fighting among themselves over feeder space.

LENGTH 5 inches. **SEXES SIMILAR** Pale gray-tan above, heavily streaked with dark brown; whitish below, heavily streaked to lightly flecked with brown; 2 white to yellowish wingbars, the lower being especially conspicuous; yellow wing and tail markings (ranging from dull, pale yellow to bright lemon yellow) especially flashy in flight; small, sharply pointed bill; notched tail. **VOICE** Song features a variety of harsh but musical rambling warbles often punctuated by a screechy *terweeeeEEE* twitter call. Calls include a harsh but musical *zereeEE*, and varied finchlike chatters and whistles.

BEHAVIORS During winter, frequently mobs bird feeders in flocks ranging from a dozen to many dozens of birds, and often forms mixed winter flocks with American and Lesser Goldfinches, and sometimes House Finches. In breeding territory, typically feeds high in the canopy and is difficult to detect. **HABITAT** Breeds in mountain forests and lower-elevation conifer woodlots, including within parks and cities; widespread in migration and winter, generally in brushy areas with at least scattered trees, especially alder, birch, and conifers. **STATUS** Common year-round resident and winter visitor; irruptive and nomadic in winter, so abundance varies annually.

Typical adult Pine Siskin

Bright yellow adult Pine Siskin

Pine Siskin range

American Goldfinch
(*Spinus tristis*)

The most seasonally dimorphic of Northwest finches, the male American Goldfinch is also the most striking in his summer breeding plumage, a conspicuous bright yellow highlighted with black and white. Come autumn, however, he molts into a much more subdued dress only to molt again in early spring back to the showy bright yellow. The female likewise changes with the seasons, but her plumage variation—from pale yellow in summer to more pale tan in winter—is not as dramatic as that of the male.

LENGTH 5 inches. **MALE BREEDING** (April–August) Bright lemon yellow with sharply contrasting black wings and forehead, and blackish tail; white undertail coverts; yellowish orange bill. **MALE NONBREEDING** (September–March) Similar overall pattern to breeding plumage, but much duller, usually warm pale tan with pale yellow throat and yellow shoulders; blotchy yellow patches in autumn; all black wings by late summer molting to distinct white or buffy wingbars on black wings in fall; increasingly yellow

by late winter, with patchy black forehead. **FEMALE** Pale yellowish overall with dusky yellowish mantle and face (breeding) or soft grayish tan overall (nonbreeding), with dark wings and prominent wingbars (thin wingbars in female Lesser Goldfinch); white undertail coverts; pale yellowish bill (breeding) or pale gray bill (nonbreeding). **VOICE** Song is a rapid, ringing, musical series of *twee* and *sweet* notes, often slowing and slurring at the end; calls include a whiny *teweeee*; a very high *pureet-peet-peet* or similar, often with rapid high chatter notes; flight call is a musical, repeated *cher-chee-chee-chee*. **BEHAVIORS** Habitually feeds on thistle, Fuller's teasel, and other stemmed seed heads, often clinging acrobatically to stems. **HABITAT** Brushy or weedy fields, fencerows and hedgerows, brushy road margins, woodland edges. **STATUS** Common year-round resident.

American Goldfinch range

Male breeding American Goldfinch

Female breeding American Goldfinch

Male nonbreeding American Goldfinch
(January)

Female nonbreeding American Goldfinch

Lesser Goldfinch (*Spinus psaltria*)

Despite the male's striking color pattern—especially flashy when he takes flight—the unobtrusive little Lesser Goldfinch is easily overlooked, even though in some areas of the Northwest it seems to outnumber the larger American Goldfinch. In much of western Oregon, Lesser Goldfinches are year-round residents that will happily spend all winter frequenting their favorite bird-feeding stations daily.

LENGTH 4.5 inches. **MALE** Bright yellow below; greenish olive to grayish olive above, with distinctive black cap; white wingbar on black wings; white edging on flight feathers; in flight, shows large, flashy white spots on wings and both sides of tail. **FEMALE** Pale tan or grayish tan to pale yellow overall, slightly darker above than below; pale wingbars; small white wing patches most evident in flight; pale yellow undertail coverts (white in American Goldfinch). **JUVENILE** Similar to female. **VOICE** Song is a complex and lengthy mixture of very high buzzing chatters, ringing *twee* notes, whistles, and brief warbles. Calls include brief chats and a

Adult male Lesser Goldfinch

Lesser Goldfinch range

Pale female Lesser Goldfinch

musical, bubbly *chee-chee-chee*. **BEHAVIORS** Easily attracted to bird feeders, especially with sunflower seeds and thistle. **HABITAT** Brushy or weedy fields, fencerows and hedgerows, brushy road margins, woodland edges, open woodlands. **STATUS** Fairly common year-round resident of western Oregon and central Oregon; uncommon year-round resident and migrant in eastern Oregon; uncommon year-round resident of Washington (primarily Klickitat County); uncommon winter resident and migrant and rare summer resident in Boise Valley, Idaho; rare elsewhere in Idaho.

First-spring male Lesser Goldfinch

Bright female Lesser Goldfinch

House Sparrow
(*Passer domesticus*)

One of the most successful introduced species in the Americas, the House Sparrow, originally from Europe and Asia, now ranges across all but the northernmost latitudes of North America. Introduced in the mid-1800s, it is supremely successful in coexisting with humans and is most abundant in urban and suburban settings.

LENGTH 6.25 inches. **MALE BREEDING** Black bib, throat, and face; pale gray cheeks; gray crown and rusty nape; pale gray belly and flanks; rufous mantle with black streaks; gray rump; black wing coverts with rusty tan edges; dark bill. **MALE NONBREEDING** Similar to breeding plumage, but paler, with black limited to face and throat; yellowish bill. **FEMALE** Mottled brown and tan above, plain buffy gray below; unmarked pale grayish tan face and head, with light eye stripe; pale bill. **VOICE** Common call is a rather dry but musical, slightly wavering *cheeeeup cheeeeup* often punctuated by a screechy *terweeeeEEE* twitter call. **BEHAVIORS** Commonly forages on city streets, sidewalks, and parking lots. Nests in cavities and crevices, including cavities and nest boxes that would normally be used by native species. **HABITAT** Cities, towns, farmlands, ranch lands. **STATUS** Abundant year-round resident.

Male breeding House Sparrow

Female House Sparrow

Male nonbreeding House Sparrow

House Sparrow range

RECOMMENDED RESOURCES

Science, Conservation, and Birding Organizations

Government Agencies

United States Fish and Wildlife Service
  Pacific Region
  www.fws.gov/pacific
United States Bureau of Land Management
  Oregon/Washington Region
  www.blm.gov/or/index.php
United States Bureau of Land Management
  Idaho Region
  www.blm.gov/id/st/en.html
United States Forest Service
  Pacific Northwest Region
  www.fs.usda.gov/r6
United States Forest Service
  Intermountain Region
  www.fs.usda.gov/r4
United States Forest Service
  Northern Region
  www.fs.usda.gov/r1
National Parks Service
  www.nps.gov
British Columbia Ministry of Environment
  www.gov.bc.ca/env
British Columbia Ministry of Forests,
  Lands, and Natural Resource Operations
  www.gov.bc.ca/for
Idaho Department of Fish & Game
  www.fishandgame.idaho.gov
Oregon Department of Fish and Wildlife
  www.dfw.state.or.us
Washington Department of Wildlife
  www.wdfw.wa.gov

Birding and Conservation Organizations

National Audubon Society
  www.audubon.org
American Ornithologists' Union
  www.americanornithology.org
American Birding Association
  www.aba.org
The Nature Conservancy
  www.tnc.org
Nature Conservancy of Canada
  www.natureconservancy.ca
Western Rivers Conservancy
  www.westernrivers.org
Ducks Unlimited
  www.ducks.org
British Columbia Field Ornithologists
  www.bcfo.ca
Oregon Birding Association
  www.orbirds.org
Idaho Birds
  www.idahobirds.net
Washington Ornithological Society
  www.wos.org
Western Field Ornitholigists
  www.westernfieldornithologists.org

Northwest Bird and Birdwatching Festivals

British Columbia
Brant Wildlife Festival (Parksville), March/April
www.brantfestival.bc.ca
Wings over the Rockies Festival (Invermere), early May
www.wingsovertherockies.org
Creston Valley Bird Festival, early May
www.crestonvalleybirds.ca
Meadowlark Festival (Penticton area), mid-May
www.meadowlarkfestival.bc.ca
Skagit Valley Bird Blitz, late May
www.hopemountain.org
Manning Park Bird Blitz, late June
www.hopemountain.org
Fraser Valley Bald Eagle Festival, late November
www.fraservalleybaldeaglefestival.ca

Idaho
Hagerman Bird Festival, February
www.hagermanbirdfestival.com
Dubois Grouse Days, mid-April
www.grousedays.org
American Falls Birding Festival, early May
www.afbirdingfestival.us

Oregon
Winter Wings Festival (Klamath Falls), mid-February
www.winterwingsfest.org
John Scharff Migratory Bird Festival (Burns), mid-April
www.migratorybirdfestival.com
Ladd Marsh Bird Festival (La Grande), early May
www.friendsofladdmarsh.org/bird-festival-2

Mountain Bird Festival (Ashland), late May
www.klamathbird.org
Dean Hale Woodpecker Festival (Sisters), early to late June
www.ecaudubon.org/?alias=ecaudubon.org/woodpeckers
Oregon Shorebird Festival (Charleston), early September
www.fws.gov/refuge/Bandon_Marsh/

Washington
Upper Skagit Bald Eagle Festival, late January or early February
www.skagiteagle.org
Port Susan Snow Goose and Birding Festival, late February
www.snowgoosefest.org
Othello Sandhill Crane Festival, late March
www.othellosandhillcranefestival.org
Olympic Peninsula BirdFest, late March/early April
www.olympicbirdfest.org
Washington Brant Festival (Semiahmoo), mid-April
www.wabrant.org
Wings over Water Northwest Birding Festival (Blaine), mid-March
www.wingsoverwaterbirdingfestival.com
Grays Harbor Shorebird and Nature Festival, late April/early May
www.shorebirdfestival.com
Leavenworth Spring Bird Fest, mid-May
www.leavenworthspringbirdfest.com
Puget Sound Bird Fest (Edmonds), mid-May
www.pugetsoundbirdfest.com
Washington State Wenas Creek Audubon Campout, Memorial Day Weekend
www.wenasaudubon.org
Ridgefield Bird Fest, mid-October
www.ridgefieldfriends.org/birdfest

Male Gadwall

FURTHER READING

Ackerman, Jennifer. 2016. *The Genius of Birds*. New York: Penguin Press.

Adams, George. 2013. *Gardening for the Birds*. Portland, Oregon: Timber Press.

Beadle, David and James Rising. 2003. *Sparrows of the United States and Canada*. Princeton, New Jersey: Princeton University Press.

Beadle, David, Jon Curson, and David Quinn. 1994. *Warblers of the Americas: An Identification Guide*. New York: Houghton Mifflin.

Birkhead, Tim, Bob Montgomerie, and Jo Wimpenny. 2014. *Ten Thousand Birds: Ornithology Since Darwin*. Princeton, New Jersey: Princeton University Press.

Burleigh, Thomas. 1972. *Birds of Idaho*. Caldwell, Idaho: The Caxton Printers, Ltd.

Cramer, Deborah. 2016. *The Narrow Edge: A Tiny Bird, an Ancient Crab, and an Epic Journey*. New Haven, Connecticut: Yale University Press.

Darke, Rick and Douglas W. Tallamy. 2014. *The Living Landscape: Designing for Beauty and Biodiversity in the Home Garden*. Portland, Oregon: Timber Press.

Erickson, Laura. 2009. *The Bird Watching Answer Book: Everything You Need to Know to Enjoy Birds in Your Backyard and Beyond*. North Adams, Massachusetts: Storey Publishing.

Haupt, Lyanda Lynn. 2001. *Rare Encounters with Ordinary Birds: Notes from a Northwest Year*. Seattle, Washington: Sasquatch Books.

Hayman, Peter, John Marchant, and Tony Prater. 1991. *Shorebirds: An Identification Guide*. New York: Houghton Mifflin.

Heinrich, Bernd. 2016. *One Wild Bird at a Time: Portraits of Individual Lives*. Boston: Houghton Mifflin Harcourt.

Howell, Steve N. G. 2010. *Peterson Reference Guide to Molt in North American Birds*. Boston: Houghton Mifflin Harcourt.

Howell, Steve N.G. and Jon Dunn. 2007. *Peterson Reference Guides to Gulls of the Americas*. Boston: Houghton Mifflin Harcourt.

Lederer, Roger. 2016. *Beaks, Bones, and Bird Songs: How the Struggle for Survival Has Shaped Birds and Their Behavior*. Portland, Oregon: Timber Press.

Marshall, David (editor). 2006. *Birds of Oregon: A General Reference*. Corvallis, Oregon: Oregon State University Press.

McFarland, Casey and Shannon David Scott. 2010. *Bird Feathers: A Guide to North American Species*. Mechanicsburg, Pennsylvania: Stackpole Books.

Mlodinow, Steven G., Bill Twiet, and Terence R. Wahl (editors). 2005. *Birds of Washington: Status and Distribution*. Corvallis, Oregon: Oregon State University Press.

Pyle, Peter. 1997. *Identification Guide to North American Birds, Part I: Columbidae to Ploceidae*. Point Reyes Station, California: Slate Creek Press.

Pyle, Peter. 2008. *Identification Guide to North American Birds, Part II: Anatidae to Alcidae*. Point Reyes Station, California: Slate Creek Press.

Rappole, John H. 2013. *The Avian Migrant: The Biology of Bird Migration*. New York: Columbia University Press.

Rich, Jeffrey. 2016. *The Complete Guide to Bird Photography*. Amherst, Massachusetts: Amherst Media.

Roth, Sally. 2009. *Backyard Bird Secrets for*

Every Season. Emmaus, Pennsylvania: Rodale, Inc.

Sibley, David Allen. 2009. *The Sibley Guide to Bird Life and Behavior.* New York: Knopf.

Strycker, Noah. 2015. *The Thing With Feathers: The Surprising Lives of Birds and What They Reveal About Being Human.* New York: Riverhead Books.

Swanson, Sarah and Max Smith. 2013. *Must-See Birds of the Pacific Northwest.* Portland, Oregon: Timber Press.

Thompson III, Bill, Julie Zickefoose, and Kenn Kaufman. 2006. *Identify Yourself: The 50 Most Common Birding Identification Challenges.* Boston: Houghton Mifflin Harcourt.

METRIC CONVERSIONS

Inches	Centimeters		Feet	Meters
¼	0.6		1	0.3
⅓	0.8		2	0.6
½	1.3		3	0.9
¾	1.9		4	1.2
1	2.5		5	1.5
2	5.1		6	1.8
3	7.6		7	2.1
4	10		8	2.4
5	13		9	2.7
6	15		10	3
7	18			
8	20			
9	23			
10	25			

Temperatures

degrees Celsius = 0.55 × (degrees Fahrenheit - 32)

degrees Fahrenheit = (1.8 × degrees Celsius) + 32

To convert length:	Multiply by:
Yards to meters	0.9
Inches to centimeters	2.54
Inches to millimeters	25.4
Feet to centimeters	30.5

ACKNOWLEDGMENTS

Without the support of numerous excellent photographers, compiling this field guide would have been impossible, and we sincerely thank each of them for their contributions herein, including Nagi Aboulenein, Bob Armstrong, Mark Carmody, Ken Chamberlain, Jared Clarke, Tom Crabtree, Kevin Feenstra, Zia Fukuda, Brandon Green, Tom Grey, Jim Hardman, Marlin Harms, Tobias Hayashi, Tom Johnson, Michelle Lamberson, Steve Mlodinow, Julio Mulero, Christian Nafzger, Neil Paprocki, Dennis Paulson, Owen Schmidt, Harry D. Sell, Doris Sharrock, David Speiser, Terry and Kay Steele, Jason Stemple, Colin Talcroft, Lyn Topinka, Steve Tucker, Alan Vernon, Glenn Walbek, Michael Woodruff, and Mike Yip. Please see the photographer biographies at the back of this book.

Additionally we wish to thank content editor Hendrik Herlyn for his superb and professional editing, which added immeasurably to the accuracy of the species accounts herein. And finally, sincere thanks to the ever-helpful staff at Timber Press, especially Juree Sondker and Julie Talbot for expertly steering this project from inception to completion.

PHOTOGRAPHY AND ILLUSTRATION CREDITS

Photography

All photographs are by John Shewey with the exception of the following:

Nagi Aboulenein, page 249 bottom. Nagi is a computer engineer in the Portland area. He keeps himself busy with birding and wildlife photography.

Bob Armstrong, page 228 left. Bob Armstrong has coauthored books and articles, worked for the Alaska Department of Fish and Game, and been an associate professor for the University of Alaska.

BC Parks, page 32 bottom

Marc-Andre Beaucher/Creston Valley WMA, page 33 bottom

Tim Blount, pages 29, 155 top, 165 top, 168 bottom, 260 bottom, 269 top right, 273 second from top, 279 bottom, 297 right, 317, 349 top, 358 right, 433, 434 top, 441 bottom, 459 right, 461 bottom, 473 top, 478, 489 top, 504 bottom

Mark Carmody, page 226 top. Native to Cobh, County Cork, Ireland, Mark Carmody (www.markcarmodyphotography.com) is the coauthor of three books about Irish birds.

Ken Chamberlain, page 129. Ken loves birding and exploring rural areas of Oregon, as well traveling to Latin American, including Mexico.

Jared Clarke, page 123 left. Jared Clarke, www.birdtherock.com, guides birders in Newfoundland, providing clients with an array of customized tours and trips.

Tom Crabtree, pages 67 bottom, 87 bottom, 198, 213 bottom, 222 right, 234 bottom, 258 top, 266 bottom right, 267, 268, 269 upper left, 281 top, 327 left, 377 left, 437 bottom, 439 top, 471 top, 483 bottom. Tom has been a member of the Oregon Bird Records Committee for over a quarter century, and has led tours throughout North America for the Portland Audubon Society and Field Guides, Inc. His photography has captured over 425 species in Oregon alone.

Kevin Feenstra, pages 151, 202 right, 203 bottom. Kevin Feenstra's love for photography emerged from his years working as a fishing guide. He shoots a variety of birds, but also mammals, insects, underwater creatures, and river landscapes.

Flickr

Used under a Creative Commons Attribution 2.0 Generic license, no alterations, www.creativecommons.org/licences/by/2.0:

AG/USFWS, page 203 top

Andrea Pokrzywinski, pages 182 right, 214 top

David St. Louis, page 177 left

Don McCullough, page 221 top

Dow Lambert, page 250 top

Duncan, page 249 top

Ed Dunens, page 125 top, 276 right

Eric Ellingson, page 248 bottom

Fyn Kynd, pages 236 top right, 449 top

Greg Schechter, page 283

Jo Garbutt, page 75 top

Kaaren Perry, page 339 bottom

Linda Tanner, page 475 bottom

Mark Johnson, Bearing Land Bridge National Monument, page 197 top

Mike's Birds, page 247 top

Peter Davis/USFWS, page 83 bottom

Peter Massas, page 231 bottom

Peter Pearsall/USFWS, page 242 bottom

Ray W. Lowe/USFWS, pages 133 bottom, 245 bottom

Robert Britt, page 309 bottom

Ron Knight, page 196 bottom

Seabamirum, pages 76 top right, 470 right

Tim Lenz, pages 167 center, 368 top

USFWS, page 334 left

Used under a Creative Commons Attribution-ShareAlike 2.0 Generic license, no alterations, www.creativecommons.org/licenses/by/2.0:

Dominic Sherony, page 308 bottom

Gregory "Slobirdr" Smith, pages 127 bottom, 286 top

Hans de Grys, page 294 top

Used under a Creative Commons Attribution-NoDerivs 2.0 Generic license, no alterations, www.creativecommons.org/licenses/by/2.0:

Kelly Colgan Azar, page 451 left

Wplynn, page 331 bottom

Yankech gary, page 486

Used under a Creative Commons Public Domain 1.0 Generic license, no alterations, www.creativecommons.org/about/cco:

Alan Schmierer, pages 171 top, 221 bottom, 442 top, 494 bottom left

Zia Fukuda, pages 99 bottom, 199 bottom, 291 bottom, 298. Zia Fukuda works as a Spotted Owl technician, using the opportunity to photograph this threatened species and spread educational awareness.

Brandon Green, pages 220 top, 332 left, 398 left, 411 right, 467 right, 481 bottom, 517 top. Brandon served as photo editor for the journal *Oregon Birds* from 2011 to 2014. His photos have been published in the *New York Times* and various publications from the Cornell Lab of Ornithology.

Tom Grey, pages 52 top, 118 top, 121 bottom, 122 bottom, 126, 178 top, 194 bottom, 197 bottom, 204 top, 215 center, 224 center, 226 left, 228 top, 229 bottom, 241, 254 bottom, 255, 257 top, 259 center, 261 left, 261 top right, 261 bottom right, 271 bottom, 295 right, 327 right, 346, 361 bottom, 501 right, 508. Tom Grey's bird pictures have appeared in *Birding*, *Birder's World*, and the Museum of Natural History's *Birds of North America*, among other publications. His website, www.tgreybirds.com, is an excellent reference for bird identification.

Jim Hardman, pages 138 top, 146 top, 148 bottom, 172 bottom, 213 center, 296, 313, 326 bottom, 328 top, 248 bottom, 355 bottom, 357 top, 394, 395 bottom, 406, 421, 426, 455 left, 480, 511 top, 513. Since 2010, Jim Hardman has avidly pursued songbird photography throughout the continental United States and Hawaii. His other subjects are nature, landscapes, and especially wildflowers.

Marlin Harms, pages 41, 119. Veteran birder and conservationist Marlin Harms has done photographic work for several land conservancies in various projects, most of which were successful in acquiring properties for the public. www.flickr.com/photos/marlinharms.

Tobias Hayashi, pages 81 top, 123 right, 124, 125 bottom, 127 top. Australian photographer Tobias Hayashi, www.tobiashayashi.com, loves finding and photographing seabirds and terrestrial orchids. He holds degrees in botany and zoology from the Australian National University.

Tom Johnson, pages 161 bottom, 295 left, 309 top. Birding guide and expert photographer Tom Johnson is a regional editor for *Birds of North America* and a columnist for *Birding* Magazine.

Michelle Lamberson, pages 34 top, 180 top, 189, 202 left, 218 bottom, 224 bottom, 225, 357 bottom. Michelle enjoys capturing birds in action, particularly in flight. Her favorite subjects are hummingbirds, raptors, and shorebirds—and pretty much any other bird she finds. Her photos can be viewed at www.flickr.com/photos/vitrain.

Steven Mlodinow, pages 310, 381 bottom. Steven has served as editor for two regional editions of *North American Birds* and has authored and edited several books. He is one of the few photographers to successfully capture the Black Swift in flight.

Julio Mulero, pages 49 bottom, 54 top, 72 bottom, 86 top, 114 bottom, 122 top right and left, 166 top, 186, 240 bottom, 242 top, 248 top, 252 bottom, 265 top, 271 top, 272 top, 281 bottom, 282, 303, 307 top, 311, 355 top, 359 bottom, 376 right, 386, 388 left, 445, 446, 485, 489 bottom, 509, 519 bottom, 520 bottom. Julio Mulero is on a quest to photo-document all the birds that breed regularly in the American Birding Association area (essentially all of North America north of Mexico). To date, he has photographed more than 600 species; see his images at www.pbase.com/juliom.

Christian Nafzger, page 101

Neil Paprocki, pages 227 bottom, 388 right, 512 bottom. An animal behaviorist, raptor biologist, photographer, and filmmaker, Neil Paprocki helped cofound a small wildlife conservation filmmaking nonprofit, Wild Lens Inc. In 2013 he codirected his first film, *Bluebird Man.*

Owen L. Schmidt, pages 120 upper right, 199 top, 209 left, 232, 260 top, 262 bottom, 280 bottom, 320 left, 360 top, 382 right, 506 right. For many years the editor of the journal *Oregon Birds*, Owen L. Schmidt has been photographing birds, or trying to, since long before the advent of digital imagery.

Harry D. Sell, page 494 right. Harry D. Sell has captured more than 100,000 images of birds in his home state of North Carolina.

Doris Sharrock, pages 48, 68 right, 112 top, 160, 178 bottom, 209 right, 211 top, 215 bottom, 224 top, 228 bottom right, 273 third from top, 297 left, 318 right, 349 bottom. Doris Sharrock (www.pbase.com/dlilly) has developed a knack for catching amazing bird scenes, including spectacular in-flight shots.

Shutterstock

Abi Warner, page 218 right

Adam Fichna, page 98 top

Agustin Esmoris, page 155 bottom

AlexussK, page 278

Andrew M. Allport, page 287 left

Anotherlook, page 208

Bildagentur Zoonar GmbH, pages 49 upper right, 492

BMJ, page 158 bottom

Bonnie Taylor Barry, pages 460, lower left and right, 464

Brian Lasenby, page 505

Bruce MacQueen, page 252 top

Charles Brutlag, page 422 bottom left

Chris Hill, page 165 lower right

Cindy Creighton, pages 254 top, 380, 490 top

Critterbiz, pages 58 bottom, 106 bottom left, 305

Daimond Shutter, pages 490 top, 285 bottom

David Spates, page 411 bottom left

Dennis W. Donohue, pages 51 bottom right, 111 top, 211 bottom left, 373 bottom, 470 top left

Dmytro Pylypenko, pages 81 bottom, 244

Don Mammoser, page 50 top right

Eduard Kyslynskyy, page 301 top

Elliotte Rusty Harold, pages 207 top, 274 bottom left, 277 center

Erni, page 53

feathercollector, pages 171 bottom, 302 top right

Feng Yu, pages 100 bottom right, 272 bottom right

Gallinago_media, page 382 left

Gerald A. DeBoer, pages 181 left, 216 bottom left, 217 bottom, 277 bottom

Glen Gaffney, pages 12, 523 bottom right

Glenn Price, pages 82 center, 185 top, 192, 496 top

Gregory Johnston, page 156 top

Gualberto Becerra, page 353

Herman Veenendaal, pages 91 bottom left, 145

Ian Maton, pages 93 bottom, 226 lower right, 231 top, 239, 240 top, 299 left, 325, 335 bottom left

Ifstewart, page 131 bottom left

Ivan Kuzmin, page 210 bottom

Jausa, page 153 bottom

John E. Heintz Jr., page 38 right

John L. Absher, page 507 bottom

Joseph Sohm, page 51 top

Kajornyot, page 215 top

KellyNelson, page 520 top

Keneva Photography, pages 93 top, 107 top

KOO, page 85 bottom

Laura Mountainspring, page 340

Lijuan Guo, page 33 top

Maciej Olszewski, page 75 bottom

Maksimilian, page 426 top

Marco Barone, page 116 top

Marcus R, page 179

Mark Bridger, page 291 top

Martha Marks, pages 136 bottom right, 339 top, 408, 471 bottom, 481 top right

Matt Knoth, page 138 right

Menno Schaefer, pages 109 bottom left, 371 top right

Michael G. McKinne, pages 435 top, 460 top left

Michal Kocan, page 229 upper right

MVPhoto, pages 70 bottom, 200 left, 206 right, 257 bottom left, 287 top right, 330 bottom right, 334 right, 337 bottom left, 352 top right, 448, 496 bottom, 522

Natalia Paklina, pages 61 bottom, 512 bottom

Neil Burton, page 207 bottom

Norman Bateman, pages 211 bottom right, 404

Pacific Northwest Photography, page 466 bottom right

Shutterstock (continued)

Paul Reeves Photography, pages 59 top, 77 bottom left, 80, 82 top and bottom, 84, 90 top, 169 top, 175 top, 200 right, 206 left, 212, 219 bottom, 253 top right, 258 bottom, 279 top, 322, 343, 344, 356 bottom, 375 bottom, 379 top, 416, 422 right, 424, 434 right, 447 top, 476, 493 top, 503 bottom

Paul Sparks, page 289

pcnorth, page 359 top right

Peter Schwarz, pages 154 top right and lower left, 159 top

Phoo Chan, pages 274, top right, 152, 166 bottom left, 176 bottom left

Pictureguy, page 164 left

Pr2is, page 167 top

Randimal, page 253 bottom left, 511 bottom

Randy Hume, page 523 top

rck_953, pages 69, 181 right, 369 top, 371 bottom

Richard Fitzer, pages 148 top, 277 top

Rob Christiaans, page 374

Rob McKay, page 164 right

Robert L. Kothenbeutel, pages 66 right, 73, 86 bottom, 92, 104 top, 105, 108, 168 top, 177 right, 195, 419, 497 top, 506 left

Rocky Grimes, page 109 top

Sari ONeal, page 88

Smishonja, page 154 top left

Soru Epotok, page 175 bottom

Stephen Bonk, page 213 top

Steve Byland, pages 67 center, 196 top, 275 bottom, 503 top, 524 left

Steve Collender, page 299 right

Steve Estvanik, page 246 top

Steve Jamsa, pages 51 bottom left, 98 bottom

Steven Blandin, page 323

Stubblefield Photography, pages 434 bottom left, 435 bottom, 447 bottom, 449 bottom, 452 bottom, 459 left, 470 bottom left

Takahashi Photography, pages 321 bottom, 407

Timothey Kosachev, page 519 top right

Tobyphotos, page 302 left

Tom Reichner, pages 87 top, 96, 97 top, 99 top, 106 right, 319, 356 top right, 373 top

Tony Campbell, page 369 bottom

Tony Mills, page 493 bottom left

Tory Kallman, page 121 top left and top right

Vikki Hunt, page 253 bottom right

visceralimage, pages 270 left, 235, 236 bottom

Wesley Aston, pages 170, 304 right

Wildphoto3, page 90 bottom

Wolfgang Kruck, pages 272 left, 276 left

xpixel, page 379 bottom

David Speiser, page 491 bottom. Images by David Speiser (www.lilibirds.com) have been featured at New York City's Central Park Zoo, the Royal Botanical Gardens of Ontario, in various birding publications, and even on a U.S. postage stamp.

Terry and Kay Steele, pages 102, 103, 107 bottom, 153 top, 154 bottom right, 165 left, 345, 378, 462 left. Terry R. Steele is a frequent and popular presenter on wildlife and photography topics. Kay Steele established Chintimini Wildlife Rehabilitation Center in Corvallis, Oregon, and is the other half of Terry Steele Nature Photography (www.terrysteelenaturephotography.com).

Jason Stemple, pages 139, 143, 144 right

Colin Talcroft, pages 230, 259 top and bottom, 263 top, 264. Colin is a fine art photographer, printmaker, and collage artist. He is the creator of Sonoma County Bird Watching Spots (www.colintalcroft.com) and the blog Serendipitous Art (http://serendipitousart.blogspot.com).

Lyn Topinka, pages 22, 60 right, 67 top, 70 top, 71 left, 72 top, 138 bottom, 233 bottom, 265 bottom, 413 top, 491 top, 499, 523 bottom left. Lyn Topinka began photographing birds in 2007 and spent the entire year at the Ridgefield National Wildlife Refuge. She and her camera now roam the Columbia River corridor. Her work is on display at www.northwestbirding.com.

Steve Tucker, page 263 center. Gull expert Steve Tucker is from Oakland, California. His rants and diatribes appear regularly at Bourbon, Bastards and Birds (www.seagullsteve.blogspot.com).

USFWS/Camas National Wildlife Refuge, page 28 top

Glenn Walbek, pages 444, 161 top. Glenn Walbek photographs a wide variety of bird species and displays many of his images at www.pbase.com/gwalbek and https://gwalbek.smugmug.com.

Michael Woodruff, pages 54 bottom, 120 left, 120 bottom right, 174, 182 left, 227 top, 294 bottom, 304 bottom left, 318 left, 342, 450 top left, 512 top. Michael Woodruff purchased his first big lens for bird photography the age of 13. He has spent a significant amount of time abroad, and his diverse photography is on display at www.flickr.com/photos/nightjar.

Mike Yip, page 128

Illustrations

Range maps by Tim Blount and Anna Eshelman.

INDEX

ABOUT THE AUTHORS

Lifelong birding enthusiast **John Shewey** is a veteran freelance writer, author, editor, and professional outdoor photographer with credits in dozens of magazines, ranging from *Birdwatching* to *Portland Monthly*. John has photographed birds from the mountains of Alaska to the jungles of Central America to the islands of the Caribbean. His website, www.birdingoregon.com, chronicles many of his travels in rich photographic detail, including bird-photography expeditions to the most remote corners of his home state of Oregon. It also provides a repository for all things related to birdwatching in Oregon. In 2012, John found and photographed only the third Ruby-throated Hummingbird ever confirmed for the state of Oregon.

Noted expert **Tim Blount** has birded extensively in the United States and Europe. Living in Oregon's Harney County, the most renowned birdwatching destination in the Northwest, Tim is a director with the Oregon Birding Association and executive director of the Friends of Malheur National Wildlife Refuge. Tim's website, www.harneybirder.com—the most-visited birding website in the Northwest—details bird sightings in southeast Oregon and serves as a wellspring of information for birders visiting the region. Rich in photography, it is an important resource for birders researching species found in the Northwest. Tim has guided numerous groups and individuals in Oregon, sharing his love of birds with beginners and experts alike.